GLOBAL RESEARCH ON DRAINAGE IN AGRICULTURE

Global Research on Drainage in Agriculture

An Annotated Bibliography, 1960-1986

S. K. Gupta M.Sc., Ph.D.

I. C. Gupta M.Sc., Ph.D.

CONCEPT PUBLISHING COMPANY, NEW DELHI - 110015

ISBN 81-7022-026-2

First Published 1987

© S.K. Gupta and I.C. Gupta 1984

Published by
Naurang Rai
Concept Publishing Company
H-13, Bali Nagar,
NEW DELHI-110015 (India)

Printed in India by
New Gian Offset Printers
495, DDA Complex, Shahzada Bagh
DELHI-110035

Foreword

Drainage of agricultural lands is one of the important elements that stands out in the strategies for better utilization of land and water resources. We need only mention the transformations of Mesopotamia plains once considered as granery of the world, now a salty desert to emphasize the importance of drainage. More recently the waterlogging problems and the accompanying evil of soil salinity in many developed and developing countries has focussed attention on the importance of maintaining the water budget in irrigated lands. The times when reclamation and management were treated as of academic interest to the scientist have changed and world demographic pressure obliges us to attempt to cultivate all types of soils and to achieve best level of production in each of the hectare under plow. It is clear that in the years to come irrigation engineers will have to pay increasing attention to the evolution of integrated irrigation and drainage practices truly reflecting the spirit of the adage that irrigation and drainage go together.

Though the practice of drainage is probably as old as the history of agriculture, it is only in the last century that work on drainage practices currently in vogue, was started. It is only in the last few decades that enormous amount of publications arising out of the drainage projects have appeared in the scientific journals, proceedings of symposium/ seminars etc. and in the form of books/bulletins and monographs. Unfortunately in developing countries not many experiments have been conducted on the improved on-farm drainage practices, though the FAO and UNDP experts have emphasized that much of the information already available on the improvement of agricultural lands through drainage could be made use of for improving production in various parts of the third world.

The information on drainage studies is lying scattered and the need for collation has been apparent to all. Any constructive efforts in this direction are really laudable and require encouragement from every quarter. The attempt to collect this information and decision to publish it is a right step in this direction. With the realization of the need of drainage in agriculture, people engaged in various disciplines and at various levels will require this information and therefore, this volume containing both theory and practices of drainge, and performance studies on drainage machinery and materials will be of utmost value to these persons. The citations in respect of drainage for forestry indicate that word agriculture is used by the authors to encompass related areas as well which is a happy augary.

The annotated bibliography entitled **Global Research on Drainage in Agriculture**, which contains around 1500 abstracts of papers on the subject is a timely publication as world wide attention is drawn to drainage in agriculture at this time. The field of bibliogaphy is vast and dynamic one and to prepare a annotated bibliography on a specialized topic has required considerable efforts on the part of the authors. Drainage is really an interdisciplinary area of research, development and practice and I find in the authors Dr. S.K. Gupta, Drainage Engineer and Dr. I.C. Gupta, Soil Scientist, a team effort which has been accomplished well. I understand that while working at the Central Soil Salinity Research Institute, Karnal, the authors got an opportunity to work in the field of salt affected soils and saline waters, the management of which is almost impracticable unless drainage practices are adopted. Their attempt to venture into the past literature and zeal to work in this direction leading to the compilation of such a volume should be interesting to all those engaged in a wider area of work connected in any manner with soil and water resources and agriculture. I am sure that this book will be a handy ready reference to all such persons and therefore deserves to be on the shelves of each and every library of agricultural and engineering organisations.

R. Ghosh
Chairman (Retd.)
Central Water Commission
New Delhi-110017

Preface

With the ever increasing population in the world which is going to rise by 30 per cent by the turn of the century than what it is today, agricultural scientists have to face the challenge for meeting the food requirements. Total land area of the world is 13,150 million hectares out of which 3,190 million hectares is potentially arable but only 1,406 million hectares is actually cultivated. Thus, about 56 per cent of the potentially arable land is lying uncultivated due to several constraints. Most of the uncultivated lands commonly classified as waste lands falling under wind erosion, water erosion or waterlogging/salinity are caused by the lack of water, excess of water and mismanagement of irrigation water respectively. Introduction of irrigation has been the key approach throughout the world for increasing the agricultural production.

Introduction of excessive water into the irrigation systems without due consideration of drainage with the intention of sustaining the green revolution has been threatening several regions of the world with desertification through waterlogging and increasing soil salinity. History is witness to the fact that many civilisations have flourished whereas many civilisations have doomed, both due to irrigated agriculture. The difference lies only in the drainage. Alone in India, about 12 million hectares of canal irrigated land has already become waterlogged and another 25 million hectares are threatened by the increasing salinity of the soil which is slowly but positively making the land unfit for cultivation. Challenge lies in the prevention of waterlogging and salinity by providing drainage and management of water under new irrigation systems being developed now in potentially arable lands.

The knowledge and experience gained in one country can be translated into useful practice in another country with similar problem of the drainage in agriculture. It is with this objective that the available information in the world has been compiled in the form of annotated bibliography.

We hope that the annotated bibliography will prove immensely useful for the scientists, agriculturists and planners. We would like to add that inspite of our best efforts, we do not claim that the entire available literature on the subject has been covered. In the end we crave the indulgence of our wary readers in the omissions and errors in which we must have fallen inadvertently and solicit their valued suggestions towards improving the contents.

S. K. Gupta
I. C. Gupta

Contents

1 Global Research on Drainage in Agriculture

AART,R.VAN. Aspects of drainage and reclamation in the lower Mesopotamian Plain - A problem analysis. In 1st Scientific Conf.,Scientific Res. Foundation, General and Agri.Sci.,Baghdad, Iraq,1972:312-28.

Problems of waterlogging and salinization in the lower Mesopotamian Plain and their control by drainage and subsequent desalinization,reduction of alkalinity and cropping are described. Outlines are given for future land reclamation involving a change in areas under existing crops, which include wheat,barley,rice,green gram (Vigna radiata) and sesame,and the introduction of new crops. An increase in summer cropping and the introduction of more cash crops, the intensification of winter cropping, the incorporation of oilseed crops into the rotation and the promotion of fodder production are among the most important recommendations.

2

AART,R.VAN. Drainage of irrigated lands. Proc. Int. Drain. Workshop, Wageningen, Netherlands (1978), 1979, 25:50-61.

Drainage of irrigated lands, its economics and recommendations arising out of 9 papers presented in the workshop have been reported.

3

ABDUEV,M.R. Improvement of hardly reclaimable solonchaks in Azerbaijan. Int. Symp. New Develop. Field Salt Affected Soils, Cairo,(1972),1975,493-8.

The report presents the main points of the results obtained in a major experimental work on washing hardly reclaimable solonchaks in Azerbaijan (16 versions of production experiments) carried out against the background of chemical amelioration, drainage (with inter-drain spacings of 200m), and flume-type feeders. Gypsum, its combination with manure, organic mineral wastes, an acidifier (wastes from oil refineries and chemical plants), and spent sulphuric acid have been tested in various doses from 10 to 80 t/ha. Most efficient and economically advantageous methods of improving hardly reclaimable solonchaks have been revealed.

4

AFANAS'EV, N.I. Water supply for crops on drained soils in White Russia. Pochvovedenie, 1969, 9:157-9.

During 1966-68, the moisture deficit under barley, potatoes and perennial herbage on drained bog and peatbog soils was 55-160mm, 60-130 mm and 80-110 mm, respectively. Irrigation increased barley and herbage yields.

5

AGARWAL, M.C., MALIK R.K. Drainage coefficients for different agroclimatic locations of Haryana. J. Res. HAU, 1983, 13: 522-30.

For estimating drainage coefficients, rainfall probability analysis using more than 22 years of rainfall data in Haryana was made for 1-4 consecutive days maximum rainfall on 5,10 and 20 years return period basis. For medium, light and very light textured soils, no drainage problems were encountered for 1-day rainfall and 5-years return period basis. However, in heavy soils of different locations, the drainage coefficient values for 1-day rainfall and 5-year return period basis ranged from a minimum of 3mm/day at Hissar to a maximum of 55 mm/day at Karnal. There were no drainage problems for 4-day rainfall in the state.

6

AGARWAL, M.C.; SINGH, J.; SINGH, R. Water balance studies

of rising water table areas in Haryana. Trans. Indian Soc. Desert Technol., 1979, 4:1-5.

In central and western parts of Haryana State, the underground water is of poor quality. Introduction of canal irrigation in these areas has led to rapid rise in ground water table and in certain pockets it has already risen to within 3m from the surface. Contrary to expectations, the rise in ground water table due to recharge from good quality irrigation and rain water has not shown corresponding improvement in ground water quality due to the presence of large quantities of soluble salts in the soil profile. In such areas, the danger of soil salinization due to high water table cannot be prevented unless certain quantity of brackish ground water is pumped out and water table is prevented from rising within 3m depth. The problem can be tackled by minimizing recharge from irrigation losses and making conjunctive use of groundwater. In this paper, an attempt has been made to assess quantitatively the amount of recharge taking place from various sources and extent to which this recharge can be reduced by improving irrigation system efficiency. The quantity of groundwater which must be pumped in each of these problem districts to check the groundwater table rise has also been estimated.

7

AGEYEV, B.M.; MAMDOV, F.F. Variations in degree and chemical properties of soil salinization under passive drainage. Int. Symp. New Develop. Field Salt Affected Soils, Cairo, (1972), 1975, 605-9.

In the experimental area situated in the middle of the Terter river drift cone within the Karabakh Plain, test plots have been singled out in the centre of inter-drains spaced at 200, 400 and 600m intervals and having a transversal and longitudinal arrangement. The salinization dynamics and the chemical properties of the soil salts to a depth of 5m were observed from 1961 to 1969. Under the conditions of a sloping plain with subterranian pressure water being present and without active interference in the soil (washing and irrigation), the drainage operation is unstable. In these conditions, the soil is not reclaimed but only retards the progress of salinization.

8

AGRICULTURAL DEVELOPMENT AND ADVISORY SERVICE, U.K. Drainage problems in hill and upland areas. Ministry of Agriculture, Fisheries and Food, Drainage Leaflets, ADAS,1978,13, Revised, 9p.

Upland areas are usually well served by small natural streams or minor watercourses and provided these are maintained the farmer need normally only be concerned with the minor ditches and infieled drainage. Where ditches present problems,piping may be contemplated. However, with underdrainage, permeable backfills may be necessary. A system of removing excess water by small surface channels (moorland gripping) often provides a cheap and simple drainage operation.

9

AGRICULTURAL DEVELOPMENT AND ADVISORY SERVICE, U.K. Draining through a growing crop. Ministry of Agriculture, Fisheries and Food. Drainage Leaflets, ADAS,1978,17, Revised,8p.

Although some crop damage results from draining through a young standing crop, there are many advantages in carrying out this work under optimum conditions in the spring.

10

AGRICULTURAL DEVELOPMENT AND ADVISORY SERVICE, U.K. Field drainage record plans. Drainage Leaflets, Ministry of Agriculture, Fisheries and Food, Drainage Leaflets, ADAS 1976, 19, 8p.

A guide for the farmer, describing the uses and basic requirements of drainage record plans, the type of information to be recorded and methods of storage.

11

AGRICULTURAL DEVELOPMENT AND ADVISORY SERVICE, U.K. Land Drainage Service. Development work in regions 1978. Reference Book, Ministry of Agriculture, Fisheries and Food U.K, 1980,339,31 p.

Brief summaries are given of the following investigations: the effect of field drainage on soil water regime, soil moisture conditions for effective subsoiling and moling, drainage economics, damage to mole channels by subsequent subsoiling, percentage of fines in permeable backfill, watertable, fluctuations in response to drainage, hydraulic conductivity values for drainage design purposes, moling without permeable backfill, effect of soil moisture at the time of mole drainage. flow to well points in a peat soil, time of drainage installation on open cast coal sites, drainage problems affecting orchards on red soils (Henford area), effects of mole drainage on permanent pasture, determination of drainage status by use of aerial photographs, monitoring peat shrinkage after drainage on peat soils, tunnel ploughing for dewatering peat soils,watertable levels in undrained pulverised fuel ash in reclaimed clay pits, border dyke irrigation of water meadows.

12

AGRICULTURAL DEVELOPMENT AND ADVISORY SERVICE, U.K. Pumped field drainage. Ministry of Agriculture, Fisheries and Food. Drainage Leaflets, ADAS, 1978. 14, Revised, 8p.

Factors affecting the feasibility of pumping selection and design of pump, maintenance and costs are outlined.

13

AHMAD,N. Soil Salinity in west Pakistan and means to deal with it. Salinity Probl. Arid Zones, Proc. Teheran Symp.,(1958), 1961,117-25.

Subsoil water-table conditions, causes of the rapid rise of the water table in the canal regions, salts in subsoil water; salts in river waters, removal of salts by leaching,difficulties of successful leaching in west Pakistan, open drains, tile drains, drainage by pumping, hydraulics of tubewells, have been discussed in the paper.

14

AHTI, E. Estimation of the influence of forest drainage on flooding. Suo, 1975, 26:9-13.

If the runoff curve corresponding to a certain hydrograph is known for a drained peatland area with a 40m ditch spacing theoretical runoff curves corresponding to the same hydrograph can be constructed for any alternative spacing. During a long spell of rainfall with a constant rain intensity. an equilibrium is achieved in which surface runoff from a drained strip equals the amount of precipitation in unit time. The time required for achieving the equilibrium is independent of rain intensity and directly related to the ditch spacing. Consequently, theoretical surface runoff curves corresponding to rainfalls with constant intensity can be constructed if the equilibration time is known.
The most difficult process is to separate the drained area-runoff from the total runoff curves. Further, it may be laborious to determine an average value for the drained parts of a catchment area.

15

AKTSUKA, K.; HAKOISHI, T.; KONNO, T. Effect of tile drains on the properties of heavy clay soils. Res.

Bull. Hokkaido Nath. Agric. Exp. Stn.,1965, 88:38-44.

Drains 1m deep at intervals of 12 m made plowing possible immediately after the snow melted. Sowing was advanced by up to one month. Growth and yield of oats increased. The effect on free Fe++ extended to the upper C horizon. Water-soluble organic matter markedly decreased in the A horizon. Bulk density and large pore spaces increased, pore-size distribution indicated that aggreggate formation was accelerated. Tile-drainage is recommended where the structure of B horizon is well developed and permeability is fairly high.

16

AKAZAWA, T.; CHIBA, T.; SAKUMA, T.; UMEDA, Y. The efficient method for drainage on clayey lands-subsoil breaking on clayey upland fields in north Japan. 10th Cong. Drain. Athens, 1978, 4:415-37.

This report is concerned with the distribution and properties of clayey soils in northern Hokkaido, changes in them due to underdrainage and subsoiling (the latter carried out to prevant subsoil hardening after drainage), factors influencing the rate of work, effect of subsoil breaking on soil physical properties, comparison of working methods, durability of the effects, yield increases, and combination with other methods.

17

AKHMEDOV, A.P. Role of drainage systems in soil reclamation in the Mugan-Salyany Steppe. Pochvovedenie, 1976, 6:93-8.

The data have been presented on the long term operation of collector-drainage system and on the leaching of saline soils over an area of 1,25,000 ha. Deep horizontal drainage proved to be very effective in salinization control in areas where ground-waters with a high mineral content and having no natural outlet were close to the surface.

18

AKPARANTA,S.E.; SKAGGS, R.W.; SANDERS,D.C. Drainage requirements for sweet potato at harvest. J. Am. Soc. Hort.Sci., 1980, 105:447-51.

Sweet potato lines with varying resistance to flood damage were grown in a sandy loam and a silt loam soil and subjected to water table draw-down rates of 5, 20, 35 and 100 cm/day for 7 days prior to harvest. Decay increased with time after harvest and decreasing drawdown rate. 'Julian' had the longest post-harvest keeping quality for the various drainage rates applied and there was no difference in keeping quality between soils. 'Centennial' was intermediate, showing wet soil damage with lower drainage rates and the damage was higher on the silt loam than on the sandy loam soil. 'NC 257' showed high wet soil injury for all drainage rates with greater injury in the finer textured soil. Decay of the roots was greater in 1976 than in 1975 for all genetic sources. pH, titratable acidity and soluble K were well correlated with decay in 1975 but not in 1976.

19

AKSENOV, P.I. Moisture accumulation in soil by deep mole drainage. Vestn. S.Kh. Nauki,1965, 2:97-101.

On autumn-plowed land in the central Chernozem zone, mole drainage at a depth of 70 cm (the average depth to which the soil freezes) increases moisture accumulation in the deep soil layers throughout the cold season and decreases evaporation during the vegetative period. Mole drainage decreases runoff and soil erosion.

20

ALDABAGH, A.S.Y.; BEER, C.E. Economics of increased mobility from the drainage. Trans ASAE, 1975, 18: 116-21.

The effects of tile drainage on soil trafficability and the consequent economic benefits from increased mobility of agricultural machinery were determined for two sites on poorly drained silty clay loam.

21

ALDRICH, S.N. Environmental aspects of sulfur dioxide treatment of iron-mangagese in pipe drain. In Water Management For Irrigation and Drainage, Proc. Spec. Conf. ASCE. Irrig. Drain. Div., ASCE, New York, 1977, 1:121-5.

The technique of removing iron-manganese deposits from the drain pipe by sulphur dioxide treatment, though effective, eventually affectsthe biological communities at the down-stream from points of effluent release. It is environmentally compatible where favourable conditions exist for controlling the rate of discharge and arrange for proper dilution of the effluent. A dilution factor of 50 to 1, water to effluent is suggested for the study area.

22

ALESSANDRELLO, E.; CONCARET, J.; GUYOT, J.; PERREY, C. Water movement in drained hydromorphic silty sols lessives. Comptes Rendus Seances Acad. Agric. Fr., 1976, 62:364-73.

Drainage conditions, after the laying of 6 cm pipes at 70 cm depth and 10m spacing in a silty hydromorphic sol lessive, are discussed with the help of hydrograms, moisture profiles and porosity data.

23

ALIMOV, A.K. Conditions of formation and differentiation of drainage from the Mugon-Salyany steppes. Pochvovedenie, 1975, 7:113-9.

The proportion of irrigation water, groundwater, and confined water in the collector-drainage systems of this area was determined by hydrochemical methods, and that of salts from the zone of aeration and from ground and confined waters in the salt discharge was established by the balance method.

24

ALLEMAN, M.; VIGNERON, J., subsurface drainage as means to reclaim saline and alkali soils. Irrig. Drain. Paper, FAO, 1972,16:5-15.

The studies pertain to the location in France, geographical department of Gard, commune of vauvert, the so called "petite camargue" (delta area). The drainage by mole drains in saline and alkali soils of fine texture and with a high lime content is feasible. The drainage is improved by enveloping the pipes with some filtering material; drain-laying techniques must be adapted to the texture of each soil. Considering the relatively high annual rainfall in the region (598mm), drainage combined with sprinkler irrigation, with doses closely approximating the crop requirements, results in adequate salt removal. Desalination does not automatically lead to sodium desorption by the clays. Plots should be arranged in such a manner so as to ensure runoff of excess surface waters. Cropping intensification should be approached carefully.

25

ALLEN, R.R.; MYERS, V.I. Draining area underlain by highly permeable shallow sand aquifer. Trans. ASAE, 1969, 12:29-31.

Field and laboratory simulation on an electric resistance network analog of drainage in a Raymondville clay loam to a Willacy fine sandy loam soil underlain by a highly permeable thick sand aquifer at the 3 m depth was studied. The drainage system spaced at 198 m and 1.8-2.4 m depth was installed on a 8 ha plot. Flow from the tiles placed at 2.4 m depth was substantial but had limited effect on the water table. The field and analog simulation indicated that underlying aquifer supplied 70-75 per cent of the total tile flow. Successful water-table control might be achieved through subsurface drains planned as an area unit rather than isolated individual system.

26

ALPERS, R.J.; SHORT, T.N. Clay draintile and its use. Proc. Drainage for Efficient Crop Production Conf. Chicago, Illinois, 1965, 46-8.

Manufacture of clay drain tile from raw materials to finished product has been discussed. Mechanical handling of a clay tile for transportation and installation has been reported and ways to save on labour suggested.

27

ALSOULI, S.S.; DAVIES G.M.; FARNWORTH, J. The effect of height of water table on the growth and development of two alfalfa cultivars. Publication, College of North Wales, Bangor and Ministry of Agriculture and Water, Saudi Arabia, 1977, 83, 6p.

In a glasshouse study at Bangor, UK in 1973-74, lucerne cv.(a) Hasawi and (b) Europe were grown on sandy soil in containers in which water-table depths of 13,38,64 or 81 cm were maintained, and cut 9 times at monthly intervals. Mean DM Yields were 630, 807, 864 and 1004 g/container for water-table depths of 13,38,64 or 81 cm, resp. and average 974 and 684 g/container for (a) and (b) resp.Root and top dry wt., CP yield, taproot length and plant persistance increased with increase in water table depth.

28

ALVAREZ LOPEZ, A.; ARIAS HERNONDEZ, A.; LEGARDA BURBANO L. Study of a drainage system for small areas in Pasto Narino (Colombia).An. Edafolo. Agrobiol., 1978, 37:895-906.

A drainage system for a 6483 m^2 area in Pasto, Colombia was designed on the basis of topography, direction of water flow, groundwater profiles and hydraulic conductivity of the soil. The system comprised open ditches in connection with collector and secondary channels within the drained area. A layer of stones and one of straw acted as a filter in the closed channels. The low hydraulic conductivity of the soil necessitated close spacing of the secondary ditches. Data from observation wells indicated that drainage lowered the mean water table depth by 0.234 m to 0.363 m, during maximum rainfall-periods.

29

AMER, M.H. Field investigation and analysis of siltation through tile drains in the delta area U.A.R. *7th Cong. Irrig. Drain. Mexico.*, 1969. Q 25 R 14:181-92.

Actual measurement of silt and clay deposits through tile drains, crack spacings and slope of the line has been determined for a 3,000 m, of tile drain laterals laid machanically without filter in the Delta, U.A.R. The nature of the subsoil that caused silting and that already moved into the system has also been determined. New empirical equations that are based on basic relations of variables involved in sediment movement are obtained from the field data. The equations can be used to predict the soil deposit through tile drains for similar areas in the Delta and Upper Egypt. Analysis showed also that a minimum slope of 0.10 per cent and a maximum of 0.25 per cent for laterals are satisfactory under conditions of the irrigated lands in U.A.R.

30

AMER, M.H.; ABDELDAIEM, S.Y. Use of critical path method in planning tile drainage network in the Delta area, U.A.R. ICID Sp. Sess. Varna, 1972, 55-78.

The critical path method (CPM) being used in planning and scheduling only the different tasks of field tile drainage work to be provided in UAR in about 950,000 feddan area is illustrated. A computer programme for the solution of CPM network is made. The printed programmes provide information for each activity, the earliest time, the latest time, the earliest finish, the latest finish, the total float and the critical path. The analysis is used for replanning and scheduling all tasks of tile drainage network according to calender dates to be set for the starting of the work. (1 feddan=0.42 ha).

31

AMER, M.N.; LUTHIN, N. Proposed drain spacing equation and an experimental check with the Hele-Shaw model. Proc. Int. Soil Wat. Symp. Praha, 1967,397-422.

An equation for the spacing between horizontal drains is derived by extending Kirkham's exact steady-state solution to transient conditions. The new equation gave better results than did the earlier equations.

32

AMERICAN SOCIETY OF AGRICULTURAL ENGINEERS. Drainage Materials. ASAE, Michigan USA, Sp. Pub. SP-01-67, 35p.

The publication discusses various drain tile, pipe or tubing, envelope (filters) and stabilizing materials commonly used in agricultural drainage. A bibliography on drainage materials is provided at the end.

33

AMI, S.R.; BROUGHTON, R.S.; SHADY, A.M. Designing and installing subsurface drainage laterals less than 100 mm in diameter. Can. Agric. Engng.,1978, 20:16-9.

Seventy five mm diameter plastic drain tubing can be used successfully for subsurface drain laterals in Canada. The use of 75-mm dia tubing can result in definite economic savings and in easier handling, storage and transportation. It is particularly advantageous on rolling land with high grades and in soils with poor hydraulic conductivities where smaller spacing between laterals can be used. During installation, care in grade control has to be exercised especially with grades less than 0.25 per cent. Awareness of the limitations of 75 mm tubing is necessary to avoid exceeding the hydraulic capacity. Hydraulic capacity data for corrugated plastic drain tubes from 59-100 mm inside dia are presented. Tables giving drained and maximum length of subsurface drain laterals for drainage coefficients of 9 and 12 mm/day, hydraulic gradients of 0.1, 0.2, 0.3, and 0.5per cent and drain spacing ranging from 9-60 m are given.

34

ANANIAN, A.K.; MARTCHIAN, S.M.; MHITARIAN, G.M.; SIMONIAN, G.A.; KARAPETIAN, S.G. Research on the design of vertical drainage for sodic solonchaks of the Ararat Plain. Agrokem.Talajt.,18 (Suppl.):139-58.

The test plots each 0.66-0.8 ha in area were located across from the initial, middle and terminal portions of the depression. The depression was forming as water was being pumped continuously from a deep well. The effective range of the well reached 500 m owing to the topography of the depression where the ground water level had dropped to a depth of 2 m from the surface. Horizontal drains upto 1.5-2 m deep were laid along the boundary of all the plots at 100 m space. All the plots had observation holes to evaluate chemistry and water head in the covering layer and low-head aquifer before, during and after washing irrigations as well as during land development for the purpose of farming. The soda solonchaks were washed with 1 per cent solution of sulfuric acid in order to obtain good results in a comparatively short period. The total mineralisation of water pumped from the test plot was equal to 1.4 g/1 while the content of CO_3, HCO_3 and Cl and other components were not unduly high. In order to completely wash a meter layer of soil, it was necessary to have 90-120t of sulphuric acid and 40-50 thousand m^3 of water per ha. In order to remove one ton of salts, 360-400 m^3 of water was required. During washing, roughly 15-20 per cent of salts were removed through horizontal drains and the rest by vertical drainage.

35

ANDERSON, G.; MURISON, J. Draining red volcanic soil for avocade culture. Banana Bull., 1975, 39:20-1.

Avocados grown in red volcanic soil on the east coast of Australia are liable to waterlogging and root rot (*Phytophthora cinnamoni*). The natural conditions may be improved by a soil drainage system which involves constructing mounds to facilitate surface drainage and installing underground drains where the subsoil water is close to the surface.

36

ANDERSEN, S.A. Experiments and experience with drainage methods and drainage materials in Denmark. Z. Kult. Flur. 1983, 24:223-37.

The effectiveness of different pipe and filter materials with trenchless and traditional drainage methods was investigated on various soil types. On ochreous soil types the effectiveness of trenchless and traditional drainage methods was similar. Use of a voluminous filter material showed no advantage on these soils. On non-ochreous, unstable soil types, however, the use of a voluminous filter meterial not liable to blockage with deposit materials gave much better results. It was necessary to develop a suitable, voluminous, factory-wound filter, or to change the drainage method, so that loose filter materials could be utilized under trenchless drainage.

37

ANDRITTZ, M. Daudiya drainage pumping station (Iraq). ICID Bull., 1965, 14:55-7.

The Daudiya drainage pumping station erected for the purpose of maintaining the ground water table at the proper level by pumping excess surface and sub-soil water from an irrigation drainage system into the river Tigris is described and its peculiar characteristics discussed.

38

ANDRIYAUSKAITE, E. Effect of drainage on hydrophysical properties of clayey and loamy soils. Pochvovedenie, 1961, 3: 89-95.

In the central Lithuanian plain, drainage lowered the ground water level, improved filtration, decreased soil moisture and increased aeration. The effects of the distance from the drains on these properties are tabulated.

39

ANJANEYULU, B.S.R. Drainage improvement for reclamation and preventing waterlogging in some soils of Rajasthan Canal areas. Proc. Symp. Waterlogging Causes Measures for Improv. New Delhi Cent. Bd. Irrig. Pwr. Pub., 1972, 118:43-8.

In the Anupgarh branch command area of Rajasthan canal, there are extensive areas affected by salinity and alkalinity, due to lack of surface and subsurface drainage. To improve subsurface drainage, vertical mulches (holes of 10 cm dia) of sand, organic and sand-organic combination with gypsum were made upto a depth of 2.5m. The maximum response to each treatment was observed when the vertical mulches of sand and organic matter were made # 100/ha and those of sand and organic combination # 50/ha. Gypsum addition to the combination increased the response. Soils were reclaimed within 6 months resulting in bumper crop of rice.

40

ANONYMOUS. Decade of research. CSSRI, Karnal, 1979, 186p.

The research work done at Central Soil Salinity Research Institute, Karnal during first ten years of its establishment, has been presented in different disciplines, viz. soil survey, soil physics, soil chemistry, soil microbiology, agronomy, genetics and plant bredding, plant physiology, drainage, water conveyance and application, hydrology, use of saline water in irrigation, extension education and production economics. A list of publications of the institute has also been appended.

41

ANTER, F.; OMER, M.; EL-HADY, O. The influence of water table levels on some chemical properties and plant growth. Egypt. J. Soil Sci., 1977, 17:169-82.

The fluctuation of the water table was measured during one year in plots of clay soils at the Agricultural Experimental Station, Shalakan, using piezometers. The imperfectly drained soil (annual mean depth of water table 72 cm) contained high levels of soluble salts, particularly in the surface layer. The well-drained soils(163 and 167 cm to water table) contained very low levels of soluble salts, while the moderately drained soil (114 cm) contained some salt accumulation, particularly in the surface layer and zone of groundwater fluctuation (110-120 cm). Yields of cotton and maize increased with increasing depth to the water table.

42

ANTHEUNISSE, J. Preliminary observations on the breakdown of coir in soil. Pl. Soil, 1979, 51:145-51.

The decomposition in soil of coco fibres, used as a wrapper of plastic drain pipes, was investigated. Unaffected and partly decayed fibres were washed and placed on cellulose-containing agar plates. Unaffected coir rarely showed microbial growth while around the partly decayed fibres fungi generally developed. A number of these were isolated and identified as cellulose-decomposing fungi. After 6-8 months aerobic incuba-tion, an increase in lignin and nitrogen content, and a decrease in sugars, were detected in partly decayed coir samples. A laboratory method is described for studying the breakdown of coir under conditions more or less equal to those in the field.

43

ARMSTRONG, A.C. The design of in-field drainage systems in the English Fenlands. In Polders of the world. Vol. 1 Polder projects. Land and Water Management Aspects. Int. Inst. Land Reclam. Improv. Wageningen, 1982, 433-42.

The current practices for the design of in-field drainage systems within England and Wales are presented, with special reference to the Fenlands. The Fens are entirely dependent upon arterial drainage for the maintenance of agriculture. Problems specifically mentioned include soils, outfalls, ochre, sub-irrigation needs, peat wastage, high value crops, and siltation.

44

ARMSTRONG A.C. A digest of drainage statistics. Tech. Rep. Field Drainage Experimental Unit, UK, 1978, 78/7, 73p.

Tables are presented which record tecnhical data relating to all underdrainage schemes in England and Wales. The tables report these data for the period 1.4.71 to 31.3.78. Examples of some of the uses of these data are presented.

45

ARMSTRONG, A.C. The effect of drainage treatments on cereal yields: results from experiments on clay lands. J.Agric. Sci. U.K, 1978, 91:229-35.

Crop yield results from 3 long-term drainage experiments in central England are analysed and show that grain yield increase of upto 1 t/ha of a winter wheat can be attributed to the effect of drainage. Mole drainage is clearly the preference treatment on heavy clay soils. There is also a suggestion that drainage permits more utliization of applied N.

46

ARMSTRONG, A.C. The interaction of drainage and the response of winter wheat to nitrogen fertilizers: some preliminary results. J. Agric. Sci. U.K., 1980, 95:229-31.

The yield data of wheat on drained (15, 30, 60m spacing) and undrained plots with secondary treatments of nitrogen levels indicated that drainage increased yields over the undrained plots. It is pointed out that the benefit of crop yield as a consequence of drainage can be largely attributed to the more efficient use of nitrogenous fertilizers. More research on this aspect is needed.

47

ARMSTRONG, A.C.;RYCROFT, D.W.; WELCH, D.J. Modelling waterable response to climatic inputs-its use in evaluating drainage designs in Britain. J. Agric. Engng. Res., 1980, 25:311-23.

The Van Schilfgaarde transient water table model is shown to be applicable to the drainage of heavy clay soils in the English context. The model is calibrated against a site in Leicestershire and then subjected to 10 years climatic record representative of contrasting areas in Bedfordshire and

Devon. In both cases, the clear superiority of mole drains over drains alone at 20 m spacing is demonstrated.

48.

ARMSTRONG, A.C.; SHAW, K.; WILCOCKSON, S.J. Field drainage and nitrogen leaching: some experimental results. J. Agric. Sci. Camb., 1983, 101:253-5.

It was observed that in the cockle Park drainage scheme area nitrogen loss and discharge rates are similar for drained and undrained soils. The authors emphasize that blame of high concentration of nitrate nitrogen on modern agricultural practices in general and land drainage in particular is not supported by the data obtained in this study area.

49

ARMSTRONG, A.C.; SMITH, Y. Soil series and drainage practices. Tech Bull Ministry of Agriculture, Fisheries and Food, UK, 1977, 77/5,, 28p.

The dominant drainage design practices for 13 of the more commonly occurring soil series in England and Wales have been identified from the computerised records of drainage grant applications, and the resulting contrasts and similarities between divisions examined. It was possible to identify those soils on which there is a uniformity of drainage treatment and those on which there is a wide variety of practices. It has thus been possible to identify research priorities where important soils have widely differing drainage treatments.

50

ASINGR, J. Economics of the use of plastics in drainage. Ved. Pr. Vyzk. Ustavu Melior. Zbraslavi., 1973, 13:89-100.

Labour productivity, when laying plastic pipe by

mechanical or manual methods, was greater than when laying clay tiles but the costs of drain construction were not reduced in proportion. Agricultural production and labour costs were similar on plots drained by clay tiles and plastic pipes.

51

ASINGR, J. Effectiveness of drainage construction by excavation with the TKBD-120 machine with adapter. UTVIZ-Melior., 1978, 14:35-43.

The effectiveness of drainage construction using the TBKD-120 machine is assessed by indices of productivity of labour and drainage construction costs. The basic data have been collected in the course of drainage construction with the TBKD-120 machine adpater (drainage plough for excavating trenches) and from data provided by enterprises owning the machine. The construction with TBKD-120 is considerably more effective than construction with machines commonly used (ETC-202 and GP-589-200). Labour productiviy increases by 100 to 130 percent and costs drop by 27 to 29 percent with the use of clay or flexible PVC pipes.

52

ASINGR,J. The effectiveness of drainage in mountainous and sub-mountainous regions. UVTIZ, Melior., 1980, 16:41-6.

Research results show that drainage costs in mountain and sub-mountain areas of Czechoslovakia range between kcs 15 503 and kcs 32 683 per ha and additional reclamation costs arising from drainage from kcs 2297 to kcs 23927. Drainage doubles or trebles crop yields per ha and doubles gross production. As prime production costs for most products are higher than farm prices the net effect is a loss or a very low profit. The indices of economic effectiveness, therefore,indicate that drainage is not economical in these regions, however high the increase in crop yields,may take place.

53

ASSEED, M.; KIRKHAM, D. Depth of barrier and water table fall in a tile drainage model. Soil Sci. Soc. Am. Proc., 1966, 30:292-8.

A laboratory model having a transparent photographable front and containing glass beads has been used for determining under idealized conditions, the influence of a barrier on tile drainage, in regard to both rate of fall of the water table and the measured discharge. Photographs of the water-table and the measured discharge as it varies with time are raw experimental data to derive charts of dimensionless drainage parameters. The results showed that changes in depth of a shallow porous medium markedly influenced drainage. A critical depth occurred when the barrier is at depth equal to one fifth of drain spacing; then the barrier had negligible influence on rate of fall of water table and on drain discharge. When steady rainfall (Glycerol drops) was applied. the water table arch at equilibrium was more curved for higher rates of rainfall. Graphs of rainfall intensity versus water-table height were straight lines both for a position over the tiles and midway between tiles. Data were compared with theoretical work, and a good agreement was obtained.

54

AVNIMELECH, Y.; DASBERG, S.; HARPAZ, A.; LEVIN, I. Prevention of nitrate leakage from the Hula Basin, Israel: a case study in watershed management. Soil Sci., 1978, 125: 233-9.

Drainage of the Hula Basin led to the exposure of about 2000 ha of organic soils. Corollary to a high subsidence rate, extensive nitrification and leaching of nitrates became evident, endangering the quality of the water in Lake Kinneret (Sea of Galilee). The problem is being solved by (a) improving the drainage system and reducing water flow through the basin; (b) maintaining a relatively high water table during the summer, thus minimizing the depth of the oxidized layer; (c) inducting denitrification through the use of sprinkler irrigation; and (d) selecting crops that reduce the accumulation of nitrates in the soil.

55

AWAN, N.M.; O'DONNELL, T. Moving water tables in tile-drained soils. J. Irrig. Drain. Div. ASCE., 1972, 98:459-77.

Results of a Hele-Shaw model of non-steady tile drainage system for falling water table and for the case of rising water table have been analysed and reported. The results are converted to standard dimensionless forms and compared with several analytical and numerical methods reduced to the same dimensionless forms. Drainage characteristics in both the cases have been found to be independent to the depth of the impermeable bed below the drain axis once that depth is 1/8 or more of the spacing between drains.

56

AYAZI, M. Drainage and reclamation problems in the Garmsar area. Salinity Probl. Arid Zones, Proc. Teheran Symp.(1958), 1961, 285-90.

Investigations are carried out on drainage problems in the field in desert areas, comprising 1,100 ha. Salt content of the soil to a depth of 3 m, ranged from 0.45 to 3.0 per cent, which decreased with depth. Four investigation lines were established and piezometers placed on the lines at a depth of 2 m and at intervals of 300 m. Auger holes were dug besides each piezometer to a depth of 3 m, to assess the soil profiles. The Kanats which locally work as underground tiles were installed for the under-ground drainage. In the district, the cost of land which has long been reclaimed is about 5-8 times higher than that of land reclaimed recently.

57

AYRES, Q.C.; DANIELS SCOATES, A.E. Land Drainage and Reclamation. McGraw-Hill Book Company Inc. New York. 1969, 496 p.

The book deals with surveying, drainage, reclamation and soil conservation planning in agricultural lands. The book is divided into 31 chapters out of which 5 chapters deal with surveying, 19 on drainage and reclamation and 7 on soil conservation and general aspects including introduction. The book can serve as a good text book to undergraduate courses.

58

BAGLEY, G.R. Drainage for increased crop production and a quality environment. 3rd Natn. Drain. Symp. ASAE, Michigan, 1976, 3-5, 13.

The role of drainage in agricultural production is outlined and its benefits enumerated. The potential for drainage to protect land from soil erosion and water pollution is discussed and the hazard of eliminating all wet-lands is placed in perspective. Present drainage research needs are also outlined and described.

59

BAIER, W.; ROBERTSON, G.W. Climatic estimates of average probable irrigation requirements and of seasonal drainage in Canada. J. Hydrol., 1970, 10:20-37.

This paper demonstrates the use of published information on average and probable irrigation requirements which were esti-mated by means of a meteorological budgeting technique from 30- year daily climatic records. A seasonal drainage estimator model requiring as input only monthly values of precipitation and potential evapotranspiration, was developed from the date for 42 stations and verified on the data for another 17 stations. An irrigation water balance is presented for 11 stations across Canada.

60

BAILEY, A.D. Drainage of clay soils in England and Wales. Proc. Int. Drain. Workshop, Wageningen, Netherlands, (1978). 1979, 25:220-42.

There is an estimated drainage need on 2.6 m. ha of agricultural land in England and Wales. Of the work currently being carried out 75 per cent is on clayey soils. In the clay soils of low conductivity, it is concluded that the only effective way of draining the subsoil is by secondary treatments of mole drainage or subsoiling over permanent pipe systems. These two operations are defined, and the conditions under which they are likely to be successful discussed. In terms of current drainage design practices on these soils there is a large traditional factors, related to the particular part of the country. A more scientific approach is advocated, but due to the fact that most schemes are small (7 ha), there is an economic limit to the amount of pre-design survey that can be reasonably carried out. The options available are discussed, and a design philosophy suggested.

61

BAILEY, A.D. A review of drainage machinery in England and Wales. Soil Water, 1978, 6:3-4, 6.

Drainage problems in the UK related to watertable, impermeable subsoils and springs, types of machinery available, the distribution of drainage problems and machinery in the 8 regions of England and Wales, machine capability in relation to scheme size and trench hazards, and machine and drainage costs are discussed.

62

BAILEY, A.D.; TRAFFORD, B.D. Drainage construction techniques in England and Wales. 10th Congr. Irrig. Drain. Athens, 1978, 5:119-36.

Most of the drainage work in England and Wales is in the drier east and is concerned with intensive arable farming. The most common drainage problem is an impermeable clay subsoil. Details of the wide range of drainage machines available are given. Most work is carried out by continuous trenchers but backater excavators are used in hilly areas or rocky soils and the trencheless method is becoming well established. The advantages and disadvantages of all machine types are discussed. Operator skill is one of the most important factor Capital and working costs for drainage machines are given. The critical factors in unit cost/m are the costs of pipes and gravel backfill. The major research needs are identified as evaluation of trencheless machines· application of vibrating techniques to draught reduction; and techniques for reducing the use of gravel backfill and improving its handling.

63

BAILEY, A.D.; TRAFFORD, B.D. Latest subsurface drainage techniques in England and Wales. 10th Congr. Irrig. Drain Athens, 1978, 4:353-67.

Most of the drainage work in England and Wales is in the drier east and is concerned with intensive arable farming. The most common problem is an impermeable clay subsoil. Plastic pipes account for less than 20 per cent of work because clayware pipes are readily available at low prices. Filters are not widely used largely because alluvial fine sand soils are not widespread Gravel or similar coarse material is used as backfill in about 60 per cent of all work, mainly to make a hydraulic connection between mole drains and permanent drains. Mole drainage is important for heavy clay soils of low hydraulic conductivity.

Typically present drains with gravel backfill are laid at 40-60 m spacing with the moles drawn over these drains at right angles and at spacing of 2-3 m. Recent work on soil fissuring patterns with mole drainage or subsoilers suggests that effective improvement in subsoil permeability at higher depths is difficult to achieve.

64

BAINS, S.S. Possibility of using rain water for salinity control through surface drainage on ditch irrigated ridged soils. Proc.Symp. Wat.Mgmt. Udaipur, (1966), 1968, 243-6.

Effect of rain water has been studied on salinity of ridge slopes under alternate ditch irrigation system on a saline soil with EC_e of 8.0 dS/m in the surface layer. Alternate ditch irrigation has been found to be a very effective system of irrigation to flank salts to the sides. Rain water has been shown to wash the salts from the unirrigated slopes thereby suggesting the possiblility of using unirrigated ditches as surface drains during monsoon season.

65

BAKER, J.L.; CAMPBELL, K.L.; JOHNSON, H.P.; HANWAY, J.J. Nitrate, Phosphorus, and sulfate in subsurface drainage water. J. Environ Qual., 1975, 4:406-12.

Annual losses of phosphorus, SO_4-S, and NO_3-N from tile-drained cropland ranged from 0 to 0.04, 0 to 32, and 0 to 93 kg/ha, respectively, being very dependent on the amount of water lost. Because of low concentrations of phosphorus, losses in subsurface drainage water were insignificant compared with losses in surface runoff. Concentrations of SO_4-S and NO_3 were seemingly inversely related. Tile drainage water with consistently high NO_3-N content relative to surface runoff ($>$10 ppm even under the low-fertility management of this study, 224 kg/ha of N over 5 years) is believed to be responsible for the high NO_3-N content sometimes found in a river draining central Iowa.

66

BAKER, J.L.; JOHNSON, H.P. Impact of subsurface drainage on water quality. 3rd Natn.Drain Symp. ASAE, Michigan, 1976, 91-8.

Surface drainage of wet areas alters the time and route by which excess precipitation reaches surface waters. This influences both quantity and quality of runoff and drainage water. The paper attempts to review the effect of subsurface drainage on the quality of water draining from cropland, primarily with respect to the inorganic ions namely NH^{+}_{4}, NO^{-}_{3} and PO^{-3}_{4} Pesticides and sediment are con⁻idered. Qualitative and quantitative data for surface runoff, tile drainage, and base flow are compared to explain quality patterns of stream draining primarily agricultural land.

67

BAKER, J.L.; JOHNSON, H.P. Nitrate-nitrogen in tile drainage as affected by fertilization. J. Environ. QuaL., 1981, 10:519-22.

The volumes and NO_3-N contents of tile drainage from two plots cropped with maize in rotation with either oats or soybeans during 1974-1978 were measured to determine the effect of differential N fertilization on NO_3-N leaching losses. The plots were fertilized with N at rates of 100 or 250 kg/ha in 1974 and 90 or 240 kg/ha in 1976. In 1978, both plots were fertilized with 90 kg N/ha. The NO_3-N contents of tile drainage from these plots had been established in a previous 4-year study in which both [illegible] received 112 kg N/ha in 1970 and again in 1972. The ratios of NO_3-N concentrations in daily samples from the two plots before differential fertilization were constant and near unity. After differential fertilization, concentrations for the higher fertility plot exceeded those for the lower fertility plot for extended periods by a factor of two after the 1974 fertilization, and by a factor of four after the 1976 fertilization. In 1974, there was a delay of about 2 months (100 mm of flow) before the effect of surface fertilization was observed in the tile drainage; a similar delay was observed in 1976. Although relative NO_3-N concentrations for the higher fertility plots eventually decreased with time after the last differential fertilization, crop samples taken 3 years later still showed the effect of the higher level of fertilization. Overall, concentrations and losses for the plot receiving 90-100 kg N/ha every-other year averaged 20 Mg/ml and 27 kg/ha; for the plot receiving the most fertilizer, the values were 40 Mg/ml and 48 kg/ha.

68

BAKHATI, H.K.;KEWAISY, H.A.; ADBEL-MALIK, S.H; ELDIN, F.E.S. Evaluation of tile drainage at Mehaier Mousa farm. 3, cotton yield as affected by different treatments of field tile drainage. Agric. Res. Rev., 1975, 52:13-9.

In field trials in 1971, cotton cv. Giza 69 was sown at 3 sites on land drained either with concrete or PVC drains each spaced in lines 10, 30 or 40 m apart at a depth of 1.35 m. Water table levels were monitored throughout the growing season. On the least favourable site, seed-cotton yields were 4.16 and 2.49 kantar/ac higher for the 10 and 40 m drain spacings, respectively and on the most favourable site 6.57 and 6.17 kantar/ac higher than for undrained control crops. Increases in drain spacings resulted in delay in cotton maturation; a drain spacing of 20 m was therefore recommended. There was no clear difference between concrete and PVC drains in their effect on crop yield.

69

BAKHATI, H.K.; KEWAISY, H.A.; EL-MALIK, S.H.A.; EL-TOHAMY, M.F.; EL-ABASSERY, M.A.; EL-SHABASSY, A.I.; MITKEES, A.I. Evaluation of tile drainage at Mehallet Mousa Farm. Influence of tile drainage on the fluctuations of water-table level and clover yield. <u>Agric. Res. Rev.</u>, 1974, 52:29-44.

In field trials in Egypt during 1971-2, field drainage more than doubled the yield of Egyptian clover (*Trifolium alexandrinum*) cv. Miskawi which yielded about 12.5 DM/ha cut before drainage. Results obtained from the 2nd and 3rd cuts showed that the type of tile (PVC or concrete) and the tile spacing had no significant effect on yield of clover.

70

BAKKER, A.C. Disposal of drainage effluent in the Sheppartion region. <u>In</u> Hydrogeology of the Riverine Plain of South East Australia. Proc. Symp. Griffith, Australia, 1977 (Storrier, R.R., Kelly, I.D.Ed.). Austr. Soc. Soil Sci. Inc. 1978, 247-58.

A case of the re-cycling of saline drainage effluent is discussed. Some results are presented which show plant response to salinity of irrigation water, and to the inter-action of salinity and waterlogging. The importance of lowering the groundwater level for salinity control in the Goulbrun Valley is discussed with emphasis on the heavy soils, where pumping from aquifers is the only succesful method of drainage. The resulting effluent can only partly be discharged into the drainage system but this is permissible under strictly defined limits at the end of the irrigation season. The only alternative to this is an external disposal system but this is not anticipated in the foreseeable future. The effect of the resulting increased salt-load in the irrigation water on the plant and soil has been examined. Related aspects which need further investigation and careful monitoring are particularly those which affect the long term use of the disposal method such as maintenance and soil permeability and salt balance.

71

BAKR, H.M.A.; ELATTAR, H.A. Leaching and drainage practices in some soils of Egypt. Int. Symp. New Develop. Field Salt Affected Soils, Cairo, (1972), 1975, 553-60.

Six field experiments were carried out on saline soils in Abid (HC 0-100 cm, 24.5 cm/day), Gharb Tira and Gharb Elmansour (HC 0.2 cm/day, 60-100 cm) land reclamation projects. The soils represent some lactustrine deposits of the northern Nile Delta. Intermittent leaching saved almost half the amounts of leaching water per unit EC compared to continuous leaching. Ponding water for 4 days increased about 90 percent of soluble salts in the preleached surface layers and 3 to 45 per cent in the medium saline layers. The amount of leaching water to decrease EC by 1 dS/m decreased from 667 to 248 mm in unploughed soil to soil ploughed and subsoiled. The addition of gypsum reduced amount of leaching water from 417 to 231 mm in plowed soil and from 248 to 94 mm in soil plowed and subsoiled. Constructed moles at 40 and 65 cm depths were more effective than surface plowing and 40 to 65 cm depths of subsoiling. Also, the moles were more effective at their lower depths. Sand filled moles were more effective than the unfilled moles. The leaching water-requirement was proportional to the salt content but inversely correlated with the soil hydraulic conductivity.

72

BLASHEV, L.S. Indication of the intensity of drainage of bogs from the plant cover. Sov. J.Ecol., 1961, 11:154-9.

A method is presented for determining the intensity of drainage of bogs from drainage indices, which are determined for individual plant species as a ratio of the frequency of occurrence on drained bogs to that on undrained bogs of the same type. The indices were determined for a large number of grassy and grassy-mossy lowland bogs in the Ukraine. The possibility of determining the intensity of drainage of the whole bog mass from the same objective indicators is demonstrated, as well as comparing the ecological series of communities which form on drained bogs.

73

BALI, Y.P. Some drainage problems and drainage reclamation practices in India. Proc. 6th NESA Irri. Pract Sem. Jordan, 1966, 165-9.

The main causes of drainage problems and accompanying hazard of waterlogging, salinity and alkalinity are frequent flodding, seepage from irrigation canals and insufficiency of drainage facilities. The monsoon type of rainfall further causes seasonal fluctuations in the groundwater table. The pilot drainage reclamation projects for farm drainage at Ranbirpura (Punjab) and coastal saline lands at Umrath (Gujarat) have been outlined. These projects would investigate drainage and reclamation through open and tile drains with and without gypsum.

74

BARKER, M.G. A drainage investigation on a clay soil. J. Roy. Agri. Soc., 1963, 124:50-9.

Two areas were tile-drained at different intensities by different spacing of drain lines. A third area was undrained. Yields of winter wheat from the three areas for 1953-1962 showed the benefits of drainage on heavy land, but gave no indication of the effect of drainage intensity. Inadequate drainage affected crop growth seriously in late winter and early spring. The undrained area, after several years, showed no evidence of permanent adverse effect on the ability of soil to produce crops.

75

BARNI, N.A. Response of soybean (*Glycine max* (L) Merrill) to flooding during the reproductive period. *Agron. Sulriog*, 1978, 14:215-25.

In a field trial in 1973-4 soybean cv. Santa Rosa were flooded to a depth of 10 cm for 10 or 20 days beginning at early flowering, 10 days later or at early pod filling or 10 days later. Flooding during flowering decreased yields to a greater extent than flooding during pod filling. Flooding decreased numbers of pods and seeds/plant and 100-seed wt. Flooding during flowering increased the height of the first pod and if flooded for 20 days caused stems to become spongy.

76

BAROZZ, D.L. Maintenance of the surface drainage network in Italy. *7th Congr.Irrig. Drain. Mexico*, 1969, 5:321-32.

The problems of the mechanisation of maintenance operations for silt clearance, weed eradication and of chemical weed eradication are reported for irrigation and drainage networks. Experiments to find out the most optimum time for cleaning operations are elaborated.

77

BAS, M.S. Agricultural utilization potential of the swamp area in the Ege University Agricultural Research Farm. Ege Univ. Ziraat Fakult. Dergisi, 1979, 16:149-52.

The swamp area investigated consisted of 15,3250 ha most of which was inundated during winter. Drainage requirements for reclamation and leaching requirements to keep the soil salinity at 4 dS/m are discussed. As an alternative to this, the area could be used partly for field fish production and partly for production of pasture and field crops. In the later case, a drainage system to protect the crops from the high water tables in winter is necessary.

78

BAUMANN, H.; DIEDERICHSEN, H.; MANN, G.; SCHLEICH, C. Effect of drain projects on the water yield of the inland-Eider-basin in Schleswing-Holstein. Kult Flurb., 1980, 21:279-87.

In the inland-Eider-basin (693 km^2), which is completely artificially drained, the discharge morphology as a result of channel improvement and drainage in the period 1965-1967 was investigated. A comparison of monthly means and frequencies of discharges has established that the frequency of peak water flow has not been increased significantly. Channel improvement and drainage have in fact equalized discharge in this boggy bottom land.

79

BAZIN, F. Ground water level control in drained land. 5th Congr. Irrig. Drain. Tokyo, 1963, Q17 R 12, 193-7.

Draining lands likely to undergo settlement or the deterioration of soil physical properties or fertility may need close check on water-table. Mobile weirs at places where level fluctuation is permissible or some degree of temporary submersion is allowed may be useful. Constant level gates are recommended for areas where more or less constant water levels are required.

80

BECK, L.A. Agricultural drainage water reuse potential. Water Management for Irrigation and Drainage. Proc. Spec. Conf. ASCE Irrig. Drain.Div., ASCE, New York, 1977, 2:45-50.

Drainage problem of San Joaquin Valley in California (USA) is discussed in relation to the disposal of drainage water. Several alternative arrangements for use of agricultural drainage waters have been made including cooling of fossil fuel or nuclear power plants through these waters.

81

BEDIENT, P.B.; HUBER, W.C.; HEANEY, J.P. Drainage density as index of watershed development. J.Irrig. Drain. Div. ASCE, 1978, 104:373-87.

Studies have been completed on watershed in Florida, and a methodology developed to predict environmental impacts using drainage characteristics of developing watersheds. The drainage density index, a measure of total length of waterways observed on aerial photographs per unit area, was computed for various watersheds in order to relate land-use changes and drainage patterns to hydrologic and water quality responses. Results indicate significant increase in drainage density over the past 19 yrs in the Kissimmee River Basin. Water quality concentrations and loading rates of total P from 7 watersheds show a significant correlation with the drainage density index. Differences in available nonpoint source loads were found to reflect only a small part of the observed differences in measured total P loads.

82

BEER, C.E.; JOHNSON, H.P.; SHRADER, W.D. Yield response of corn in a planosol soil to subsurface drainage with variable tile spacings. Res. Bull. Iowa agri. Exp. Stn., 1965, 540:563-74.

In claypan soils of southern Iowa, yields with 4.5 m tile spacing were 6 and 4 bu/acre higher than with 18 m spacing for rotation and continuous maize respectively. The average advantage of the 18 m tile spacing over untiled plots was 13 bu/acre on rotation and 1 bu/acre on continuous maize. The nature of tile discharge for the 1951-1963 growing seasons is presented. (1 bu/ac = 50.44 kg/ha).

83

BEERS, W.F.J. VAN. Computing drain spacings. Bull Int. Inst. Land Reclam. Improve. Wageningen, Netherlands, 1976, 15, 47p.

A generalized method has been presented with special reference to sensitivity analysis and geohydrological investigations. The latest developments that have taken place in the Netherlands on the subject of computing drain spacings using drainage equations, based on the assumption of steady-state conditions are summarized.

84

BEERS, W.F.J. VAN. Some nomographs for the calculation of drains spacings. Int. Instt. Land Recl. Improv., Wageningen. Netherlands, 1965, 8,48p.

The paper presents nomographs for drain spacings based on Hooghoudt and Ernst's formulae; develops nomograph for homogenous soil with impermeable layer at great depth ($D > 1/4\ L$) and a note book method based on modified formula of Ernst.

85

BELGAUMI, M.I.; RAMAJAH, R.:ACHAR, H.P. Effects of circular openings of tile on the efficiency of tile drainage system. J. Agril. Engng., 1981, 10:9-12.

A field study was conducted to evaluate the performance of clay tiles with circular openings used in Malaprabha Command Area. There were seven treatments of different circular openings of 1 cm dia whose number varied from 5 to 60 and four depths of irrigation water. The results indicate that the tile outflow increased due to increased number of openings. But the relative increase in the tile outflow and the bearing strength of tiles decreased significantly with an increase in the number of openings of the tiles. The load bearing strength decreased from 10.83 kg/linear cm in case of 6 openings to 5.83 kg/linear cm with 60 openings.

86

BELLIN, K. Accuracy of drain pipe installation-experiment, principles, requirements. Wasser Boden, 1979,31:10-4.

Accurate installation of a ~iped drain system is essential for its prologned satisfactory functioning. High-speed installation machinery requires careful monitoring of pipe laying accuracy, in order to assess machine performance. Up to now, the tolerance acceptable in control of pipe installation has not been known. A code of conditions for assessing installa-tion accuracy and recommended procedures are presented.

87

BELLIN, U. Accuracy of pipe laying-experiments, principles, demands. 10th Congr. Irrig. Drain. Athens,1978, 5:137-46.

The need for accurate drain installation to ensure prolonged adequate functioning, hydraulic suppositions for drainage discharge, practical aspects of controlling and checking the accuracy of pipe laying, sources of error in positioning, and the degree of accuracy which is reasonable to demand in practice are discussed.

88

BELTRAN, J.M. Study of drainage and filter materials in saline silty-clay soils of the Bardenas area, Spain. 10th Congr. Irrig.Drain. Athens, 1978, 4:527-41.

A ten-ha experimental field was established in 1974 in the Bardenas irrigated area. Using clay pipes with and without gravel filter and corrugated PVC pipes without filter and with gravel, barely straw, esparto fibre and coco fibre as filters, the entrance resistance to water flow and equivalent drain spacings were determined.

89

BENETIN, J.; HUSKA, D.; KOLAR, A. Calculation of the operational function for drainage regulation. 10th Congr. Irrig. Drain. Athens, 1978, 6:215-25.

In lowland regions, where drainage and supplementary irrigations are alternate, and where hydrological conditions and field configuration are suitable, the soil water regime can be governed by drainage regulation. The conditions necessary for economic drainage regulations are discussed.

90

BENGSTON, R.L.;CARTER, C.E.; MORBIS, H.F.; KOWALCZUK, J.G. Subsurface drainage effectiveness on alluvial soils. Trans. ASAE., 1983, 26:423-5.

A sub-surface drainage-runoff-erosion experiment on clay loam soil near Baton Rouge, Louisiana was conducted on four plots, of which two were subsurface drained. Corn silage was grown for two years (1980, 1981). Average yields from surface and subsurface drained plots were 82 per cent higher than those from the only surface drained plots. Surface runoff and soil loss were decreased by 16.9 per cent and 17.8 per cent respectively, through subsurface drainage.

91

BENNETT, D.R.; WEBSTER, G.R.; PATERSON, B.A.; HARKER, D.E. Drainage of an irrigated saline soil in Alberta. Can. J. Soil. Sci., 1982, 62:387-96.

A shallow subsurface drainage system effectively controlled a high water table and reduced salinity in an irrigated soil near Magrath, Alberta. Plastic corrugated tubing was installed in 1976 at depths of 1.1-1.5 m and spacing of 15 and

30 m in a moderately saline soil. During the irrigation period, the water table rose to within 0.3 m of the surface but was lowered to pre-irrigation levels within 48 hr. The watertable was maintained at, or below, the depth of the drains between irrigations. The 15- and 30-m spacings of the drain lines were effective in providing water table control in this lacustrine soil which was underlain by a coarse sand and gravel layer. Salinity levels were decreased substantially only within the surface 0.3 m soil depth. Quality of the drainage effluent remained constant throughout the growing season with only small dilution effects detected during irrigations. Barley yields increased to 3900 kg/ha in 1978, 2 yr following drainage of this saline soil which had been out of crop production for 20 yr.

92

BENOIT, G.R.; BORNSTEIN, J. Considerations for effective sloping land drainage systems. <u>Soil Sci. Soc. Am. Proc.</u>, 1972, 36:819-23.

Water balance equations for two spring and four autumn periods of 10-40 days were prepared for drainage systems consisting of surface drains 30 or 51 cm deep and 61 or 122 cm apart, with and without subsurface drainage spaced 61 or 30.5 cm apart at 102 cm depth in a poorly-drained glacial till fragipan soil with a 6 per cent slope. There was great variation in drain performance between drains during the same drainage period and between drainage periods for the same drain. Differences in performance were related to initial soil water content, for gain or loss of soil water, the magnitude of potential evapotranspiration and drain location with respect to the depth of soil above the bedrock. Greatest drainflow occurred from surface drainage located where the bedrock was closest to the soil surface but any drainage systems for wet sloping soil should also include other subsurface drains to intercept bypass flow before it reaches downslope soil surface areas.

93

BENOIT, G.R.; FISHER, K.D.; BORNSTEIN, J. Alfalfa survival-indicator of sloping land drainage effectiveness. Agron. J. , 1967, 59:444-7.

Plant-cutting data of lucerne for 1964-66 and plant counts for 1966 were taken from a 12- plot surface and sub-surface drainage experiment on silt loam typical of many sloping, hardpan, wet soils of the Northeast. Drainage effectiveness trends were established as early as the first cutting. Percent lucerne (dry weight basis) and plant-count data each gave a good indication of drainage effectiveness. Percent winterkill was significantly correlated with both soil and water content. Drainage of excess water to the content present at 0.1 bar of tension ensured negligible (10 per cent or less) winterkill.

94

BENYAMINOVICH, E.M.; SHEDER, V.R.; KOROTKOV, P.A.; OKULICH-KAZARIN, E. Irrigation and reclamation of Golodnaya steppe. Trans. 6th Congr. Irrig. Drain., New Delhi, 1966, Q.19. 28:493-505.

The paper describes the gradual and systematic reclamation development of the Golodnaya steppe, a long desert plateau situated on the left bank of Syr-Darya River in Uzbekistan; deals with the layout of the irrigation system in the area and the horizontal subsurface drainage layout for salinization control and leaching of saline soils and reviews the working of the maintenance and land reclamation services which achieved, by 1965, the reclamation of more than 300,000 ha of old and virgin lands and of more than 430,000 tonnes of cotton picking.

95

BENZ, L.C.; DOERING, E.J.; REICHMAN, G.A. Watertable management saves water and energy. Trans. ASAE, 1981, 24:995-1001.

Field-plot and lysimeter experiments on Hecla-Arveson-Fossum Association soils (sandy loams to loamy sands) show that relatively small increases in watertable depth, beyond the optimum depth of about 1.3 m, increased the irrigation requirement and thus energy use for maximum corn and sugarbeet production. Also, watertable depths shallower than optimum decreased yields. Lowering the watertable 0.2 m and 0.3 m for maximum corn dry matter and corn grain yields, respectively, increased the irrigation required by 26 cm. Similarly, irrigation requirement was increased by 32 cm for maximum production of sugarbeet root and sucrose yields when water table depth was lowered, 0.4 and 0.5 m, respectively. Thus the watertable is a valuable agricultural resource that should be maintained at the optimum depth to save both water and energy.

96

BENZ, L.C.; DOERING, E.J.; REICHMAN, G.A.; FOLLETT, R.F. Evaluation of some subsurface drainage envelopes. 3rd Natn. Drain Symp. ASAE, Michigen, 1976, 31-3.

A fabricated laboratory apparatus is described for use in testing envelope materials used with buried plastic pipe drains. Some modifications are planned for the soil loading and water-flow into the 'plastic-pipe' aspects of the experimental apparatus. Short-term flow tests showed that all envelopes tested had acceptable flow rates, ranging from 8.9 liters/hr (2.1 gal/hr) to 16.8 liters/hr and would retain the aquifer, but gravel had a slightly higher flow rate and occasionally would not regain the aquifer. Long-term tests showed that flow rates decreased with time. Flow rates into a 0.30 length of 10.2 cm diameter plastic pipe in the field, assuming full pipe flow and conditions similar to those of the laboratory test cylinder, are 10 times greater than those obtained in the test cylinder.

97

BENZ, L.C.; REICHMAN, G.A.; DOERING, E.J. Drainage requirements for alfalfa grown on sandy soils. Trans. ASAE, 1983, 26:161-64,166.

Alfalfa hay yields were evaluated as functions of water-table depth and irrigation level for a sandy soil in field plots and field lysimeters in south eastern North Dakota. The yields were normalized on an annual basis, a 4- plot year basis, and then the data from both experiments were combined to provide information for a wider range of watertable depths and irrigation levels. The results show that optimum depth to water table was 1.5 m and irrigation was unnecessary when the water table is at this depth. For alfalfa production significant energy and cost savings, estimated at $ 57/ha annually can result by omitting irrigation which can be done when a shallow 1.5 m watertable is present.

98

BERGEDALEN, J. Field trials with drainage materials. Meld. Norges Landbruk., 1977, 50, 12p.

Soil types, filters, perforation and slope of pipes were investigated at 3 sites. 2m^3 of sawdust per 100 m drain used as a filter material prevented silting at 2 sites. Where soil and slope were mor^ critical, a sawdust filter of 5 cm thickness protected against silting. 2 cm thickness of glasswool was too rigid to wrap the pipes and cracks appeared between filter and pipe, causing silting.

99

BERRY, W.L.; STETSON, E.D. Drainage problem of the San Joaquin Valley. Trans. ASCE, 1961, 126:1-11.

The paper outlines the geography and geology of the San Joaquin Valley of California with special emphasis on the factors affecting drainage; unfolds the existing drainage problems (1959) through a description of the cultural changes that have brought them into sharp focus and describes the objectives and scope of the drainage investigations in progress.

100

BERRYMAN, C. Improved production from drained grassland. Tech Bull., Ministry of Agriculture, Fisheries and Food, Field Drainage Experimental Unit, UK, 1975, 75/7, 14p.

Soil drainage in grass production is discussed with reference to poaching of grassland and the effect of soil drainage on crop N supply. Effects of drainage on sward composition are considered and effects of short-term flooding and constant water table depths on herbage yields are discussed.

101

BESEDNOV, N.A. Washing and drainage of heavy solonetz-like solonchak soils. Tr. Pochv. Inst. Dokuchaeva, 1957, 52:123-219.

The soils dominant in sulphate and having low infiltration capacity, should be washed with small amounts of water during hot weather and in the presence of a fine closed drainage network. After washing, the soil should be plowed down the slope. Winter barley should then be sown. After harvesting, the soil sould be washed again in summer and autumn and then cultivated by methods suitable for heavy soils.

102

BEYE,G. Assessment of 5 years of study on the desalinization of the Medina polder soils(Lower Casamance).Agron. Trop. 1975, 30:251-63.

Cycle processes of desalinization and resalinization, with overall loss of salt, were identified with all drainage methods(gravity, pumping station with ditches or tile drains). Desalinization was due mainly to rainfall, resalinization occurred during the dry season. Drainage by pumping to reduce the groundwater depth to 1.3 m was the most effective method.

103

BHATTACHARYA, A.K.; BROUGHTON,R.S. Optimization of subsurface drainage system. J. Agric. Engng., 15:6-10.

Waterlogging condition in farm land causes reduction in yield and loss of available working days. Sub-surface drainage is one of the most effective methods of water table control in the farm. A design of subsurface drainage system involves determination of appropriate depth and spacing of the tile drains. Apart from being hydraulically efficient, the system should also be economical. Larger depths increase cost of trenching but reduce the total length of tile requirement which reduces the cost. In this project an analysis has been done with a view to obtain a combinatioı of depth and spacing which, apart from being effi· cient, will also minimize the cost of installa tion.

104

BHATTACHARYA, A.K.; BROUGHTON, R.S. Spacing drains to maximize revenue increases.Trans.ASAE,1979, 22:351-6.

Procedures were developed to compute maize crop loss for different depths and durations of high groundwater level,and to calculate the revenue increases from subsurface drainage system designed with different spacings in soil with various hydraulic conductivities. Seventy-six years of weather data were used for the analysis. With no subsurface drainage,, the crop loss was found to be 50 per cent of the normal production. The spacing required to obtain maxmimum revenue increases was found to be sensitive to soil hydraulic conductivity, but insensitive to interest rate, amortization period and installation cost.

105

BHATTACHARYA, A.K.; BROUGHTON, R.S. Variable drainable porosity in drainage design. J. Irrig. Drain. Div. ASCE, 1979, 105:71-85.

Following field and laboratory investigations into the dependence of drainable porosity on suction in two soil types, empirical nonlinear relationships were found between them. These relationships were different for sand and clay soils. The relationships were integrated between various limits to obtain the volume of water drained from saturated soil columns for different watertable suctions. The results obtained were used to derive water table depth and transient water content relationships for soil columns of known lengths. These relationships can be used to predict water table depth from the known value of transient water content in a soil column, in the water balance approach to subsurface drainage design. The effect of depth of installation of subsurface drains on soil moisture depletion in coarser soils is examined.

106

BHATTACHARYA, A.K.; FAROUD, N; CHIENG, S.T.; BROUGHTON, R.S. Subsurface drainage cost and hydrologic model. J. Irrig. Drain. Div. ASCE 1977, 103:299-308.

A computer simulation water balance model was used to analyze the variation of daily water-table depths in the field using 27 yr weather data and soil water parameters from the Macdonald College Farm. The high water table in the study area resulted from a cool moist climate and poor natural drainage. The effects of drainage rates and drain depths on the water table were studied. A frequency analysis was done to obtain the recurrence intervals of specific water-table depths for various successive numbers of days. The net financial benefit from the system was evaluated corresponding to various chances of crop failure due to inadequacy of the drainage system during its assumed amortization period.

107

BHATTACHARYA, A.K.;SARKAR, T.K. Proposed combination of surface and sub-surface drainage system for Madanpur, Khadar, Asola and Ali Areas. Symp. Waterlogging Causes and Measures for its Prevention, CBIP Pub. No. 118, 1972, 1:138-42.

The existing drainage problems of Madanpur, Khadar, Asola and Ali areas in Delhi have been studied.The causes for the waterlogging of this one time cultivable area have been investigated. It is suggested that as a remedial measure, both surface and sub-surface drainage systems are to be installed. The selection of drainage coefffi-cient, on the basis of which the design of either system of drainage will depend has been discussed in detail and procedure for its selection has been given. Procedure for design of the sub-surface and surface drainage system has also been given. It is based on the selected value of the drainage coefficient.

108

BIANCHI, W.C.; HASKELL,, E.E Jr.; JOHNSTON, W.R. A soil drainage study in southwestern Fresno County. Hilgardia,1962, 33:171-8.

A new drainage problem is developing with the rise of a semiperched water table into the crop-root zone. The water is perched on an extensive layer of heavy textured sediments 30 m below ground surface. This perched water is part of a groundwater ridge extending 15.8 km southwards and maintained by deep percolation from irrigation although irrigation is greatly limited by dwindling groundwater supplies. The soluble salt content of this water has increased the salinity of the surface soils where the water table has reached the root zone.

109

BIGGAR, J.W.; COREY, R.B. Agricultural drainage and eutrophication. Agric. Sci. Rev., 1967, 5:422-8.

Factors affecting the transport of soil nutrients (eg. soluble N and P) into lakes and streams from rural areas, and excessive fertilization of waters with these nutrients(eutrophication), are discussed.

110

BIRD,N.A.; MCCORQUODALE,J.N. Computer simulation of tile systems. Trans. ASAE, 1971, 14:175-9.

A mathematical model is developed that would predict the performance of tile system by predicting the tile runoff hydrograph. The model accuracy was limited primarily by the accuracy of the soil moisture model used. By using many years of climatic data from the area many years of predicted water table heights and tile runoff hydrographs could be generated for statistical analysis. The performance of a drainage system for Brookston clay had been tested using the model and found comparable to the observed data.

111

BIRJARH, S.R.; RAO, B.N. Water-table profile and their importance in designing the tile-drainage system.J. Res. PAU, 1975, 12:317-26.

The water table profiles at two sites with tile drainage show that initial and boundary conditions should be considered in tile-spacing equations for sites where pumping ensures the free flow of water.

112

BISHAY, B.G.; DIERICKX, W.; BOODT, M.F.D. Conditioned sandy soils as an envelope to prevent silting up in drain tubes. J. Agric. Engng. Res., 1975, 20, 139-48.

A sand tank model experiment under ponded water conditions was carried out in order to compare the effectiveness of three filters. Flax straw of 1 cm thickness, or conditioned sand with bituman emulsion or conditioned sand with PAM solution as envelopes of 4 cm thickness was put around the drain tube. All of these treatments have prevented the silting-up completely. After 24 hr, when conditioned soil is used as a filter around the drain tube the water discharge was 1.3 times more than from the drain without filter but 1.16 times less than using flax straw.

113

BISHAY, B. G.;MOUKHTAR, M.M; ELHAMSHARY, S.; DAYEM, M.A.; ELSHABASSY, A. The effect of different drainage and amelioration techniques on the chemical and physical properties of a salt affected and heavy texture soil in Egypt. Agric. Res. Rev., 1978, 56:63-74.

The behaviour of certain soil chemical and physical characteristics under different amelioration and drainage methods were studied. The highly saline slick spots did not disappear from the soil surface of the non-drained fields after three cultivations even when gypsum was applied. Drainage alone without any surface gypsum addition improved the chemical properties of the soil. Moreover, it accelerated the process of exchange between the clay complex and the added gypsum to the top soil. Better results were obtained by ameliorating the trench backfilled with gypsum. Ameliorating the top soil increased its hydraulic conductivity while not influencing that of the deeper layers. The drainable porosity was not affected by any techniques used for applying gypsum or changing the drain spacing.

114

BITOUM. M.; KOESSNER, E.P. Tidal zone riceland drainage in Guinea, West Africa. 8th Congr. Irrig. Drain. Varna, 1972, Q 27 R 5:61-77.

Drainage of coastal belt of Guinea(West Africa) which helped in relieving excessive flooding and retention of natural balances in the soil oxidation-reduction status is discussed. Principal drains at 600 m spacing with intervening secondary drains spaced at 400 m were used as a guide in establishing the drainage pattern. Independent drainage system sections with each section having a separate outlet control structure has been the basic premise in planning the drainage system.

115

BLACK, R.D.; WALTER, M.F.; SWADER, F.N. Apparent effects of subsurface drainage on soil physical properties. 3rd Natn. Drain. Symp. ASAE, Michigan, 1976, 123-6.

During a research period of normal and below normal growing season rainfall corn, beans and wheat showed no yield response to tile drainage in upper New York State. During wet growing season corn yields were improved by tile drainage. Measurement of changes in soil permeability and bulk density due to tile drainage or crop residue management were inconclusive apparently due to the extreme heterogeneity of the soil. Surface residue management greatly affects the aggregate stability of the top 0.25 m of the soil surface. The well managed plots had 20 per cent more water stable aggregates after 9 years of treatment than did poorly managed plots. No substantial change in this difference occurred over the next 11 yrs. Management of surface residue did not affect subsoil aggregate stability. Tile drainage had no statistically significant effect on aggregate stability. No statistically significant increase of soil organic matter resulted from either residue management or tile drainage.

116

BLIDARU, V.; NICOLAU, A.;NITESCU. E.; COJOCARU, I. Model study of a drain filter for establishing their hydraulic and functional efficiency. 10th Congr. Irrig. Drain. Athens. 1978,511-25.

Using special laboratory equipment, the influence of water movement within five filter materials (graval, glass fibre, polystyrene granules, peat, glass wool) on drain discharge was analysed. Hydraulic characteristics are given for glass wool and polystyrene granules, relating them to hydraulic conductivity and resistance of the filters. Calculation of the head losses created by the drain filters is demonstrated.

117

BLINTSOV, I.K.; IPAT'EV, V.A. Mobile forms of nitrogen in drained peaty forest soils. Lesn Zh. 1974, 17:16-9.

Data are given on the content of NH_4- and NO_3 -N in drained peat-bog soils of the transitional type carrying scots pine stands in Belorussia. The content of mobile forms of N varied with season, with stand age (20 and 60 years) and with distance from the drainage ditch (10, 50 and 150m). The content of NO_3-N was highest in summer, when nitrification exceeded ammonification. In summer, the content of NO_3-N was highest in the places farthest from the drainage ditches, where tree growth was poorest. In spring and autumn, the content of NO_3-N tended to decrease with increasing distance from the ditch.

118

BLOOMSBURG,G.I. Finite element computer program to determine parallel drain spacing. 3rd Natn. Drain. Symp. ASAE, Michigan, 1976, 51-4.

A finite element computer program is used to develop dimensionless curves for determination of parallel drain spacing. The curves are a plot of the ratio of midpoint water table height to original water table height versus a dimensionless function of time. This dimensionless time is developed from drainage simlitude relationships developed a+ olor do State University. The use of these curves in a drain spacing problem is described.

119

BLOSSER, R.H. How drainage affects the economics of fertilizer use on a fine textured soil. <u>J. Soil Water Conserv.</u> 1960, 15:210-33.

Drainage rating and crop yields for corn, soybean, oats and wheat are tabulated. Whereas increasing dose of fertilizer incresed yield, at the same dose of fertilizer application yield increased with drainage rating qualitatively defined as below average, average and above average. Average corn yield at 336.3 kg/ha of N (300 lbs/acre of N) increased from about 2.47 t/ha (45.9 b/acre) to 3.58 t/ha (71 b/acre). It is concluded that to take advantage of heavy doses of nitrogen, drainage efficiency has to be improved on poorly drained soils.

120

BOEKEL,P. The significance of drainage to soil structure in loam and light clay soils and its financial aspects.<u>Bedrijfsontwikkeling</u>, 1974, 5:875-80.

A high groundwater level, especially on the light loams, in conjunction with prolonged exposure of the topsoil to winter weather, may cause serious crusting and therefore, poor growth of winter cereals and tulips and unfavourable seedbed conditions for spring-sown crops. Drainage conditions also affect cultural activities. A high groundwater level in winter and early spring delays spring cultivation and reduces the time available for cultural activities. A gradual deterioration in soil structure occurs.

121

BOELS, D. Installation methods. Proc. Int. Drain. Workshop, Wageningen, Netherlands, (1978)) 1979, 25:39-49.

Drainage installation methods and recommendations arising out of the 12 papers presented in the workshop are included.

122

BOELS, D. Methods to predict changes in hydraulic conductivity caused by drainage plows and backfilling of trenches. Agric. Wat. Mgmt. 1979, 2:11-24.

The paper reports a method to predict the reduction in hydraulic conductivity around drain pipes due to the laying technique. For the trenchless technique, from the stream lines and an assumed displacement of the soil mass, a relation is derived between distance to the plow body. The relation between hydraulic conductivity and deformation was derived from cores placed in a triaxial apparatus. When the friction angle is known, the reduction in hydraulic conductivity can be determined. For the trenching technique, the change in dry bulk density due to settling of the back-fill is derived from the relation between the change in density of loosened soils and the original density of the soil before loosening. Using a relation between density and hydraulic conductivity, the reduction in conductivity is obtained. The predicted hydraulic conductivity in the deformed zone caused by a drainage plow and that in back-filled trenches in a silt loam soil and a silty clay loam soil was compared with the hydraulic conductivity derived from measured drain discharges, heights of water table and known original hydraulic conductivities of the undisturbed soil.

123

BOELTER, D.H. Water table drawdown around an open ditch in organic soils. J. Hydrol., 1972, 15:329-40.

The effectiveness of a ditch in controlling the water level in a peatland area depends on the hydraulic conductivity of the peat material in the organic soil profile. An open ditch has accomplished the drainage of water from surface or near surface horizons of fibric moss in a typical lake filled organic soil. However, it had little influence on the water table beyond 5 m from the ditch when the water table was in the deeper horizons of moderately decomposed peat. In a soil with less decomposed peat throughout several meters at the surface, a similar ditch influenced the water table as far as 50 m from the ditch. Comparative rate of water table recession 50m from the ditch in two bogs were calculated to be $7x10^{-4}$ cm/day and 2.3 cm/day.

124

BOER,D.W. DE.; CHU, S.T. Application and economics of bi-level drainage. 10th Congr. Irrig. Drain. Athens, 1978:369-80.

Bi-level drainage is the arrangement of subsurface drain lines at two depths on an alternating basis. A theory has been developed which describes the movement of the maximum water table elevation. A computer programme was used to simulate the moovement of the water table throughout a crop year. Drain line spacings for single and bi-level drainage systems which satisfied drainage criteria were determined from the results. An economic criterion based on drain spacings was developed to determine the economic feasibility of bi-level drainage systems.

125

BOERS TH. M.; VAN SOMERAN, C.L. Design of gravel envelope for silty and fine sandy soils in Pakistan.

Proc. Int. Drain. Workshop, Wageningen, Netherlands, (1978) 1979, 25:713-31.

This paper deals with the design of a graval filter for a pilot area in Pakistan. Two criteria based on U.K. Road Research Laboratory and U.S. Bureau of Reclamation are explained. The criteria are applied to actual soil conditions, from which a range of required grading of the filter results. From available filter materials in Pakistan, gravel from Shadi Shaheed is selected as most suitable, on the basis of grading, distance from quarry to pilot area, and performance in a sand tank model. Two aspects of the execution are discussed: contamination of gravel and piping.

126

BOGDASZEWISKI, H; DUTKIEWCZ. Z. Technical and economic problems concerning application of drain pipes made of plastics, Polimery, 1968, 1:27-9

The capital costs of the tile and plastic (PVC) pipes used in drainage have been compared. Although the latter costs about 3 per cent more but nevertheless the cost is likely to reduce in view of adapting to thin pipes and the increase in their production. Some features of plastic pipes being used in other West European countries, have been described.

127

BOGGIE, R.; MILLER, H.G. Growth of Pinus contorta at different water-table levels in deep blanket peat.Forestry, 1976,49:123-31.

In an experiment in which the watertable in peat was maintained at levels ranging from 0 to 34 cm, the height growth of lodgepole pine was linearly related to both depth of water table and air volume in the surface horizons. Stem volume, tree weight and foliage area increased exponentially in response to drying of the peat. Foliar nitrogen and phosphorus levels were broadly related to treatment but, together with potassium levels, declined with time and showed no improvement as a result of the drying and increase in rooting volume that followed canopy closure.

128

BOKERMANN, R. A method for calculating the cost-saving effects of drainage on arable land. Kult. Flurb., 1980, 21:44-54.

Cost saving effects of soil drainage include facilitated tillage and an increase in the number of days on which field work can be carried out. Field working techniques may be divided into five categories according to their soil moisture requirements. The influence of drainage on each field work category was calculated. While cost saving through facilitated soil tillage is slight, the economic effects of extending field work periods are of considerable importance.

129

BOLTON, E.F.; DIRKS, V.A.; HORE, F.R. Corn, soybean and wheat yields on Brookston clay drained by plastic tubing installed by two methods at seven spacings and two depths. Can. Agric. Engng., 1980, 22:145-8.

Corn, soybean and winter wheat were grown on Brookston clay to establish yield response to subsurface drains installed at seven spacings, two depths and by two methods of installations. The closest spacing at 6.1m, gave the highest average yield from 4 years of corn, 3 yr of soybean and 1 yr of wheat averaging 1.73 t/ha more grain corn and 0.24 t/ha more soybean than the conventional 15.2 m drain spacing. The effect of drain spacing on yield was largely linear within the range of spacing from 6.1 to 24.4 m for both corn and soybeans. Spacing by year interactions were statistically significant, but did not affect the yield superiority of the 2 closest spacings. No average difference in yield could be demonstrated between drain depth at 70 cm and those at 95 cm or between wheel trencher installation and drainage plow installation. The advantage of close drain spacing was attributed to improve soil condition and tilth in spring.

130

BOODT, M.F.L.P. DE. Artificial soil stabilisation of backfill in field drainage. J. Sci. Fd. Agric.1978, 29: 417-22.

Washed gravel, used as backfill, can be replaced by ordinary soil when treated with soil conditioners such as bitumen emulsion or a solution of either polyurethane or polyacrylamide. Laboratory tests show that when the backfill is treated with these solutions, it can be put on the drains immediately. When bitumen-emulsion is used, the backfill should be dried first. Further, it is shown that conditioned light soil used as backfill significantly decreases the silting-up of the drains while conditioning heavy soils can increase their permeability by a factor of 10 to 100.

131

BORCHERT, H. Micromorphological observations on reclaimed soils. "Soil Micromorphology" Proc. 2nd int. Wk.Mtg. soil Micromorph, Arnhem,1964, 459-65.

Microscopic investigations on thin sections of degraded loess before and after reclamation showed that mole drainage increased looseness and porosity of soil as a consequence of increased clay content and faunal activity.

132

BOREK, S. Changes in properties of Blonie-Sochaczew black earth soils after drainage. Roczn. Glebozn., 1975, 26:101-40.

The favourable effects of drainage on moisture distribution, infiltration, O_2 and CO_2 diffusion through the profile and reducing conditions became more marked as the period of drainage increased.

133

BORNSTEIN, J. Ground water movement in poorly drained glacial till soil. Trans. ASAE, 1964, 7:38-41, 45.

The drainage problems of Cabot loam soil arising from surface runoff, seepage and subsurface water flows are investigated. The topsoil (20-27.5 cm) is loam and is underlain by a fine sandy or fine gravelly layer of 10-30 cm. Fragipan layer extending upto 120 cm occurs below this depth. Cabot loam soil having a slope of 5-7 percent was drained by diversion ditches at two depths (30 and 45 cm) and two spacings (60 and 120 cm) alone and in combination with 10 cm dia perforated bituminized fibre drains at two spacings (30 and 60 m.) The paper outlines further experiment set ups such as installation of piezometers and studies on ground water movement. Actual tests on drainage are under way and are likely to lead to solutions of the drainage problem of the region.

134

BORNSTEIN, J.; BENOIT, G.R. Subsurface drain and diversion discharge comparisons on a slopping fragipan soil. Trans. ASAE, 1967, 10:590-3.

The drainage studies on a fragipan silt-loam soil with surface (60-120 m spacing) and combination of surface and subsurface drains (30 - 60 m spacing) indicated that total drain discharge was greatest from the 30 m subsurface drain spacing and with surface drainage. The flow in surface or subsurface drains is not affected by each other. The minimum effective subsurface drainage treatment is with a spacing of 30 m to maintain optimum moisture conditions.

135

BORNSTEIN, J.; THIEL, T.J.; BENOIT, G.R. Characteristics of flow to diversions and subsurface drains on a sloping fragipan soil. Trans.ASAE, 1967, 10:586-9.

Research findings of a 12-plot drainage project on a Cabot silt loam three-layered glacial till soil with a uniform 4-7 percent slope are presented. From waterflow data it was evident that excess soil water was removed principally by subsurface drains and flow to diversions was primarily in the upper 75-90 cm of the soil profile, below 1.5-2.1 m the flow continued down slope. The subsurface drains flow was not affected by surface drains. Analog model studies and field observations indicate that electrical model was a good representation of prototype.

136

BORNSTEIN, J.; GARWOOD, S.; DOTY, C.W.; RYAN, W.F. Drainage bibliography from computer based data sources. Trans.ASAE, 1979, 22:529-31.

Data bases with possible references to drainage are listed, and a procedure for compiling an individually tailored bibliography to cover a specific aspect of drainage from computer based data sources on retrieval systems is described. The searching processes outlined in steps and the techniques in on line-off line and current awareness-retrospective searching are outlined. Sample computer printouts are included.

137

BORNSTEIN, J.; PRESTON, H.A.; WINANT, W.M.; BENOIT, G.R. Automated flow-recording system for field drainage monitoring- direct data compilation of surface and subsurface drain flow. J.Agri. Engng. Res., 1973, 16:31-5.

An automatic recording console has been developed to accept water storage data from remote weirs and flumes and store the data on magnetic tape. The sensing-trnasmitting-cycling-recording components can economically replace the mechanical-manual systems for converting flow events to card or tape data ready for computer analysis. A sequence of problems that had prevented operation was solved and is reported in the paper.

138

BOTTCHER, A.B.; MONKE, E.J.; HUGGINS, L.F. Nutrient and sediment loadings from a subsurface drainage system <u>Trans. ASAE</u>, 1981, 24:122-6.

Sediment, pesticide and nutrient losses were measured from a 17 ha sub-surface drainage system for the years 1976-78 using automatic sampling equipment. The monitored drainage system was installed on a nearly flat Hoytville silty clay soil with limited surface runoff due to raised field borders. Dynamic responses of the drainage system are graphically presented and discussed as they relate to field management practices and climatic variations. A comparison was also made between this system with its low surface runoff and a more normal situation with partial subsurface drainage and greater surface runoff. For the 17 ha system, runoff per unit area was substantially reduced resulting also in a similar reduction in sediment and nutrient losses.

139

BOTTCHER, A.B.; MONKE, E.J.; HUGGINS, L.F. Subsurface drainage and sediment transport model. <u>Trans. ASAE</u>, 1980, 23:870-6.

A computer model was developed, using the GASP IV simulationn language to simulate the water flow and sediment movement from a subsurface drainage system. The model used a one dimensional form of the Richard's equation and a steady state tile flow formula by Toksoz and Kirkham to express the water movement process. The particle detachment portion of the model, which was based on a force balance relationship was controlled directly by the output of the flow portion of the model. Data required by the model include rainfall, evapotranspiration, soil properties and drainage system layout. Calibration and verification were completed using data collected from a 17 ha tile drainage system located on a flat Hoytville silty clay soil. A comparison of simulated and observed results indicate that the model will reliably predict water yield, sediment yield and the sediment concentration curve. However, some difficulty in simulating the actual shape of the flow hydrograph occurred.

140

BOULTION, D. Operation and groundwater hydraulics of an experimental tile drainage system at Kerang. In Hydrogeology of the Riverine Plain of South East Australia, Proc. Symp. Griffith, Australia, 1977. (Storrier, R.R.; Kelly, I.D., Editors); Austr. Soc. Soil Sci. Inc. 1978, 229-46.

A description of the tile drainage installation in saline soils of the Kerang Region is given. The discharge hydrograph, water balance, 'apparent' specific yield, hydraulic head vs discharge relationship and equipotentials adjacent to the centre lateral are reported. Hydraulic conductivity in the depth range 0.5-10 metres has been measured using the auger hole and piezometer methods and the values obtained are reported. Soil layers to a depth of 11 meters are important in transmitting groundwater to the tile drains. A large part of the subsurface runoff is attributed to seepage of groundwater into the drained area. The amount of irrigation water applied increases, and percolation to the water table decreases, as the water demand of maize and sorghum increases during the growing season.

141

BOUMA, J.; DEKKER, L.W.; HAANS, J.C.F.M. Drainability of some Dutch clay soils: a case study of soil survey interpretation. Geoderma, 1979, 22:193-203.

Riverine and marine non-calcareous Dutch clay

soils have moderate limitations for use as grassland under existing conditions due to high ground-water levels in winter and early spring, which have traditionally been attributed to a very low saturated hydraulic conductivity (K-Sat) of the clay. Physical research which used soil morphology to define sample size and soil maps to locate test sites within defined mapping units, showed that permanent high ground-water levels were primarily due to low hydraulic gradients and not to a low K-Sat. Well-maintained tile drainage associated with low water level in the ditches in winter, resulted in much lower ground water levels in the soil, thereby reducing limitations imposed by existing conditions and improving soil suitability. The latter was specifically characterized in this study in terms of the assessment factor; 'drainage status'. Tile drainage resulted in a significant increase in K-Sat which was thought to be due to increased drying and cracking of the relatively young riverine soil. In the older marine soils, the increase of K-Sat upon tile drainage was due to deep earthworm activity. Existing suitability can thus be improved by using appropriate technology, i.e., tile drainage associated with better management.

142

BOUMA, J.; DEKKER, L.W.; VERLINDEN, H.L. Drainage and vertical hydraulic conductivity of some Dutch "Knik" clay soils. Agric. Wat. Mgmt., 1976, 1:67-78.

The vertical K-sat of a clay layer, occurring between 30 and 60 cm below the soil surface was measured *in situ* in early spring at 13 sites, using large soil columns. Gypsum was used to form a barrier around the column and K-sat values were measured with an infiltrometer in columns that were first attached and then detached from the subsoil. The procedure allows an estimate of the occurrence of large continuous pores, such as vertical worm channels. Highest values were found in tile-drained grassland, followed by grassland with surface drainage only, and by tile-drained arable land. Relatively low K-sat for the silty subsoil, rather than the (high) vertical K-sat for the clay layer, is considered to be responsible for high groundwater tables in the wet season. Undisturbed, large columns were taken to the laboratory and saturated for a period of three months to simulate prolonged swelling after a very wet season, and to measure chloride-breakthrough curves, for characterizing soil-pore continuity. The clay layer, sampled in the surface-drained grassland, showed no significant reduction of K-sat after prolonged swelling, but the one for arable land was reduced. Moreover, flow in the latter occurred through only a few relatively large, continuous pores, whereas a more heterogeneous pore system was found for the column from grassland. The already high K-sat of the clay layer in surface-drained grassland increased as a result of tile drainage. Compaction of the clay layer in tile-drained arable land reduced K-sat well below the level found in surface-drained grassland.

143

BOUMANS, J.H. Drainage calculations in stratified soils using the anisotropic soil model to simulate hydraulic conductivity conditions. Proc. Int. Drain. Workshop, Wageningen, Netherland (1978),1979, 25:108-23.

Drainage often occurs inalluvial plains, which are characterized by layered soils that have different hydraulic conductivities for horizontal and for vertical flow. The generally applied calculation models, assuming isotropic permeable soils are not suitable to simulate the drainage flow in layered soils. Drainage in such soils can be more suitably calculated by means of anisotropic soil models. The article further discusses the theory and application of anisotropy in drainage calculations, the determination of the horizontal and vertical hydraulic conductivity components, data on soil anisotropy from literature and collected in the field.

144

BOUMANS, J.H. Extended singular drainage: an interesting layout for draining irrigated lands. Proc. Int. Drain. Workshop, Wageningen, Netherlands, (1978), 1979, 25:701-8.

The singular drainage layout with laterals of a length not exceeding 250 to 300 m is the commonly used system in flat area. The system requires a relatively long system of open drains which have to be deep in irrigated areas and are as a consequence expensive in terms of construction, maintenance and loss of land. Open drains can be avoided by the composite drainage layout though this system has other disadvantages, however. Attention is drawn to another alternative, viz. the extended singular layout which like the composite system does not use open drains and offers the advantages of the singular layout.

145

BOUMANS, J.H.; HULSBOS, W.C.; LINDENBERCH, H.L.J. et al. Reclamation of salt affected soils in Iraq, Soil hydrological and agricultural studies. Int. Inst. Land Reclam. Improv. Wageningen., Netherlands, 1963, 11,175p.

The saline soils of Iraq are described; results of leaching of salt-affected soils (including the alkalinity aspects of the process) are reported, and crop yields, consumptive use, irrigation and drainage of reclaimed soils are discussed.

146

BOUMANS, J.H.; MOLEN, W.H. VAN DER. Drainage requirements of irrigated soils in relation to their salinity. Landbouwk. Tijdschr., 1964. 76:880-7.

The method of calculating salt balance for irrigated soils is discussed briefly. For cases in which the irrigation waters have a noticeable salt content, a correction is proposed. Moreover, a method is given for dealing with cases in which certain salts (mainly gypsum) are precipitated in the soil. From such balances the drainage requirements of irrigated soils can be derived. An important factor determining these requirements is seepage, which is a wide-spread phenomenon in irrigated areas.

147

BOUWER, H. Developing design requirements for parallel drains. Proc. Drain. Efficient Crop Production Conf. Chicago, Illinois, 1965, 62-4.

Development of drainage criteria depends on the intended function and design procedure of the drainage system. Such systems may be designed on the basis of the steady-state condition, the falling water table, the fluctuating water tables, or of salinity control in irrigated land. In most instances formulating the design requirment either in terms of a maximum water-table height and corresponding drainage rate, or of a certain rate of fall of the water table midway between drains, should be adequate. The "drainage criterion" describes the total drainage requirement for a given region. The "design requirement" describes the difference between the total drainage requirement and the natural drainage of a given field. Thus the "design requirement" reflects the drainage deficiency of a field to be absorbed by the drainage system. A empirical way of adjusting the design requirement to the natural drainage conditions of the field has been suggested.

148

BOUWER, H. Developing drainage design criteria. In Drainage for Agriculture (Agronomy No.17; Schilfgaarde, J. Van, Editor) Am. Soc. Agron, Wisconsin, 1974, 67-79.

Drainage criteria are evaluatedd for a steady-state condition, a falling water table, salinity control and trafficability of the soil, depending on the main functions of the drainage system. The design requirements are then found from the difference between the drainage criteria and the existing natural drainage intensity of a given field.

149

BOUWER, H.; SCHILFGAARDE, J. VAN. Simplified method of predicting fall of water table in drained land. Trans. ASAE, 1963, 6:288-91, 296.

Steady state drainage solutions are analytically or graphically integrated to predict the rate of fall (or rise) of the water table midway between tile lines or ditches. The procedure assumes abrupt drainage of pore space. A factor C is introduced to account for the non-uniformity of flux per unit area of water table if the water table changes in shape during recession. The procedure is amenable to steady-state formulas or to graphical or tabular solutions obtained by analogs, models, numerical procedures, field experiments, or others. Two integrated steady-state drainage equations (by Hooghoudt and by Toksoz and Kirkham) are compared with four solutions developed specifically for the falling water table. The transient formula by Schilfgaarde accounts both for flow convergence and changing height of the flow system and shows good agreement with the two integrated equations, which also agree mutually. The other three transient solutions have some inherent weaknesses and show poor agreement.

150

BOWER, C.A. Salinity of drainage waters. In Drainage for Agriculture (Agronomy No.17; Schilfgaarde , J. Van, Editor) Am. Soc. Agron., Wisconsin, 1974, 471-87.

Factors determining the quantity and generalised quality characteristics of agricultural drainage water and their effect on the quality of surface and ground waters are discussed.

151

BOWER, C.A.; SPENCER, J.R.; WEEKS, L.O. Salt and water balance in Coachella Valley, California, J. Irrig. Drain. Div. ASCE, 1969, 95:55-64.

It has been found that the salt balance index (output/input) for the irrigated land of Coachella Valley during 1957-1965 was highly related to percentage of land under drainage, the index becoming close to one when about half of the land became tiled and when the leaching percentage increased to about 30. At salt balance, evapotranspiration, taken as the difference between depths of applied and drainage water, agreed within about 10 per cent of that calcuated by the Blaney Criddle formula. About 10 per cent of the applied salt appears to precipitate in the soil largely as calcium carbonate, and calcium and magnesium in water replaces exchangeable sodium in soil during irrigation operations.

152

BOWLER, D.G. New aids for improving wet land. New Zealand J. Agric., 1978, 136:40-3.

The use of corrugated plastic piping for drainage and the application of laser beams to drainage and irrigation surveys are discussed.

153

BOXEM, T. Lowering the water table on fen grassland gives good results. Bedrijfsontwikkeling, 1975, 6:991-94.

In trials in 1970-5 the effect of lowerig the water level in drainage ditches from 25 to 75 cm below soil surface level was investigated in grassland on sedge peat. Deeper drainage maintained the water table at > 30 cm below the surface and the load -bearing capacity of the sward at > 6 kg/cm^2. Damage to the sward during spreading of slurry in winter was minimised by fitting double width tyres to the front of the tractor. On deep-drained plots the cows could be turned out to graze upto 1 week earlier than on traditionally-drained plots and could graze up to 3 weeks longer in autumn. The resulting improved grazing management pattern gave milk yields of 18.6-19.1 kg/cow day compared with 17.6-18.3 kg for traditionally-drained plots during the grazing season. Herbage yields at 0, 150 and 300 kg N/ha, respectively, were 12.0, 12.8 and 13.1 t DM/ha with deep drainge and 10.7, 12.9 and 13.7 t DM/ha with traditional drainage. Net yields of SE were about 12 per cent higher with deep drainage; calculations showed that this would more than offset the drainage costs.

154

BOXEM, T.; LEUSINK, A.W.F. Drainage of fen grassland. Report of a comparative investigation under farm conditions at Zegveld from 1970 to 1975. Publikatie, Proefstation voor de Rundvee-houderij, Netherlands, 1978, 11, 75p.

A 30-ha grassland farm on formerly wooded peat soil was divided into 2 parts on which the ditch water levels were maintained at 25 or 75 cm below the soil surface level. During the following 6yr (1970-75) the rainfall ranged from 147 mm above to 181 mm below the seasonal mean. Lowering the ditch level lowered the water table by about 20 cm and increased the load-bearing capacity of the sward to above the poaching threshold ($>$ 6 kg/cm^2). On the undrained land it was difficult to apply slurry (30 t/ha in 2 dressing) during the winter except when the soil was frozen. Drainage enabled N to be applied 2 wk earlier in spring and reduced the N requirement by 100 kg/ha in the last 2 yrs. Botanical composition of the sward in 1974 showed few differences between the 2 drainage levels except for a reduction in the number of moisture indicating plants and an increase in grazing season by 4 days in spring and 21 days in autumn, and in the wet summer of 1972 increased milk production by 1.5 kg/cow a day. The percentage of each farm area cut for fodder conservation was 135% at the lower water level and 120 percent at the higher, partly due to more even-grass growth and better distribution of cutting dates on the drained land. Where no N was applied, the herbage yields were 1.5 t DM/ha higher on the drained land, at 150 kg N/ha the yields were about the same and at 300 kg N/ha the undrained land yielded slightly more. Estimated gross DM yields were increased by drainage by 35 percent in the wet yr 1972 and by 11 percent overall.

155

BRANZANTI, E.C.; FALCIAI, M.; VITTORI, V. The behaviour of peach trees in soils with various systems of drainage. Frutticoltura, 1976, 38:34-42.

On flat ground with a soil permeability of 3.5 cm/hr, the trees grew and fruited best where deeply placed drains and a pump were employed.

156

BRENNER, W.; HOLTI, W. Application of drainage technology, Kult. Flurb, 1980, 21:287-95.

The most important cause of runoff problem in pipe drains and their prevention are discussed from a consideration of the design, construction and the maintenance of drainage system.

157

BRETFELD, Z. The use of two-layer drainage in the Trebon basin. Ved.Pr. Vyz. Ustavu. Melior. Praze. Zbral., 1978, 14:29-41.

In the years of 1967-1971, the effectiveness of cross and mole drainage, and of the combination of deep loosening with tile drainage, was evaluated in comparison with alone tile drainage in heavy-textured soils, waterlogged on the surface, at the experimental site at Bor. Under the soil and climatic conditions of the Trebon basin area tile drainage is more effective than the combined methods. Tile drainage shows a higher water outflow rate and gives greater benefits.

158

BROADHEAD, R.G.; SCHWAB, G.O.; REEVE, R.C. Synthetic drain envelopes and soil particle-size distribution. Trans. ASAE, 1983, 26:157-60.

Laboratory tests were conducted to evaluate the ability of the synthetic envelopes to prevent sediments from entering subsurface drains. Nine sand samples in the sediment problem range were fabricated and tested. Drain sedimentation can be prevented by selection of envelope opening size such that the envelope opening size is less than about 2.4 times the dia of the 60 percent (D_{60}) size of sand particles. For greater opening sizes, sediment movement through the envelope is a function of the imposed hydraulic gradient in relation to the failure gradient of the particular sand mixture. At opening sizes less than 2.4 times the D_{60} sand size, sediment is essentially completely restrained regardless of the hydraulic gradient.

159

BROOKS, D.R. Leaching agricultural soils. Trans. ASAE, 1966, 9: 826-7.

The leaching operations were conducted on a 30 ha parcel in the basin, typical of the entire dry lake area in south-western United States. Prior to leaching, EC_e and ESP of the soil were 9.5 dS/m and 24, respectively. 24 to 36 cm levee was built around the field. A 1.5 m deep perimeter drain ditch was constructed outside the levee and connected to an outlet drain from the field. Gypsum was applied # 12.5 t/ha. EC_e was reduced to 5.8 dS/m and ESP to 9 in surface 30 cm of the soil and EC_e to 8 ds/m and ESP to 15 in 30-60 cm depth of the soil.

160

BROOKS, R.H. Unsteady flow of ground water into drain tile. J. Irrig. Drain. Div. ASCE, 1961, 87:27-37.

A solution to the nonlinear differential euqation describing unsteady flow toward equally spaced drains above a horizontal impermeable boundary is presented. The solution is compared with field data and the numerical solution when the drain spacing was large relative to the depth of drains.

161

BROOKS, R.H.; COREY, A.T. Hydraulic properties of porous media and their relation to drainage design. Trans. ASAE, 1964, 7:26-8.

The three soil properties namely pore size distribution index, saturation at which the effective permeability of water is assumed to be zero and the capillary pressure have been discussed in relation to the variables of saturation, capillary pressure and permeabilities. The significance of these properties in drainage design is investigated and similitude requirements for modelling techniques outlined to take care of these parameters. Five diamensionless parameters are suggested.

162

BROUE, P.; MARSHALL, D.R.; MUNDAY, J. The response of lupins to waterlogging. Aust. J.Exp. Agric. Anim. Husb., 1976, 16:549-54.

Growth response of species and lines of lupin (Lupinus spp.) to short-term flooding of the root system during vegetative growth was measured in glasshouse and controlled environments. L. Luteus was consistently more flood tolerant than L.albus, L.angustifolius and L.mutabilis. Some data showed also that lines within species may vary in their response to flooding, but the differences among the lines were less marked than those among species.

163

BROUGHTON, R.S.; ENGLISH, B.; DAMANT, C.; AMI, S.L; MCKYES, E.; BRASSEUR, J. Tests of filter materials for plastic drain tubes. 3rd Natn. Drain. Symp. ASAE, Michigan, 1976,34-9.

Several filter envelop materials for corrugated plastic drain tubes had been tested in the laboratory. The full warp filter materials Reemay and spun nylon (27 g/m^2). Typer (135 g/m^2 and 118 g/m^2), knitted nylon stocking with very fine weave and polyester + 20 percent rayon all did a good job of preventing soil from entering plastic drain tubes. In each case decrease in water flow was observed with time though after periods of no flow increased drain discharge was observed. Coconut fiber matting did not provide a satisfactory filter for the fine sandy soils. Coarse concrete sand filters gave the best water flow with satisfactory retaining the soil.

164

BROUGHTON, R.S.; FOROUD, N.A. A model to predict water table depths for flat lands. Can. Agric. Engng., 1978, 20, 81-6.

A simulation model was developed to predict soil moisture content and water table depths for flat lands. The model used a water balance procedure to take into account rainfall, evapotranspiration, water table position, soil moisture content and subsurface and surface runoff for the snow-free portion of the year. The soil profile was divided into three zones to allow for changes in soil properties with depth. Weather data and measured water table depth from a field at Macdonald college were used to verify the model. The simulated water table depths agreed favourably with observed field data. Recommendations were made to use the model with long-term precipitation and potential evapotranspiration data to study the effect of drainage rates on the duration of particular water table depths.

165

BROUGHTON, R.S.; GAMEDA, S.; GIBSON, W. Field tests of some drain tube envelope materials. Adv. Drainage (4th Natn. Drain. Symp. Proc. ASAE, 1982, pub.12-82, 69-76.

An experiment on subsurface drainage was installed in fine sandy loam soil with standard corrugated polythylene drain with two slot sizes, four envelope materials (Nylon, typar, cerex, reemay and gravel) and also with no envelope material. It is concluded that Typar at 119 g/m^2, knitted nylon sock and gravel performed best in allowing water to enter and preventing fine sand from entering the drain tube. Reemay and cerex suffered from abrasion damage. No envelope resulted in filling up of drains with fine sand and even reduced slot size from 1.5 mm x 30 mm to 0.7 mm x 8 mm did not stop fine sand from entering the drain.

166

BROWNSCOMBE, R.H. Field evaluation of tile drains laid with organic blinding materials. Trans. ASAE, 1962, 5:61-3, 67.

Two organic blinding materials were tested for their performance in a tile drained field with surface 0-30 cm of friable fine sandy loam to friable silty clay loam. Straw and wood chips were used as the blinding materials. Visual inspection showed very little or no movement of mineral soil particles into or through the straw. Straw as old as 11 years was found to be moderately decayed but observed to be fairly effective. The life of straw as a filter material is expected as 15 years. Wood chips decay was slower compared to straw. The results are however, based on stable soils as these were not used on unstable sandy or silty subsoils. Therefore, these are not indicative of filtering effectiveness of the materials.

167

BRUSH, J.C. Jr. Seepage streams out of canals and ditches overlying shallow water tables. J. Hydrol., 1979, 41: 31-41.

When there is seepage from a canal or ditch overlying a shallow water table, waterlogging of the soil at distant points is a possibility. Thus, for given flow field parameters, it is desirous to be able to predict the limit of the drainage capacity which can be handled without producing waterlogging at distant points. A simple numerical scheme is presented for the solution of such problems. This is based upon the Baiocchi method and transformation and gives a quick and accurate means of obtaining the discharge and the free-surface location knowing the physical parameters of the particular canal or ditch problem. The data necessary to investigate a particular flow situation can readily be obtained from the application of this approach. The numerical results obtained for specific applications are compared with published results to similar problems.

168

BRUTON, R.; CONVERY, F.J. Land drainage policy in Ireland. Policy Res. Series, Economic and Social Res. Ireland, 1982, 4,93p.

The report on Irish land drainage policy reviews past activity in arterial and field drainage and the environmental impacts; presents underlying economic concepts; analyzes arterial drainage in some detail (with the legislative and institutional setting and a cost benefit analysis is applied by the Office of Public Works); reviews progress and expenditure on field drainage as well as economically related policy issues. Recommendations include further work on redistribution implications of drainage investment.

169

BUKAVETSKAS, A.K. Effectiveness of mole drainage for peat bogs. Vest. S-Kh. Nauki, Mosk., 1966, 7:26-34.

In a Lithuanian fen bog with peat layer 2-3 to 8 m thick, mole drainage in addition to tile drainage or open drains improved the air-moisture, temperature and nutrient regimes of the soil. Mole drainage and tile drainage increased grain yield of maize by 20 per cent, tuber yield of potatoes by 13 percent, roots and tops of sugar beet by 14 percent each, and perennial herbage yield by 18 percent compared with tile drainage alone. Mole drainage and open drains increased perennial-herbage yield by 57 percent compared with open drains alone.

170

BULGAKOVA, L.S. Volume and composition of drainage runoff from soils on clay alluvium of Permian deposits. Moscow Univ. Soil. Sci. Bull., 1980, 35:63-5.

The initial degree of waterlogging of soils had a substantial effect on the volume and composition of drainage runoff waters. Maximum volume of runoff and maximum removal of water-soluble compounds were observed with deep loosening and with mole plowing. With chemical fertilizers applied in the summer, no marked change in the chemical composition of the drainage waters was observed; with application in the fall, the concentrations of chlorides, sulfates, and magnesium increased and the concentration of Ca^{+2} and HCO^{-}_{3} ions decreased. Phosphorus and manganese were not found in the drainage waters during the entire summer-fall period and iron was not found in the fall. The ions (HCO^{-}_{3}, Ca^{-2}, Mg^{2+} $SO1_{4}$, and CP^{-}) showed maximum removal. No significant removal of K, Na, Si and Al was observed and the removal of nitrate nitrogen varied within wide limits.

171

BURAS, N.; SPENCER, J.R.; PILLSBURY, A.F. Cost of reclamation and development of saline lands. Trans.ASAE, 1961, 4:105-7.

The operation involved in the reclamation and development of saline lands is discussed and their cost indicated. It was observed that total cost of reclamation and development is dependent upon native vegetation and topography. Rough topography usually results in high cost. The cost of operations directly related to leaching operations with tile drains comes to about 15-16 percent of the total cost of reclamation and development including installation of tile drains.

172

BUREAU OF RECLAMATION, U.S. DEPARTMENT OF THE INTERIOR. Drainage Manual, USA, Bureau of Reclamation US Department of Interior, Denver, 1978, 286 p.

This book gives particular emphasis of the drainage of irrigated lands and contains chapters on (a) basic data (geology; soil characteristics; salinity and alkalinity; surface runoff) (b) field and laboratory procedures (in place and laboratory hydraulic conductivity tests; observation holes and piezometers) (c) drainage investigations (field reconnaisance; subsurface investigations; water source studies; groundwater studies; electric analog models (d) design and construction (drain spacing; interceptor drains; open drains; pipe drains; wells; drain layout (e) operation and maintenance of drainage systems and (f) special drainage problems (return flow analysis).

173

BURGHARDT, W. Development of plastic fibre drain filters. Wasser Boden., 1979, 31:14-7.

The hydraulic properties (resistance to inflow of

water) and filter characteristics of drain pipes of 50 mm dia without coating, with coconut fibre coating and polypropylene fibre coating were determined in short and long term tests. Filter pipes coated with coconut or polypropylene fibre showed similar filter and hydraulic qualities. Best results were achieved using fibres of different gauge. Filter efficiency can be improved by matching fibre gauge to soil texture. Adequate thickness of the fibre coating is required.

174

BURGHARDT, W. Investigations on the decomposition of coconut fibre filters in marsh soils. Z. Kult. Flurb., 1983, 24:43-9.

The decomposition of 60 coconut fibre drain filters was examined after they had been installed for a period of 1 to 4 years in marsh soils. Decomposition proceeded from the exterior inwards, at a rate of 1mm/year. The decomposed organic filter material was partly deposited in the drain pipe. The deposition of soil material in the filter reduced the volume of filter pores by 3.9 percent per year. Means of maintaining the filter function are discussed.

175

BURGHARDT, W. The movement of water to drains as affected by direction and magnitude of flow, drain water level, drain diameter and the use of drain filter. Z. Kult. Flurb.., 1978, 19:17-25.

In laboratory experiments, the flow of water to unfiltered corrugated PVC drains (50 and 80 mm dia), wire tube (50 mm dia) and a filtered drain (50 mm dia) was determined with water supplied from above, below or from the side, at different velocities and at different water levels in the drain. When partly filled with water, large diameter and filtered drains considerably reduced the pressure at the drain surface. The decrease in pressure increased the discharge but the effect was diminished by increasing the velocity of water flow to the drain and by raising the level of water in the drain. The wire tube did not behave as an 'ideal drain'.

176

BURGHARDT, W.; FOERSTER, P.; SCHEFFER, B. The effect of land use, management and installation technique on the durability of drain filters made from coconut fibre. Report 2. Z.Kult. Flurb., 1979, 20:11-9.

In marsh soils, destruction of the filters was greater in grassland than in arable land. Mineral fertilisers, slurry, and lime, and the ploughing up of grassland did not affect destruction. Farmyard manure increased destruction slightly, as a result of its effect on the C/N ratio of the soil. About 10 percent of the destruction was not chemical or microbiological but was caused by mechanical stress during drain laying, whether by machines that worked without trenching or by pulverizing machines with chains. The frequent occurrence of negative slopes in the drain lines indicated the importance of the laying technique.

177

BURGHARDT, W.; FOERSTER, P.; SCHEFFER, B. The influence of some soil properties on the decomposition of drain filters made of coconut fibres. Part I. Z.Kult. Flurb., 1979, 19:363-70.

The degrees of decomposition of coconut filters upto 4- years old from drains of gleyed and pseudogleyed sandy soils, calcareous and non-calcareous marsh soils and low-moor soils was measured by their content of soluble lignin. Decomposition was insignificant in the sandy soils. In the marsh soils, which are rich in clay, decomposition was related to the narrow C/N ratio, was not affected by the $CaCO_3$ content, and increased with increasing soil aeration.

178

BURKE, W. Drain investigation on bog-land: the effect of drain spacing on ground water levels. Ir. J.Agri. Res., 1961, 1:31-4.

The ground water level was lowered only to a distance of 1.5-1.8 m from the edge of 90 cm open drains in blanket bog. With drain spacings of 7.5, 15 and 30 m the areas remaining undrained were respectively 60, 80 and 90 percent. There was a maximum seasonal fluctuation of 30 cm in the water table but no permanent drainage effect.

179

BURKE. W. Effect of drainage on crop yield on blanket peat. Ir. J.Agric. Res., 1969, 8:229-38.

In a 5 year field experiment perennial ryegrass, cocksfoot (replaced by tall fescue after 2 years), potatoes and oats were grown on blanket peat with drains at 7.5 cm, 30 cm, 60 cm and 90 cm depths and 3 m apart. All crops responded to drainage and yields were higher on drained than on undrained plots. However, with grass, increasing the N fertilizer rate from 4 to 8 cwt/ac eliminated the yield reduction caused by high water tables. When 4 cwt/acre of N fertilizer was applied to oats, there was no significant difference in yield between drainage treatment, but severe lodging occurred in well drained plots. A significant difference in the yield of oats between drained and undrained plots emerged when the rate of N fertilization was reduced to 2 cwt/acre. Potato yields were significantly higher on drained than on undrained plots, due to failure of planted sets on undrained plots. (1cwt/ac = .126 t/ha).

180

BURKE, W. Effect of drainage on the hydrology of blanket bog. Ir. J. Agric. Res., 1975, 14:145-62.

The results are presented of a study of precipitation drain outflow from a drained area, and surface run-off from an undrained area, on blanket bog at Glenamoy. Data are summarised under the heading: water balance, quantities and frequencies of events, and pattern of precipitation, drain outflow and surface runoff for the period 1968 to 1973 inclusive. Some rain fell on average on 3 out of 4 days. Precipitation was almost frequent in summer as in winter, but the summer amounts were less. Drain outflow occurred on all days of the study period, while surface run-off occurred on about 1 out of 2 days with most run-off occurring in winter. The net effect of drainage was to lower the water-table and cause of progressive drying of the bog.

181

BURKE, W. Long-term effects of drainage and land use on some physical properties of blanket peat. Ir. J. Agri.Res., 1978, 17:315-22.

Measurements were made of moisture content at different tensions, air volume, bulk density and hydraulic conductivity of blanket peat samples from an undrained site and three other sites that had been subjected to different drainage regimes and land use. The sites were as follows: a forest where the peat had been drained by open ploughed ditches 0.3 m deep and 2 m apart; the same forest where tunnel drains 0.7 m deep had been installed at 2.2 m. Each site had responded differently and these differences are described and commented upon.

182

BURKE, W.; MCCORMACK, P. Gravel drains for peat. Ir. J. Agric. Res., 1969, 8:285-7.

A new type of gravel-on-polythene drains for peat and a simple plough to install it are described.

183

BURKE, W.; TAYLOR, G.S. Soil stratification and ponded flow into subsurface drains. Res. Cir. Ohio Agric. Stn., 1965, 138, 23p.

Presents report on the experiments conducted to study soil stratification effect for the underground drains simulated for boundary conditions and flow by the Laplace equation for two dimensional steady-state, laminar flow and the results have been analysed with an electronic computer.

184

BURNETT, R.; TEKRONY, R. Drainage investigations on the OAHE unit, South Dakota. In Water Management for Irrigation and Drainage. Proc. Spec. Conf.ASCE Irrig. Drain. Div., ASCE, New York, 1977, 1:22-9.

The procedures adopted for the drainage investigations in South Dakota (USA) is reported. The plan envisages construction of subsurface drainage before the problem is likely to arise and hence its effectiveness will be judged only after few years of installation. It is suggested that investigation programme must be flexible to adapt to the changing technology and future projects need.

185

BURT, T.P.; DONOHOE, M.A.; VANN, A.R. The effect of forestry drainage operations on upland sediment yields: the results of a storm-based study. Earth surface processes and Landforms, 1983, 8:339-46.

The effects of pre-afforestation open ditching on the sediment yield of an upland catchment in the Southern Pennines, England are described. Marked increases in the production of suspended sediment following ploughing caused major pollution of a local reservoir. Sediment supply prior to ditching was limited and easily exhausted, but afterwards remained plentiful, allowing sustained erosion during periods of storm runoff. Large volumes of sediment were carried into the reservoir during the winter of 1980-81. More recently exhaustion of supplies of loose material in the ditches and revegetation have led to a lower rate of sediment output.

186

BURT, M.; MUNTEANU, M. Effect of soil moisture due to the groundwater table on wheat, corn and sunflower crops. An.Inst. Cerc. Pentru Pedcl.Agrchim., 1973, 42:181-9.

Results from statistical and mathematical studies showed that a groundwater depth of 0.3 to 0.4 m was favourable for obtaining high yields of corn and wheat and to less extent of sunflower, especially on medium textured steppe and forest -steppe soils. A rise in the groundwater level to 0.2 and, especially, to 0.1 m led to a yield decrease especially for wheat and sunflower.

187

BUSCH, J. R.; LINDEBORG, K.H.; GALIMATO, G.D.; MILLINER, J.E.; ALI, M.B. Socio-economic analysis of a major rehabilitation of irrigation and water management system in eastern Idaho. Moscow, Idaho, USA; Dep. of Agrl. Engng. Idaho University, 1975. 40 p.

High water table caused by overuse of irrigation water have created physical and socio-economic problems in some areas of south eastern Idaho. In one area, the Snake River Fan of Jefferson county, the irrigation systems and management practices were investigated to determine their contribution to the high water problem. The investigation showed that the present systems are quite inefficient due to soil with high water intake rates, long irrigation runs and duplication of many irrigation canals. Engineering and economic analyses were made to determine the least cost method of rehabilitating the present system and to determine the capability of the area to support various rehabilitation schemes. Both analysis utilized linear programming models to determine optimum solutions subject to various social and legal constraints.

188

BUSONI, E.; KOWALIK,P.; SANESI, G. Effect of drainage conditions on winter wheat production. Agric. Wat. Mgmt. 1983, 7, 425-38.

To forecast the suitability of land for winter wheat, soil investigations were directed towards the evaluation of the effect of drainage conditions on productivity. Experiments were carried out on two soil types in Mugello (Italy): a well drained soil and poorly drained one. Measurements of plants atmosphere and water-table data were taken to simulate soil water contents and yield on a computer model. Computed and measured soil water contents and yields showed good agreement. On the poorly drained soil, root development was limited by the groundwater table and the final productivity, expressed as dry biomass, resulted in 50 per cent yield depression compared with that on the well drained soil.

189

BUSSER, J.W.; SCHOLTEN, J. Drainage maintenance in the Netherlands. Proc. Int. Drain. Workshop, Wageningen, Netherlands (1978), 1979, 25:424-38.

Drain clogging can have a big influence on the efficiency of the drains, both on the discharge and on the hydraulic head. For cleaning drains with a length up to 500 m in general a flushing equipment is used. It is well known that with this equipment sand cannot be removed, but only levelled. The equipment is built up by a hose and a jet. The pump at the other end of the hose causes the water discharge with a certain discharge-head. The amount of water and the discharge-head depend on the used equipment. The discussions concerned about different types of clogging materials under different circumstances and the frequency of maintenance needed. This frequency varies from once a year ;up to once every five years. In addition to that, every year the outlet pipes have to be cleaned. In practice a half to two third of the drainage systems never have been pierced or flushed, while for about 10 percent of the drainage systems just rarely this is done.

190

CALDWELL, R.L; MCPHERRON, E.L. Costs and returns of a surface water management system in Jefferson county , New York. J. Soil Water Conserv., 1971,26:76-8.

A surface water management system consisting of outlet ditches, random surface drains and land smoothing has allowed farmers on nearly level clay soils to increase yield and intensity rotations. All practices including initial cost comes to about $16.75/ha. Given normal rainfall and growing conditions , a farmer can expect to recover $117 per ha investment within 2-3 years.

191

CALEMBERT, J. Contribution to the study of drainage in the heavy soils of Famenne. Bull. Inst. Agron. Gembl., 1962, 30:249-59.

A method is described for classifying clayey grassland soils according to their drainage requirements by means of the botanical composition of the sward, moisture tension (pF 0.2 to 3.7-4.2)and hydrological conditions.

192

CALVER, D.A. Nitrate, phosphate, and potassium movement into drainage lines under three soil management systems.J. Environ. Qual. 1975, 4:183-6.

Nitrate, orthophosphate, and potassium concentrations in drainage water were determined from subsurface drains installed in shallow tilled (ST), (15cm); deep-tilled (DT), (107 cm); and deep-tilled and limed (DTL) Florida Oldsmar sand planted to citrus. The drainage water samples were collected during the period May, 1971 through July, 1972. Total discharge of NO_3-N was significantly greater from ST plots than either DT or DTL plots. The NO_3-N in the drainage water from all treatments ranged from 1-8 mg/l during the period of study. Concentration $<$1mg/PO_4/L were common on the drainage water from all plots during this period while K concentrations varied between 1 and 12 mg/l. Peak concentration and discharge of fertilizer nutrients were shown to be a function of rainfall, irrigation and timing of fertilizations. Deeply incorporated limestone applications into the sub-soil tended to increase the NO_3-N and decrease PO_4-P discharged over that from deep tillage alone.

193

CAMP, C.R.; CARTER, C.E. Sugarcane yield response to subsurface drainage for an alluvial soil. Trans. ASAE, 1983, 26:1112-6.

The yield response of sugarcane to various subsurface drainage treatments were investigated. Three spacings (6.1, 12.2 and 24.4 m) and two drain depths (0.9 and 1.4m) were included. All subsurface drainage systems controlled the water table adequately. Sugarcane and sugar yields for the subsurface drained treatments were significantly higher than yields for the undrained treatments during the wet year (1975) when the water table in the undrained treatment was very high during most part of the year. Although yield increases were obtained only in one year of three year study (wet year only), the magnitude of the increase for that year was large enough to defray the significant portion of the system installation cost.

194

CAMPBELL, K.L.; JOHNSON, H.P. Hydrologic simulation of watersheds with artificial drainage Water Resource. Res., 1975, 11:120-6.

A deterministic hydrologic watershed model that simulates the watershed discharge and soil moisture status continuously throughout the crop season was developed for drainage watersheds with depressional storage, infiltration, surface runoff, soil profile storage, percolation to the water table, sub-surface tilled drainage , soil moisture redistribution, evapotranspiration, and routing through depressions, tile mains, and the drainage ditch. The resulting outputs are daily evapotranspiration, soil moisture status in the crop root zone, and watershed discharge. The model simulated the watershed discharge and soil moisture status of the 61.2 km^2. East Fork Hardin Creek watershed near Jefferson, Lowa, for the 1964 crop season. Discrepancies between the measured and predicted discharges. were in part due to spatial variations in precipitation.

195

CAMPBELL, K.L.; ROGERS, J.S.; HENSEL,, D.R. Waterable control for potatoes in Florida. Trans. ASAE, 1978, 21:701-5

Subsurface drain tubes 18 and 36 m apart were compared with conventional water furrows 18 m apart for drainage and irrigation of potatoes. The water table returned to the desired level after rain more quickly in the subsurface-drained plot. Drains 18 m apart provided adequate water table control during irrigation; those 36m apart did not. Subsurface drains resulted in more efficient water use, with 10 percent higher tuber yield and 12 per cent less water use. Subsurface drainage permitted 12.5 per cent more land utilization and increased yield by 23.3-56.4 percent with an increased proportion of No. 1 grade tubers.

196

CANADA AGRICULTURE. Glossary of Terms in Soil Science. Deptt. of Agriculture, Ottawa, Canada, 1459, 1972

(Rev. 1976), 44 p.

Contains definition of terms associated with Soil Science. Terms commonly used in drainage are also included. All terms are arranged in alphabetical order.

197

CANNELL, R.Q. Effects of soil drainage on root growth and crop production. In Soil physical properties and crop production in the Tropics. (Lal, R.. Greenland, D.J. Eds.) John Wiley and Sons,New York, 1979, 183-97.

Response of waterlogging to crop growth is reviewed. Occurence of anaerobic conditions are responsible for injury to plants. Effects of waterlogging on root growth, root function, crop yields under constant and fluctuating water-table and root distribution are reported. Effect of drainage on soil physical properties and nitrogen suppply is discussed. For waterlogged conditions genotype which are more tolerant of anaerobic conditions should be selected.

198

CANNELL, R.Q. Crop growth in drainage experiments in the field and lysimeters in the UK. In Land drainage (Ed. Gardiner, M.J.) A.A. Balkema, Rotterdam,Netherlands, 1982, 309-14.

The results from lysimeter experiments with wheat, barley, oilseed rape and peas and early findings from field investigations indicate that, except for short periods, water tables should not be less than about 50 cm below the soil surface (though the range of acceptable depths is unknown). When the water table rises to the soil surface, drainage systems should be designed to remove this excess water within six days for winter cereals, and within two days where peas are grown. If waterlogging of winter cereals before emergence can be avoided by other meas, e.g. earlier sowing, then wider drain spacing would be adequate to avoid direct effects of waterlogging on crop growth. For these crops drainage may be more important to improvtrafficability for timely application of herbicides and fertiliser.

199

CANNELL, R.Q.; BELFORD, R.K. Crop growth after transient waterlogging. 4th Natn. Drain .Symp. Proc. ASAE, 1982, Pub.12- 82, 163-70.

The results indicate that, except for short periods, water-tables for the main crops in the United Kingdom should be maintained at about 50 cm or more below the soil surface. when the water-table rises nearer to the soil surface, drainage systems should be designed to remove this excess water within 4 or 5 days for winter cereals to avoid pre-emergence waterlogging, and within 2 days where peas are grown. If the risk of waterlogging of winter careals before emergence can be avoided by earlier sowing, or the effects lessened by heavier seed rates, wider drain spacing would be adequate to avoid direct effects of waterlogging on crop growth, because the rate of decline of the water-table would be less important. For clayey soils, it is advisable to maintain the water-table about 50 cm below the surface to minimize shrinkage compaction and structural damage. With clayey soils lowering of the water-table to a suitable depth to give both the direct and indirect benefit can be achieved economically by closely spaced mole drains draining into a piped system.

200

CANNELL, R.Q.; BELFORD R.K. Effects of prolonged waterlogging under outdoor conditions on the growth of winter wheat. Ann. Rep. Agricultural Research Council Laboratory, Wantage UK. 1978, 59-61.

Wheat cv. Maris Huntsman waterlogged for 25 days on 19 Nov. yielded 837-850 g grain/m^2 while wheat waterlogged for 120 days yielded 876g/m^2. Wheat waterlogged for 47 days on 2 Feb yielded 847-891 g/m^2 while that waterlogged for 80 days yielded 785 g/m^2. Waterlogging on 24 April for 25 days gave 917 g/m^2 compared with 937 g/m^2 for wheat grown with free drainage. Treatments had a slightly greater effect on vegetative growth and straw yields than on grain yields. The yield component most affected by waterlogging was number of fertile tillers. Tiller initiation and growth occurred at a low rate during waterlogging. Aerenchyma was formed in the cortex of nodal roots during waterlogging; such roots could grow to a depth of several cm in waterlogged soil.

201

CANNELL, R.Q.; BELFORD, R.K. Effects of waterlogging at different stages of development on the growth and yield of winter oilseed rape (Brassica napus) J. Sci. Fd. Agric., 1980, 31:963-5.

Oilseed rape, growing in a sandy loam soil in lysimeters, was subjected to waterlogging to the soil surface at different stages of growth. Brief periods of waterlogging in Dec. and Jan. (10days) or in May(10 days) resulted in slightly shorter plants but hardly affected the yield. Waterlogging in Dec./Jan. for 6 wk at 1-2°C slightly restricted leaf development and delayed flowering, but yield was unaffected. At the end of waterlogging for a similar duration between Jan. and March(mean temp.6°C) the height and leaf area of the plants was approximately halved, the number of senescent leaves were almost doubled and flowering accelerated; at harvest in July these plants were 17percent shorter and yields of seed, oil and straw were 14, 17 and 23 percent less, resp., than for plants in freely drained soil. During the prolonged waterlogging, short flashy white roots developed slightly below and on the soil surface.

202

CANNELL, R.Q., BELFORD, R.K.; GALES, K.; DENNIS, C.W.; PREW, R.D. Effects of waterlogging at different stages of development on the growth and yield of winter wheat. J. Sci. Fd. Agric., 1980, 31:117-32.

The response of winter wheat to waterlogging at different stages of growth outdoors was studied for 3 yr using lysimeters containing undisturbed monoliths of sandy loam and clay soils. The treatments were selected as the most extreme likely to occur in field conditions in Britain, i.e., waterlogging to the soil surface for up to 16 days in Oct., 42 days in Jan. and Feb. and 6 days in May. Waterlogging in mid winter was common to all experiments. Yield losses from mid winter waterlogging ranged from 15 percent when yields were heavy (10t/ha) to nil when yields were light (3-4 t/ha) after the plants had been affected by frost damage in late winter. Winter wheat was most sensitive to waterlogging after germination but before emergence. At this stage, 16 days waterlogging killed all seedlings and 6 days waterlogging depressed plant populations to 12 percent (clay) and 38 percent(sandy loam) of the control. However, there was vigorous compensatory growth in the remainder of the growing season and yield was only depressed to 82 per cent of the control. When the crop was waterlogged at any stage after emergence , plant populations were not affected, but winter waterlogging treatments usually depressed shoot numbers and, to a lesser extent, ear numbers at harvest. Shoot survival under waterlogged conditions was related to take-all (Gaeumannomyces graminis var. tritici) and was increased by waterlogging.

203

CARDENAL, J.B.; BUENO, G.C. Present subsurface drainage techniques in the Ebro watershed (Spain). 10th Congr. Irrig.Drain. Athens, 1978, 4:277-84.

The drainage scheme adopted for the new irrigable

lands in the semi-arid Ebro valley is outlined. The purpose of this scheme is to prevent or reverse soil salinization due to the rise of saline water tables. A summary of the stages in the execution of the project is given.

204

CARDOSO, J.C. Reclamation of saline soils in Portugal. Agrokem. Talajt. 1969, 16(suppl):329-35.

Halomorphic soils occur on marine alluvium, covering about 22,700 ha. The saline soils correlate well with solonchaks. Spartina stricta, Atropis convoluta and Arthroeneun perenne had the most active role on sediment fixation and on micro-relief formation before reclamation started. Possibility has been explored to reclaim the soils without irrigation. Good yields of wheat (3t/ha) were obtained in the second year of trials under non-irrigated conditions because of spring rains. The drainage system established at a 1 m depth was generally adequate, in non-irrigated conditions. Rice has been the main successful crop under irrigated conditions. A 10 cm thick mulch of sand (90 per cent of the particles with a diameter of 0.2-1mm) increased the water content of the upper 15 cm of the soil, reduced salinity and increased the yield of tomatoes. The cost of mulch application varied from about $800 to $3000(US dollars) per ha. The cost of reclamation of saline soils on marine alluvium is $ 1200 and 2400 with no irrigation and irrigatioon, respectively.

205

CARLIER, M. Recent development in drainage in France. C.R. Acad Agric. Fr., 1979, 65:535-7.

A brief summary of the history of drainage of agricultural land in France is given. Improved technology and greater awareness among farmers of the profitability of subsoil drainage have led to a great increase in its use. Research into the behaviour of drained soils has developed in recent years, and has given rise to a new drainage policy in France.

206

CARREKER, J.R.; LONG, F.; DANIELS, J.M. Drainage properties of some soils in the Atlantic Coast flatwoods. Trans. ASAE, 1967, 10:337-9.

The Atlantic Coast flatwood soils are sedimentary and range from sand on the low ridges to clay on the more level areas with some organic soils in the lowest part. Climate and soil data indicated that the drainage problem of these soils can be tackled by a combination of surface and subsurface drainage. The surface drainage should include major outlets, field drains and land grading. The augmentation of system with subsurface drains may be made to permit wider spacing of the field drains. The subsurface drains spaced at 12, 24 and 36 m will drain the 30 cm depth of the C horizon in 1.08, 4.3 and 9.6 days.

207

CARSON, W.M.; WATTS, K.C.; SARWAL, S.N. Galerkin's method in Agricultural Engineering. Can. Agric. Engng., 1979, 21:125-30.

This paper discusses a powerful method for obtaining explicit solutions for Agricultural Engineering problems which involve differential equations. The use of Galerkin's method is demonstrated by solving five sample problems of increasing complexity including a falling water table case of subsurface drainage.

208

CARTER, C.E. Drainage parameters for sugarcane in Louisiana. 3rd Natn. Drain. Symp. ASAE, Michigan, 1976, 135-8.

Three water table experiments which rep: ented various drainage situations were conducted at

Baton Rouge, Louisiana, during 1967 through 1973 to evaluate the response of sugarcane to drainage and to evaluate two crop-related parameters, y and t needed in transient subsurface drain spacing equations. Parameter y is the height of the water table above the drain lines midway between them after time t. Parameter t is the time in days that can be allowed for the water table to decreese from, its initial level y_o to depth y. These experiments were conducted on 0.004 ha concrete bordered plots equipped to maintain the water table at any level from the soil surface to 1.5 m below. Cane and sugar yields were increased by subsurface drainage. The number of crops, harvested from one planting were also increased from the normal 3 to 5 . Poor drainage during sugarcane's dormant season, December through March, adversely affected yields from the next crop, thus indicating a special need for drainage during this time. The value of y for sugarcane in transient subsurface drain spacing equations should provide a water table, in silty clay soils, at least 1.2 m below the soil surface for best results. The value for t for sugarcane in silty clay soil should not exceed 7 days during the growing season. A smaller t value may be needed if the probability of rainfall during 7-days period is high.

209

CARTER, C.E.; CAMP, C.R. Subsurface drainage of an alluvial soil increased sugarcane yields. Trans. ASAE, 1983, 26:426-9.

An experiment was conducted on a 10 ha tract of commerce silt loam in St. James Parish, Louisiana to determine the response of sugarcane to subsurface drainage. Cane yields from a drained tract were compared to those from an undrained and it was observed that cane yields were higher in 3 out of 4 years of study in the drained plots. The value of the increased cane yields was enough to pay for a subsurface drainage system at existing cane prices and drain installation costs.

210

CARTER, C.E.; FLOYD, J.M. Subsurface drainage and irrigation for sugarcane. Trans. ASAE, 1973, 16:279-82,84.

Water management practices, particularly subsurface drainage, increased annual sugarcane yields from 24 to 62 percent over those from check plots and increased the number of crops grown after a single planting from the normal three to five. No response to irrigation was recorded. Sucrose content of the cane juice decreased as cane yields increased. Extra cane produced by the treated plots more than compensated for lower percent sucrose; consequently, sugar yields per ha were higher from the treated plots.

211

CARTER, C.E.; IRVINE, J.E.; CAMP, C.R. A system for managing water in fine textured soils Trans. ASAE, 1983, 26:767-71.

A water management system which compensates for erratic rainfall was tested on silty clay and silt loam soils in Louisiana. The system that was tested consisted of bordered areas in which subsurface drains were installed to remove water during wet period. The same system was causing the water table to rise into the rootzone. The system tested on $40m^2$ and 0.4 ha bordered plots was found successful.

212

CARTER, C.E.; Mc DANIEL, V. ; BENGTSON, R.L. Potential for subsurface drainage in the lower Mississippi Valley. 4th Natn. Drain. Symp. Proc. ASAE, 1982, Pub. 12-82:23-33.

High water tables are a serious problem in the lower Mississippi Valley. These water tables be lowered by subsurface drainage. Sugarcane, soybean, wheat and corn silage all responded favourably when the water tables were lowered with maximum annual yield increases of 33,39, 89 and 73 percent, respectively. With only one exception, the value of the average annual crop yield increase due to drainage indicated that subsurface drainage systems could be amortized in a short period; thus a great economic potential for subsurface drainage exists in the lower Mississippi Valley.

213

CASSELMAN, T.W.; GREEN, V.E. Jr.; Plastic line mole drains in organic soil. Proc. Soil Crop Sci. Soc. Florida, 1971, 31:12-5.

The drains, laid at spacings from 3 to 24 m in a shallow organic soil, had no significant effect on the water table level due to seepage from surrounding areas, and the drain spacing had no effect on the growth of maize.

214

CAVELAARS, J.C. Composing a drainage pipe line out of sections with different diameters. Proc. Int. Drain. Workshop, Wageningen, Netherlands(1978), 1979, 25:402-13.

Formulae and graphs for hydraulic design of drain pipe lines, assuming varied flow and full-flowing pipes, are only valid if the entire pipe line consists of one and the same diameter. If a drain is to be composed of sections with different pipe diameters, then a correct design can be obtained by using each diameter till a certain percentage only of the drainable area (or drain length) as derived from the formulae and graphs. Percentage to be taken for practical design purposes are: 85 percent in case of 3 or more different diameters.

215

CAVELAARS, J.C. Subsurface field drainage systems. In

Drainage Principles and Applications. Vol. IV Design and Management of Drainage Systems. Intern. Inst. Land Reclama. Improve. Wageningen, 1974, 16:1-65.

The design, constuction and maintenance of the various types of field drains are described.

216

CENTRAL BOARD OF IRRIGATION AND P(WER. Proc. Sym. Waterlogging Causes and Measures for Prevention, New Delhi, 1972, No. 118,I, II and III: 175p, 170p, 88p,.

Proceedings of the symposium on waterlogging run in three parts. First part includes 18 papers on causes of waterlogging (3); water-table fluctuations and its interpretation(7); design criteria for drainage channels and their maintenance(6) and cropping patterns for waterlogged areas without amelioration (2). The second volume contains 18 papers on waterlogging problems, their effect and remedial measures. Waterlogging problems in different command areas of India and Srilanka are presented in 12 papers. The third volume contains the discussions on the papers presented at the symposium.

217

CERVANTES, A.L. Results of a field experiment on reclamation of saline soils. Bol. Azuc. Meseic., 1959, 116-17:19-21

Tile drainage of wet, saline clayey soils of pH 8.2-8.4 increased sugarcane yield about 3 times, improved soil structure and decreased soil pH to 7.2-7.6.

218

CHACHAVA, Yu. N. Water regime and aeration of waterlogged soils in the Kolkhida low land. Pochvovedenie, 1976, 4:125-32.

Data on the dynamics of saturation and effective aeration of the waterlogged clayey Gley soils of this lowland are presented and methods for improving them are discussed. Convex profiling of the soil surface promotes drainage of excess surface water and some what improves aeration of the plow layer.

219

CHALKIAS,, N. The design of tile drainage systems. Bull. G. C. I. D. 1965, 17: 5-22.

The criteria for designing tile drain systems have been discussed. A nomograph to work out the size of the tile has been drawn. An illustrative example of designing a sub-drainage system, has also been incorporated.

220

CHAO, C.C.; CHENG, Y.S. Effect of tile drainage on rice soils in Suchow district, Kiangsu. Acta Pedol. Sinica, 1978, 15:187-93.

On a slighty acid silty clay loam derived from Lacustrine deposits, utilized under a three-crop annual rotation (rice-rice-winter crop usually wheat) wheat growth was often retarded by wetness of the surface soil as a result of high groundwater levels, which normally occurs during a short rainy period in spring, when daily rainfall often exceeds 50 mm. By laying drainage tiles at 110 cm depth, with intervals of 9 m between the tiles, the groundwater was kept at lower levels, so that under the levels <50 cm below the soil surface lasted only 1-2 days. Experiments with P^{32} in the form of KH_2PO_4 showed that as a result of this treatment, the actively absorbing roots reached about 40 cm depth, which was much deeper than those in neighbouring check plots.

221

CHAUDHARY, T.N.; BHATNAGAR, V.K. PARIHAR, S.S.

Drainage requirements of crops: response of maize and wheat to soil submergence and depth of ground water. Soil Conserv. Digest, 1974, 2:8-14.

Maize and wheat plants are able to adapt to conditions of poor drainage for 1 and 3 days, respectively. To ensure optimum crop production, therefore, excess water should be drained from the root zone within this period. To prevent soil salinization, increase in the storage zone for rain water, and ensure utilization of water and nutrients by roots from deeper depths, lowering of ground water to 120 m depth is desirable. Further increase in water table depth would necessitate larger and more frequent irrigations. The depth of water table should be modified according to the salt concentration of the ground water. A leaching schedule whenever conductivity exceeds 5 to 6 dS/m would be desirable.

222

CHAUDHARY, T.N.; BHATNAGAR, V.K.; PARIHAR, S.S. Growth response of crops to depth and salinity of groundwater and soil submergence. I-wheat (Triticum aestivum L) Agron. J. 1974, 66: 32-5.

Response of a semidwarf wheat to four depths (60, 90, 120 and 150 cm) and three salinities (EC 0.5, 2.9 and 5.2 dS/m) of groundwater was studied in lysimeters filled with silt loam during 1970-71 and 1971-72. The plants were irrigated with water of EC 0.5dS/m when the lysimeters with a 150 cm water table depth indicated a tension of 0.5 bar at 75 cm depth. Groundwater of EC 0.5 dS/m at 60 and 90 cm depths gave highest yields. Comparable increase in the salinity of groundwater caused greater reduction in yield with shallow water tables than with deep water tables. It is indicated that critical depth of groundwater, for optimum crop production, would vary in relation to its salinity. Growth and yield response of wheat to soil submergence was investigated in another set of lysimeters with water table at 60 cm. Soil submergence up to two days did not affect growth.

It decreased yield gradully upto 5 days and sharply thereafter. Prolonged soil submergence reduced protein content in the grain as indexed by N content. To ensure optimum grain quality and production, excess water must be drained from the root zone within 3 days of submergence.

223

CHAUHAN, H.S.; RAM, S. Surface and sub-surface drainage hydrology. Symp. Waterlogging causes and Measures for its Prevention. CBIP pub. no. 118, 1972, 1:19-28.

Hydrological studies were conducted on a combined system of surface and subsurface drainage at three drain spacings of 35, 25 and 15 m for two given replenishments of 2 cm/hr lasting for 4 hr and 2 cm/hr lasting for 9 hr. The subsurface drains were laid at a depth of 1 m. The experiment was located in the humid region of Tarai area at Pantnagar. Narrow spaced subsurface drains resulted in higher steady discharge rate and total volume removed, considered per unit area of the plot than the drain at farther spacing. At a rainfall rate of 2 cm/hr and rainfall depth of 8 cm, created through sprinklers, the tiles removed 4.92 to 30.25 percent of the applied precipitation as the tile spacing decreased from 35 to 15 m. The combined system removed 38 to 59.75 per cent of applied precipitation. Such findings based on data of larger periods can be utilised for estimating drainage coefficients for a given area. U.S.SCS recommendations for average drain spacing and depth based on textural classification can possibly be utilized for working drain spacing after experimental verification for a given area.

224

CHENG, Y. Drainage of paddy rice and its significance. Acta Pedol. Sin, 1983, 20:215-24.

Drainage increased the percolation rate of paddy soils by about 10 mm/day and gave yield responses of 10-20 per cent.

225

CHERNENOK, V. YA. Water balance of the aerated zone during drainage of bottomland peat bogs. Vest S-Kh. Nauki Mosk, 1969, 8:110-4.

In 1963-67, in the central bottomland of the river Yakhrom, evapotranspiration during the vegetative period drained by closed drains at depths of 1.7-2.0 m or 0.8-1.2 m exceeded precipitation by 69-114 per cent, depending on the crop and drain depth. Evapotranspiration was less with the deep than with the shallow drains.

226

CHIDLEY, T.R.E.; GOODWILL, I.M. A compact watershed model system. J. Hydrol., 1975, 24:155-69.

A framework for implementing watershed simulation and drainage network studies on a small computer is described. The method is illustrated by reference to a catchment model built for a large catchment in Botswana.

227

CHIENG, S.T.; BROUGHTON, R.S.; AMI, S.R. Graphical solutions to drainage equations. Can, Agric. Engng., 1981, 23:91-6.

Multiple correlation graphs for solving Hooghoudt's drain-spacing equation for homogeneous soil were developed. Graphs for obtaining Hooghoudt's equivalent depth of flow below the drains for seven commonly used subsurface drain pipe were constructed. A new multiple correlation graph for determining the required size and maximum allowable length of drain pipe is presented.

228

CHIENG, S.T.; BROUGHTON, R.S.; FOROUD, N. Drainage

rates and water-table depths. J.Irrig. Drain. Div. ASCE, 1978, 104:413-33.

Precipitation and evapotranspiration data for 27 years from three stations in the Ottava and St. Lawrence Lowlands have been used together with a hydrologic model of the drainage process to obtain the duration of particular water table heights to be expected. A computer was used to make day-by-day calculations of expected water table heights for a range of conditions including drainage coefficients of 3-15 mm/day, drain depths of 1, 1.4 and 1.8 m, drainable porosities of 0.03-0.11, and available soil moisture in the root zone of 60-151 mm. A statistical frequency analysis of the output was made. The results were prepared in tabular and graphical form to show the effect of drainage rate on water table height. By using a criterion for the number of days, the water table may be at or less than a particular depth, one can select a design rate.

229

CHIENG, S.T.; BROUGHTON, R.S.; KOK, R. Computer-aided drafting of subsurface drainage plans. Can. Agric. Engng., 1980, 22:137-43.

A systematic approach to preparing spot elevations and contour topographic maps from field survey data by using a high-speed computer plotter has been made. Computer programme in FORTRAN language was written to design subsurface drainage system from assigned design parameters and to draw plans of those designs by means of a computer-controlled plotter. The new computer aided design and drafting approach reduces the time and cost in drafting subsurface drainage plans to about 1/5 of that needed for the conventional manual method.

230

CHILDS, E.C. The achievements of drainage theory in relation to practical needs; taking stock. Trans. 9th int. Cong. Soil Sci., 1968, 1:349-56.

A discussion of the theoretical-practical advances in drainage during the last 30 years, has been presented.

231

CHILD, E.C. The physics of land drainage. In Drainage of Agricultural Lands (Luthin, J.N. Ed.). Am. Soc. Agron. Madison, Wisconsin, 1957:1-78.

A chapter outlines briefly the salient relevant structural features of soil constituents, dependence of the static physical properties of soil upon the prevailing moisture content and the influence which these properties have upon properties other than physical. Laws of soil water movement and application of these laws in the particular circumstances viz. under drainage situations are presented to lead to the formulation of drainage problem in precise mathematical terms.

232

CHILDS, E.C. A treatment of the capillary fringe in the theory of drainage. II. Modifications due to an impermeable sub-stratum. J. Soil. Sci., 1960, 11:293-304.

When a drainage system copes with both artesian and precipitated water, the streamline dividing the two flow zone passes through a stagnation point in the medial vertical plane parallel to the drain lines. The approximation is made that the water-table and capillary fringe boundary may be regarded as those appropriate to precipitated water only, in the presence of an impermeable bed passing through the stagnation point. While the absolute heights may be in error it is shown that the partition of the capillary fringe about the water-table location in the absence of a fringe is given with only small error and this partition is calculated from hodograph theory with the aid of Edsac II for combinations of three values of precipitation rate, seven values of fringe thickness, and nine values of depth of impermeable bed. By interpolation, curves are presented to show the partition for combinations of three selected precipitation rates, five depths of bed and continuous variation of fringe thickness. It is shown that for high precipitation rates the greatest fringe thicknesses are accommodated almost wholly above that surface which would have been the water table in the absence of a fringe, while for the lowest rates of precipitation and zero bed depth only 30 per cent of the fringe is so accommodated, the remainder is below.

233

CHILDS, E.C.;YOUNG, E.G. A study of some three-dimensional drainage problems. Soil Sci., 1961, 92:15-24.

The Dupuit-Forchheimer approximation has been adopted to deduce relationship between rainfall, soil hydraulic conductivity and the geometry of the drainage system for three dimensional systems comprising arrays of pumped wells and rectangular meshes of major and minor drains. The theoretically deduced relationships are tested by experiments with drainage systems set up in the laboratory and also by comparison with numerical solutions for particular cases obtained by relaxation method. Agreement between theory and observations is satisfactory.

234

CHISCI, G.C.; ZANCHI, C. Preliminary results of drainage in a hilly district. An. Inst. Sperim. Stud. Difesa. Suelo, Firenze, 1971, 1:167-91.

Subsurface drainage of a clayey-loamy soil in the Florence region by PVC tubes increases the flow of surface water and groundwater, and can cause temporary saturation of the rooting zone at the end of winter. Drainage decreases runoff and erosion on soil tilled and sown in autumn, and increases dry-matter yield from grass/legume swards.

235

CHOWDHURY, P.K.; ANJANEYULU, B. Drainage by partially penetrating recharge wells in a leaky aquifer. J.Hydrol., 1978, 39:51-7.

Recharge wells can be considered as a possible solution for the drainage of waterlogged areas which have no gravity outlets. A knowledge of the capacity of the recharge wells is useful in estimating the number of wells needed to provide drainage relief to a particular area. Laboratory experiments were conducted on a sand model simulating a leaky aquifer underlying an unconfined aquifer. The unconfined aquifer in the model was overlain by topsoil which was kept ponded to simulate a waterlogged condition. The recharge characteristics of a well partially penetrating the leaky artesian aquifer were studied under different conditions. Based on experimental data, empirical equations are proposed for predicting the recharge rate.

236

CHU, S.T., Water table response to a sequence of recharge. Water Resource.Res., 1977, 13:738-42.

Water table behaviour under subsurface drainage conditions was analyzed by considering water table fluctuations in response to a sequence of recharge events. The derived solution is applicable to a nonhomogenous soil profile. An iteration procedure was introduced to illustrate the application of the theoretical results. Agreement between theoretical analysis and field data from a drained plot at Redfield,South Dakota was shown to be adequate.

237

CIESLINSKI, Z.; ROGUSKI, W. Moisture content in soil and productivity of grasslands on alluvial soils on peat in the Vistula valley drained with ditches and drain pipes. Wiadom. Inst. Melior. Uzytkow Zielon., 1974, 11:125-59.

Pipe drains at 42- m centres were found to be more effective than ditches for grassland drainage on peat soil. The data on soil moisture are given. Grassland at Blota containing *Festuca pratensis, Dactylis glomerata* and *Agropyron repens* given 2 hay cuts/year in 1965-9 yielded 6.67 t/ha without fertilizer, 7.47 t/ha with 80 kg P_2O_5 and 120 Kg K_2O and 10.33 t/ha with PK and 120, 150 or 200 kg N/ha. Percentage herbs and weeds in the hay declined on all plots, whereas that of legumes increased on unfertilized and PK plots and declined where N was applied. In trials at Plutowo in 1965-9, pasture grassland containing *Poa pratensis, Festuca rubra* and *Poa trivialis* cut for hay yielded 6.69 t/ha (2 cuts) without fertilizer and 9.57 t/ha with 120 kg N (in 2 dressings), 80 kg P_2O_5 and 120 kg K_2O/ha. Corresponding yields for a nearby *Festuca Pratensis/Dactylis glomerata* meadow sward were 8.54 and 12.01 t/ha. Data are given on the content of N, P, K, Ca, Mg, Na, Cu, Mn and Mo in hay from all sites.

238

CIESLINSKI,Z. ; WANKE, A. Effect of agro-reclamation measures in connection with technical measures in Poland's conditions. 10th Congr. Irrig. Drain.Athens, 1978, 151-4.

The effects of ploughing with a subsoiler (30+12 cm), deep ploughing (50-70 cm), soil loosening and mole drainage (50-70 cm) combined with tile drainage on some physico-hydrological and biological properties of heavy soils are discussed.

239

CLEMENS, J.; KIRK, A.M.; MILLS, P.D. The resistance to waterlogging of three Eucalyptus species-Effect of waterlogging and ethylene releasing growth substance on *E. robusta, E.grandis* and *E. saligna*. Oecologia,

1978, 34:125-31.

Seedlings of E.grandis were more resistant to waterlogging than those of E.robusta and both were more resistant than E. Saligna. This result, unexpected in view of the natural habitats of the 3 species, was related to the ability of the resistant plants to form adventitious roots at or just below the water surface. The ethylene-releasing growth substance 2- choreethylphosphonic acid (ethephon), applied to non-flooded plants, was used to study the role of ethylene in the response to waterlogging (it produces symptoms similar to those of waterlogging). There was no difference between the species in their response to ethephon. Plant height and total dry weight decreased linearly with increasing stem ethylene production.

240

COINTEPAS, J.P. Saline water irrigation and drainage in Tunisia. Cah. Pedol. Orstom, 1965. 3:299-305.

Importance of deep drainage under saline-water irrigation is discussed with reference to various locations in Tunisia. Drainage depth is determined by plant type and soil physico-chemical characteristics.

241

COLE, G. Land drainage in England and Wales. J.Inst. Water Engrs. Scient., 1976, 30:345-67.

The paper reviews and summarises developments in design, construction and maintenance of land drainage systems since 1954.

242

COLES, E.D. Some notes on drainage design procedure. Proc. S.Afr. Sug. Techn. Assn., 1968, 189-99.

Drain spacing formulae were considered suitable for designing subsurface drainage systems. The various formulae can be solved directly, but graphs have been included to simplify the solutions of different equations.

243

COLLINS, H.J.; CHRISTOPH, F.; REGNER, F. Indirect connection of lateral pipe drains to main pipe drains by filter material in drain trenches. 10th Congr. Irrig. Drain. Athens, 1978, 247-59.

Hydraulic tests have been conducted on a laboratory model, scale 1:1, in which laterals are connected to the main drain indirectly by means of a gravel filter. Comparisons were made of corrugated PVC pipes differing in position and type of slots and in water outlet areas. Various gravel filters, and the influence of the height of the lateral above the main pipe were tested. Results with this model are encourging.

244

COMMONER, B. A system approach to evaluate nutrient and pesticide transport from artificially drained cropland. Environ. Mgmt., 1979,, 3:123-32.

An inventory of artificially drained croplands in the USA is presented. The results of studies to determine losses of nutrients, sediment and pesticides in drainage water are summarised. Losses are extremely variable. Average losses of nitrate through subsurface drainage and surface runoff are 17.6 and 11.6 kg/ha/year, respectively. Total phosphorus is lost mainly through surface runoff (average 1.3 kg/ha/year). Sediment yields average 505 kg/ha/year. Losses of the herbicides atrazine and dicamba average 5 per cent of amounts applied, while losses of the insecticides aldrin and heptachlor are less than 0.01 per cent of amounts applied. An interdisciplinary system approach, utilizing a computer simulation model for predicting nutrient and pesticide load in sediments and water from artificially drained cropland, is described.

245

CONCARET, J. et al. Agricultural drainage, Theory and Practice. Chambre Regionale d'Agriculture de Bourgogne, Dijon, France, 1981, 509 p.

This is a multi-author book prepared with the collaboration of research workers, lecturers, field engineers, technicians and extension officers. Five sections and four appendices deal with the techniques and applications of drainage, equipment and material, drainage networks and their maintenance, including pedological and hydrological background information, historical and legal aspects, and methods of soil analysis in relation to drainage problem.

246

CONSTANTINIDES, N.J. Field irrigation and drainage in the Mouria and Agoulinitsa lowlands, Greece. _8th Congr. Irrig. Drain. Varna_., 1972. Q 27 R 28. 439-44.

The paper deals with the irrigation and drainage techniques to be used in the newly reclaimed Mouria and Agoulinitsa lagoons. A sprinkling irrigation system is under construction and its several details are included. Drainage of the lagoons may be effected by tertiary drains at a minimum depth of 1.8 m spaced at 200 m. For new reclamation spacing may be reduced to 40-100 m depending upon the texture. Suggested cropping patterns and anticipated economic results are given.

247

COOKE, G.W. A review of the effects of agriculture on the chemical composition and quality of surface and underground waters. _In_ Agriculture and water Quality. Ministry of Agric. Fisheries and Food, London, _Tech. Bull_., 32:5-57.

A review paper on effect of agriculture on the

quality of surface and ground water. The chemical composition of drainage water in Britain and other countries of the world in relation to agriculture practices, has been reviewed.

248

COOPER, T. Practical land drainage. Leonard Hill Co, London. 1965, 162 p.

The book contains 21 chapters divided in five parts. Three chapters deals with basic information on drainage, 5 chapters on old drainage system, 4 chapters on tile drainage, 4 chapters on mole drainage and 5 chapters on special situatons.

249

COREY, A.T. Pumped outlets for drainage systems. Trans. ASAE, 1981, 24:1504-7.

The storage capacity of sumps for pumped outlets of subsurface drainage systems often can be reduced in size and cost by using the disposal tile for supplementary storage. The practice is feasible where the disposal tile can be placed at an elevation below the grade of the collector tiles so that its periodic submergence does not interfere with drainage. This paper describes procedures for the design of sumps when additional storage is available in the disposal tile.

250

COULBOIS, P. Working and maintenance of the drainage ditches in the swamps of Charente-Maritime. 7th Congr. Irrig. Drain. Mexico, 1969, Q 25 R 7:73-86.

Analysis of the principles that fix the calculations of drains and ditches; their working and the principal means for maintenance are described.

251

COULBOIS, P. Drainage of clayey soils in Charente estuary. 8th Congr. Irrig. Drain.Varna., 1972. Q 27 R 19:287-96.

Heavy and saline soils of Charente estuary are difficult to drain with traditional open drains. Underground drainge with PVC pipes at different spacing and depths are being examined with application of soluble calcium salts to improve hydraulic conductivity. The paper presents the preliminary findings of this study.

252

COURTNEY, J.V. Gravel tunnel drainage in Ireland. Soil Water, 1979, 7:12-4.

A mole drainage system which forms as gravel filled channel is described. The advantages of the gravel filled system over conventional mole drainage include prevention of structural collapse of the channel in unstable soils and of deterioration by erosion and siltation, and increased flexibility in terms of soil moisture content during installation. A comparison of the gravel tunnel system with mole and conventional pipe systems on a silty clay soil showed that both mole and gravel tunnel drains were superior to simple pipe drains in lowering the water table and increasing grass yields. The grvel tunnel system was stable than the conventional mole system.

253

COVIAN, R.G. Experiences in Spain with subterranean drainage as means of saline and alkali soil reclamation. Irrig. Drain. Paper FAO, 1972, 16:31-61.

The various operations carried out in the Guardalquivir Marismas for desalinization of the soils through drainage and leaching (both rain and irrigation water) have been described. With an efficient drainage net work, the soils in the area can improve, becoming salt-free and alkali free vertisols without any regression in their improvement. Crop yields increased as salinity decreased, reaching the same level as adjacent normal soils, in 5 or 6 years.

254

COVIAN, R.G. Subsurface drainage in Spain, technical and construction aspects - critical judgment of mechanical drainage systems preferred in Spain. 10th Congr. Irrig. Drain, Athens, 1976, 4:175-81.

An appraisal is given of the use of trenching and trenchless machines as illustrated by drainage schemes in the Guadalquivir marshes and the Alto de Aragon. Particular attention is paid to the costs involved.

255

COVIAN, R.G. Subsurface drainage in Spain, technical and construction aspects - present subterranean drainage techniques in Spain. 10th Congr. Irrig. Drain. Athens, 1978, 5:261-77.

Drainage projects undertaken in the Gaudalquivir marshes and the Alto de Aragon are described in some detail in order to illustrate current drainage techniques in Spain. The costs associated with various methods are also compared.

256

CRECY, J. De. Structural behaviour of heavy soils and drainage systems. In Land Drainage (Gardiner, M.J.Ed) A.A. Balkema. Rotterdam, Netherlands, 1982, 73-84.

Soil structural changes in relation to drainage are discussed.

257

CROS, P.; DEVILLERS, J.L. Quality of coiled PVC drain pipes required with modern laying methods. 10th Congr. Irrig. Drain. Athens., 1978, 4:137-46.

This report underlines problems of fragility, longitudinal shock resistance, flexibility, suppleness, coefficient of hydraulic resistance to entry, free flow coeffcient and light-induced ageing.

258

CROS, P.; GUYON, G.; HERVE, J.J. Experimentation with agricultural drainage. 10th Congr. Irrig. Drain. Athens., 1978, 4:295-310.

The objectives, techniques and treatment of data in field drainage experimentation are presented.

259

CROS.P.; LECHAUX, H.; WOLSACK, J. Drain laying quality survey. Proc. Int. Drain. Workshop.,Wageningen, Netherlands, (1978), 1979, 25:590-8.

The efficiency of field drainage demands particularly careful work. A survey of the laying quality has been recently held in France. The present paper gives a quick description of data handling and processing and then analyses the respective influence of various parameters, among which those relative to soil and machine appeared to be most important.

260

CROS, P.; LECHAUX, H.; WOLSACK, J. Parameters influencing drain-laying quality. 10th Congr. Irrig. Drain. Athens., 1978, 4:33-42.

A compaign for monitoring the accuracy of installation of subsurface drains has recently been conducted in France. The influence of factors relating to the soil and to the equipment used is discussed.

261

CRUZ, J.B.; DE,L.A.; BALMASEDA, A.G. The Ebro delta drainage project. 8th Congr. Irrig. Drain. Varana,

1972, Q 27 R 22: 325-48.

A proposed change from rice growing to horticulture in Ebro delta which is low lying marshy area calls for drastic lowering of water table. The drainage system consists of sub-dividing each polder into a number of secondary units served by secondary pumping stations pumping the water from wells into which a number of subsurface collector drains discharge; leading the discharge to main open drains and to main pumping stations. The subsurface collector drains receive the discharge from subsurface in field drains. The PVC laterals in sandy and silty soils will be prewrapped in nylon tissue, the best filtering material judged from studies so far. The discussion on the whole system as a whole are presented.

262

DAGAN, G. Linearized solution of unsteady deep flow towards an array of horizontal drains. J. Geogphys. Res., 1964, 69:3361-9.

An approximate method for predicting water-table fluctuations caused by a variable recharge is presented. It is assumed that the drains are at the water table and that the drained permeable stratum is deep, homogenous and isotropic. The main approximation of the flow system is the linearization of the boundary condition on the free surface. In this way the problem is reduced to the solution of a Volterra integral equation of the first kind which is solved by a numerical technique. The results are presented in a dimensionless form in tables and graphs.

263

DAGAN, G. Spacing of drains by approximate method. J. Irrig. Drain. Div. ASCE, 1964, 90:41-6.

The paper studies steady seepage of rainfall through soil toward tile drains and drainage ditches by approximate method; replaces nonlinear condition for velocity potential on unknown free surface by linear condition in vicinity of drain and by horizontal solution by superposition of complex potentials; solves horizontal flow by ellipse equation; combines two solutions to derive formulas and curves for spacing of tile drains, at and below water-table, and ditches in both isotropic and nonisotropic soils; presents numerical applications.

264

DAGAN, G. Steady drainage of two-layered soil. J. Irrig. Drain. Div. ASCE., 1965, 91:51-64.

The paper investigates by an approximate method the steady seepage of rainfall through a two layered soil towards drains. In a zone adjacent to the drain line arises the boundary condition on the free surface whereas in the remaining region assumes the flow essentially horizontal. Thus derives linearlized solution by superposition of complex potentials in a semi-infinite strip and presents formulae for drain spacing. The use is illustrated by numerical examples.

265

DAGAN, G. Unsteady deep flow towards drains in anisotropic soils. J. Geophys. Res., 1965, 70:837-45.

The drainage of deep anisotropic soil layers was investigated by an approximate method similar to that for isotropic soil. The main approximation of the flow system is the linearization of the boundary condition on the free surface. The problem is transferred from the anisotropic medium to an isotropic one. In the transformation, the drain shape becomes elliptic. The potential distribution and the movement of the free surface caused by recharge are determined by conformal mapping. The mathematical problem is reduced to the solution of a Volterra integral equation of the first kind which is evaluated numerically. Results are given in tables and graphs.

266

DAHMEN, E. Drainage and irrigation requirements: an integrated approach. In Polders of the world. Volume 1. Polder projects. Land and Water Management Aspects. Int. Inst. Land Reclam. Improv., Wageningen, Netherlands, 1982, 443-54.

This paper describes an integrated approach to establish design criteria with emphasis on the determination of an optimum drainage modulus for lowland rice. The method used is a daily water balance computation, simulating water management and cultural practices at field level. Computed crop yields are then analysed for variability over the years and an optimum drainage modulus is determined, incorporating both farmers' benefits and overall project economics. Additional possibilties briefly touched upon are the estimation of required storage capacities in case of impeded drainage, and the analysis of the operation procedure and efficiency of the irrigation network.

267

DAHMS,P.; SOWA, E.; HILLIGER, D.; HEIM, S. Methods for measuring valocity, volume and pressure head of flow in drainpipes in relation to entry of water. Arch. Acker.Pflan. Bodenk., 1978, 22:89-96.

An outline is given of measuring methods applied during investigation on the hydraulic pressure in pipes and of the groundwater in systems, for groundwater regulation. An electronic water level indicator suitable for field use attains the required accuracy of ± 1mm. Current meters and salt velocity methods are suitable for measuring the velocity and the volume of flow in drainpipes during the entry of water. Portable measuring devices in the drain are used in combination with piezoelectric gauges. These methods were calibrated on a laboratory unit of half-shell drains. It was possible to record minimum velocities of 0.006 m/sec and flow volumes of 0.03

L/sec respectively.

268

DAI, T.X.; LI, X.L.; ZHANG, L.T.; YOU, W.R.; SUN, C.L.; LIU, W.Z.; XIAO, Z.H. Experiments on the reclamation of saline soils by well drainage in the coastal areas of Shandong Province. Acta. Pedol. Sin., 1980, 17:255-66.

In experiments during 1973-78 in coastal areas of Shandong Province, irrigation water was conducted from the Yellow River and pumping wells in connection with open ditches were constructed for drainage. Soil permeability in the experimental area was poor (about 0.7 m/day). Compact clay layers of 2-7 m thick occurred at a depth of 10-20 m below the surface, which impeded water percolation from the upper part of the solum. The yield of a pumping well of 40 cm dia and 34-38 m depth (constructed with a cement conglomerate wall 5 cm in thickness) was only 5-10 m^3 of water per hour;drained water mostly originated from underground water lying at 10 m below the land surface. The efficiency of a single well was confined to an area about 30-60 m in diameter. More recent observations have shown that pumping wells should be situated above the compact clay layer, i.e. not exceeding 10-20 m below the surface in most cases. Distances between the pumping wells should not exceed 120 m. The use of large pumping wells of about 70 cm diameter or vacuum tube pumping in connection with the drainage seemed to be feasible.

269

DAISHEV, T.I.; SANNIKOV, G.P. Some problems connected with operations of drainage system. ICID Bull., 1968-69, 17-18:15-20.

Studies on silting of drainage systems have shown that to avoid silting it is necessary to increase the minimum slope of drainage lines upto 0.004-0.006 in order to avoid reverse slope and depressions. Gradient of seepage head at gaps reach 30-35 but 3 cm from the gap they are less than unity. So the zone of dangerous piping does not exceed 3-4 cm. Factors affecting ice clogging of drainage system are discussed. Use of herbicides to remove undesirable plants from drains is recommended by placing them on the bottom of the drain in spring. This use of herbicides reduces expenses and increases labour productivity.

270

DAMOUR, L.; DUPRAT, J.; JANIN, J.L.; ROLLAND, L. Drainage of heavy soils in the marshes of Charente Maritime. 10th Congr. Irrig. Drain. Athens, 1978, 4:63-89.

Charente Maritime contains 80,000 ha of coastal marshes on very heavy soils which are being drained for conversion into arable land. This report describes the conditions of the marshes, the present state of the hydraulic network, and field studies undertaken to determine appropriate drainage techniques and farming systems.

271

DANNOWSKI, R.; QUAST, J. Comparative studies of steady-state geohydraulic models for drain spacing determination. Arch. Acker Pflan. Bodenk., 1983, 27:559-66.

The principles underlying the use of steady-state drain spacing equations for characterizing systematic pipe drainage in homogenous, isotropic aquifers are analysed. Variants are calculated to enable a comparison of the simpler equations of Guyon (1978), Hooghoudt (1940), Aver'yanov (1957), Dagan (1964), and Ernst (1956) to be made with the two dimensional solution according to Toksoz and Kirkham (1961).

272

DARBISHIRE, P. The development of drainage

machinery. In Dryland, Saline Seep control. Proc. 11th Int. Soil. Sci. Soc. Congr., Canada, 1978: 4.30-4.46.

History of drainage machinery since "Cata" who is credited as being the original 'drain brain' to present day laser technology is reported.

273

DARGAN, K.S.; SINGH, O.P.; GUPTA, I.C. Crop Production in Salt Affected Soils. Oxford & IBH Publishing Co., New Delhi, 1982, 276p.

The book first of its kind, ever written by a team of scientists of three different disciplines, viz, agronomy, agricultural engineering and soil science, contains 14 chapters discussing principles and practices of crop production in salt affected soils, comprehensively. The recent work done mainly in India has been reviewed. The reclamation and management has been presented in 7 chapters viz, land development, hydrological aspects, irrigation practices, drainage and leaching, chemical amendments, electrical and thermal methods, agronomic and cultural practices. Crop production practices have been outlined for rice, wheat. barley, bajra, sugarbeet, sugarcane, cotton, berseem, shaftal, para grass, Karnal grass and dhaincha. Economic aspects of crop production have also been discussed.

274

DARRA, B.L.; JAIN, N.; SINGH, H. Saline and alkali problems in soils of Matunda region of Chambal Command area. Indian J. Agric. Sci., 1970, 40:518-25.

In the waterlogged Chambal command area, the water table has crossed the critical limit of 300 cm after the inception of canal irrigation. Fluctuations of high water table with the opening and closing of the canal deteriorated the soils considerably. The fertility was quite poor. A few measures are suggested to improve the physico-chemical structure of the soil. The provision of field drainage is essential to check up further deterioration of the soils.

275

DAS, G. Correction of sub grade gravel as filters for drains. J. Agric. Engng. Res., 1973, 18:333-41.

A theoretical procedure has been proposed to modify the sub-standard pit run gravel materials available on river beds as filters for subsurface drains by adding to it another gravel sample of known properties. The proposed method was experimentally verified and found to be satisfactory.

276

DASS, P.; MOREL - SEYTOUX, H.J. Subsurfacae drainage solutions by Galerkin's method. J. Irrig. Drain.Div. ASCE, 1974, 100:1-16.

Use of Galerkin's method to solve the non-linear partial differential equation of flow through drains in a aquifer is demonstrated. The non-linear partial differential equation is first transformed to a system of non-linear ordinary differential equations by Galerkin's method and the non-linear system so obtained is then solved by a finite difference approach. The finite difference scheme adopted was found to be stable for almost all values of d/H ratios (depth to impermeable layer below the drain d; to depth of water table above the drain H).

277

DAVIDCHENKO, N.N. The filtration coefficient in the design of horizontal drainage on irrigated land. Vest. S-KH. Nauki., 1973, 1:67-73.

The paper presents a computer model by which the drainage coefficient can be calculated following stochastic analysis of the field determined variables.

278

DAVIDYUK, K.S. Effect of normal and controlled forest drainage on the surrounding area. Lesovod Agrolesomelior., 1979, 53:27-33.

Studies on Ukraine showed that draining of bogs significantly affects the ground water level of the dry land (outside the bog) from 200-250 m from the drainage ditches. Controlled drainage with sluices considerably reduced the extent of this effect compared with normal open drainage dithes. Growth reduction of neighbouring forests was much less for dry (usually more elevated) than for moist site types. It is recommended that drainage ditches be no less than 100 m from adjoining moist site type, and that bogs of less than 20-30 ha surrounded by such sites should be left undrained.

279

DAVIS, S.; WILLARDSON, L.S.; SPENCER, J.R.; MULDER, D. Drain envelope performance in a sandy soil. Proc. Natn. Drain. Symp. Chicago, Illinois, 1971, 8-10, 17.

Three tile materials clay, concrete and plastic with three envelope materials Pea gravel, Pitrun (sand from the beach) and screened pit run in which fines less than 1.19 mm have been removed were tested for a sandy soil. The results show that any combination of tile and envelope will work satisfactorily provided installed carefully. The tile lines without envelope failed indicating that dirty envelope material can cause failure of the system.

280

DAYEM, M.S.A.; BISHAY, B.G.; SAID, N.F. Economic analysis of improving a salt affected heavy soil by means of tile drainage and amendements in Egypt. 10th Congr. Irrig. Drain, Athens., 1978, 4:127-35.

Barley, maize and cotton were planted in a two-year rotation under drain spacings of 10 and 20 m, with and without gypsum. The gypsum was applied to trench or topsoil only, or to both. The economic analysis was based on both the benefit-cost and the increments of benefit/increments of cost ratios. The most economic project was tile drains of 10 m spacing accompanied by a surface application of gypsum.

281

DAYEM, M.S.A.; EL-GIZAWI, M.G.; BISHAY, B.G. A field study to control water table in low permeable salt affected soil in Egypt. 10th Congr. Irrig. Drain. Athens, 1978. 4:101-34.

Hydrological conditions in a salt-affected heavy soil in the Nile delta were determined under undrained conditions and with two drain spacings, combined with the application of gypsum either to all the topsoil or to the trench backfill only. The water-table under drained conditions was in most cases horizontal with an increasing gradient only in proximity to the backfill. The use of gypsum had only a limited influence on water table drawdown, although its use in the trench was more effective in promoting flow to the drains than its general addition to the topsoil.

282

DEBOER, W. D.; BUCKS, A. D.; LEMBKE, D. W.; WIERSMA, L.J. Envelope performance in a coarse-silt base material. Proc. Natn. Drain. Symp. Chicago, Illinois, 1971, 5-7,9.

Field and laboratory investigations on drain envelope materials are presented. The results indicate that 7.6 cm gravel envelope performed as satisfactorily as a 15.2 cm gravel envelope. A field study on the performance of a combination of gravel and glass fibre material envelope indicated that the performance is satisfactory and without any problem.

283

DEER, H.J., MANN, W.F. Jr. Bedding poorly drained sites for planting loblolly and slash pines in southwest Louisiana. USDA Forest Service Res. Paper, Southern Forest Experiment Stations, 1977, SO-134, 5p.

On 6 sites of different soil type in Louisiana, loblolly (*Pinus taeda*) and/or slash pine (*P.elliottii*) were planted (a) on unprepared sites, (b) after thorough flat discing or (c) after flat discing and formation of beds of max. ht. 20 cm and 2.4 m wide. Data are tabulated on survival, ht., and fusiform rust incidence upto 10 yr old. Treatment (c) (bedding) significantly increased ht. in some trials, but the overall effects were not large; (b) was nearly as effective especially with *P.taeda*. It is suggested that intermittent beds ('pimple mounds') might improve drainage more effectively.

284

DEKKER, K. The discharge capacity of corrugated plastic drain pipes, determined by field data. Proc. Int. Drain Workshop, Wageningen, Netherlands(1978) 1979, 25:414-23.

Analysis of field data on discharge and hydraulic heads from corrugated pipes indicate that a hydraulic rough turbulent flow in the pipe allows a good explanation for the measured hydraulicc heads losses. The roughness coefficient, apparently is related to the Reynolds number Re: $=7.0\ R_e^{-0.5}$ for clean and well-maintained pipes. Field data indicated that for badly maintained and partly silted drainlines the co-efficient 7.0 may increase to about 10 to 14. On the basis of the derived flow equation this implies that a silting percentage of 22 per cent to 30 per cent can be responsible for this increase. The latter figure can be explained when a silt layer of 13 mm is present on the larger diameter pipes, a reduction in drainable area is found which is dependent on the applied pipe diameter. The data introduced in this way have been worked out in a nomograph that can be used for design purpose.

285

DEMPSTER, J.I.M.; STONER, R.F. Methods of operating tubewell projects. 7th Congr. Irrig. Drain. Mexico, 1969, Q25 R11:135-50.

Advantage of tubewell drainage are listed which include its extreme flexibility. The paper outlines various methods of operating a tubewell field and the economic implications. Drainage for saline and fresh ground waters has been discussed under separate heads. Aspects of seasonal and annual variations in drainable surplus, changing irrigation demands, effect on operational sequences and temporary and long term over pumping aspects are discussed.

286

DENNIS, C.W. Field drainage-specification for permeable backfill. J. Agric. Engng. Res., 1982, 27:529-35.

Data are presented to cover mechanical strength, size and fine contents of the permeable backfills while the possible complications of toxicity and solubility are briefly discussed. It is demonstrated that a more detailed, performance based , specifications may now be written which covers the majority of materials encountered.

287

DENNIS, C.W. The failure of a pipe drainage system in an organic soil and subsequent remedial measures. Tech. Rep. Field Drainage Experimental Unit U.K. 1978, 78/3, 11p.

A pipe drainage system laid in a predominantly organic soil failed within three years. The diagnosis and attempts to restore performance are described. Hydraulic conductivity measurements and piezometric investigations indicated that the layer within the immediate vicinity of the pipe is restricting the flow since the ability of the soil to transmit water was adequate. Detailed physical examination revealed substantial ochre deposits within and around the pipe which were almost certainly the cause of failure. Attempts to restore performance by prolonged sub irrigation, sulphur dioxide treatment, drain rodding and water jetting all met with negligible or, at best, only partial success. Redrainage of the site is to proceed cautiously. Most of the area will be left in its present unsatisfactory state while the rest is to be used to compare and contrast the performance of simple deep ditches, mole drainage and three different types of pipe drainage.

288

DENNIS, C.W.; CROOTE, S.G. The minimum size of permeable fill used with mole drainage. Agric. Wat. Mgmt., 1977, 1:143-53.

Laboratory work involving the full scale simulation of flow from a mole drainage channel into a trench filled with permeable material and thence to the pipe drain at the bottom of the trench is described. Four sizes of permeable fill, two trench widths, two pipe types and a wide range of flow rates were used in the investigations. Provided that there was reasonable vertical separation between the mole channel and pipe the flow could be divided into three distinct phases namely: entry from the mole channel, near vertical flow within the permeable fill and entry into the pipe. The most limiting of these was the abililty of the fill to accept water from the mole channel. Variation of hydraulic head with flow rate and permeable fill size at critical points, of the system is presentedd in graphical form. It is suggested that permeable fill with a mean diameter of 5 mm is the minimum size acceptable under U.K. conditions and that this result will not be affected by the presence of a small quantity of fine material. The hydraulic conductivity associated with such materials, in the absence of fines, is approximately 2500m/day. Application of the data to other design conditions is also possible.

289

DERBIN, V.P.; ORLOV, B.F.; TARAKANOV, A.M. Using the MTP -32A machine with truncated -cone rotary cutters in forest drainage. Lesn. Khoz. 1978, 2:42-4.

An account is given of trials in the Archangel region with three rotary drain-cutters, each in the form of a truncated cone with base diamter 2000, 2100 and 2300 mm; the cutters are 1600-1700 mm long and can cut and renew ditches 140-160 cm deep(with sides of slope 0.81-0.83). The rate of work is 1300m^3 per (7 -hr) shift or 44.4 sec. per m^3. Details are tabulated on rate of work, costs and performance. The rate of work is 3.7 times that of the TE-3M excavator.

290

DEVILLERS, J.L.; GUYON, G. Application of modern concepts to drainage projects.C. R. Acad. Agric. Fr., 1979, 65:815-20

The choice of drainage method (ditches, or underground channels associated with subsoiling or mole draining) should be governed by pedological characteristics and groundwater depth. Precautions may be necessary to prevent the blockage of pipes by fine sediments, iron salts or plant roots and may include the use of filters, and the stabilisation of soil by the application of amendments(gypsum, lime). The optimum drain spacing for efficient drainage is then calculated in relation to the hydraulic conductivity of the soil.

291

DEVILLERS, J.L.; GUYON, G. Choice of a field drainage treatment.Proc. Int. Drain. Workshop Wageningen, Netherlands, (1978), 1979, 25:165-79.

This paper is a general review of water excess situations occurring in France and briefly describes the different field drainage treatments recommended by the Ministry of Agriculture . Special atttention is paid to the determination of the origin of the excess water and the need for pedologic and hydraulic soil data.

292

BEVILLERS, J.L.; MONTIER, J. Drainage and farm economics. In Land Drainage (Gardiner, M.J. Ed) A.A.Balkema, Rotterdam, Netherlands, 1982, 123-30.

The use of simulation models is discussed.

293

DHANAK, G.G.; SHALIGRAM,V.K. Reclamation of saline lands, special problems of low areas along the sea coast.Trans, 6th Int. Congr. Irrig. Drain., New Delhi, 1966, Q 19 R 16:219-53.

The Gujarat state forming 6.1 percent of the area of the Indian Union has coastal line extending to 1,829 km along which are flat areas consisting of saline lands, waste lands and low lying lands, having different degrees of salinity in them, most of them tending, to become barren and unproductive. A pilot polder project was undertaken to obtain information on problems like the construction details involving civil works pertaining to bunds, inlets, sluices, the most effective process of utilizing available water and evolving special methods of cultivation. The state has another problematic area known as the Little Rann of Kutch which is a vast stretch of salt affected barren land covering a total area of 3,818km^2. In the saline lands in Surat and Broach districts, leaching with sweet water is necessary, after the ingress of water is stopped by putting bunds. In the "Ghed" area, training of rivers providing by outlets to the river to carry flood discharges, and leaching are necessary. The Bhal lowlands require, primarily, a well laid-out drainage system to remove the excess waters that accumulate during the monsoons, and sufficient quantities of waters for reclamation of the salt affected lands. The Datha area also requires a planned drainage system which has now been prepared. Pot culture experiments have indicated that the saline soils of the Little Rann of Kutch can be leached and made free from the harmful salts with a water application of 170 cm. It is shown that agricultural crops can be grown successfully. It is estimated that requirements of water would be 40.68 million ha-cm to reclaim nearly 2,40,000 ha of land.

294

DHRUVANARAYANA, V.V. Drainage of salt affected soils in India. Proc. Indo-Hungarian Sem. Mgmt. Salt Affected Soils Karnal, 1977, 160-76.

For working out the drainage specifications, the salt affected soils of India may be grouped as :(a) alkali soils of Indo-Gangetic alluvial plains, (b) coastal saline soils of West Bengal, Andhra Pradesh, Maharashtra and Gujarat,(c) acid sulphate soils along the coast of Kerala and (d) heavy saline soils in the inland tracts of Chambal, Tungabhadra, Nagarjuna Sagar, Nizamsagar and Kadana command areas. These groupings are based on the broad similarities of hydrologic and soil characteristics. This paper briefly outlines these and recommends the general practices of drainage that can be adopted in their reclamation.

295

DHRUVANARAYANA, V.V.. Drainage requirements of salt affected soils. Ann. Arid. Zone, 1980, 19:259-63.

The drainage characteristics of salt affected soils from different locations are tabulated. Three tier system has been described for reclamation of alkali soils by surface drainage method which consists of (a) collection of part of the rainfall in the crop land, till such time and depth as will not be harmful to the crops,(b) directing the excess rainwater into the dugout ponds of sufficient storage capacity located in the lowest portion of the catchment(farm area), the stored water is utilized for irrigation during dry spells, and(c) the remaining excess water is then led into the regional drainage channels. The sub-surface drainage is not necessary in Karnal region as water table is declining at the rate of 30 cm/yr and also because of impervious nature of alkali soils.

296

DHRUVANARAYANA, V.V. Improvement in land utilisation resulting from drainage. In State of the Art, Pub. No. 2, ICID, New Delhi, 1981, 185-247.

A review paper containing 40 references. The suggestions have been given for sustaining high yields of irrigated crops for a long period at a global scale through rehabilitation and improvement of existing drainage schemes. The measures have been suggested for improving water management in irrigated agriculture and reduce the losses from irrigation. The steps to improve drainage and salt leaching in the irrigated land have been outlined.

297

DHRUVANARAYANA, V.V. Reclaiming alkali soils-engineering aspects. Bull 6, CSSRI, Karnal. 1980 ,61p.

The basic data required for reclaiming alkali soils are: infiltration rate, hydraulic conductivity, water table depth and flucuations, quality of ground water, rainfall distribution and runoff. Subsurface and surface drainage have been discussed. Water management practices have been described and hydrologic factors have been evaluated.

298

DHRUAANARAYANA, V.V ; KAMRA, S.K. Subsurface drainage

for saline soil reclamation. In Polders of the world. Vol III ; Late papers. Int.Inst. Land Reclm. Improve. Wageningen, 1982, 12-21.

An area of inland saline soils near Bidaj was selected for reclamation by construction of open seepage drains of 200 m spacing, leaching the salt by ponding and adopting suitable crop rotations. An experiment with open drains at 15 to 25 m spacing was also started on barren land. This system reduced flood losses, soil salinity and groundwater levels and increased the cultivated area and crop yields. Improvements were greater with the more closely spaced drains.

299

DHRUVANARAYANA, V.V.; PANDDEY,, R.N.; GUPTA, S.K. Drainage of alkali soils. J. Indian Assocn. Hydrologists, 1977, 1:21-8.

The design of drainage systems for alkali soils is conventionally linked with sub-surface drains and permeabillity improvement through application of amendments and this is considered relatively more difficult than for saline soils. Excess soluble salts, render them nearly impervious. Good quality groundwater is available at 3 to 5m below ground level. High water table conditions used to prevail in the rainy season upto 1972. Subsequently, there has been a general decline in the water levels due to the increased withdrawals of ground water. The rainy season is also characterised by the occurrence of two to three heavy storms frequently intervened by continuous dry spells. Field experiments conducted on alkali soils at Karnal, since 1971, have indicated that (a) sub-surface drainage is neither feasile nor necessary and that (b) a three tier system of surface drainage would be useful to tackle the alternating problems of water stagnation and drought condition that prevail during the rainy season.

300

DICKY, G.L. SCS drainage design procedure. In Water Management for Irrigation and Drainage. Proc. Spec. Conf. ASCE Irrig. Drain Div. ASCE, New York, 1977, 2:1-6.

The drainage design procedure as followed by soil conservatioon service USA is explained in 10 steps. The problem of disposal of water is highlighted in reference to San Joaquin valley and suggested that problem should be tackled at the earliest possible.

301

DIELEMAN, P.J. Drainage knowledge and research needs in developing countries. Proc. Int. Drain. Workshop, Wageningen, Netherlands, (1978), 1979, 25:620-32.

Short and long term research strategies for developing countries are discussed. It is observed that present status of knowledge and technology is generally adequate to meet short-term needs of drainage design and implementation in developing countries. The development of new and advanced theories and methodologies is not considered to be of immediate concern. Short term research priorities should be directed towards constraints in the transfer of information. Constraints realized are primarily in the adaptation and acceptance of technology and concepts at the level of farming community as well as to higher decision-making levels Long -term research needs cover the most aspects of drainage investigations, rationalization of designs and implementation. It is also observed that effective research programmes at the local level call for national as well as regional and interregional coordination, guidance and support.

302

DIELEMAN, P.J. Drainage of sandy soils, In Sandy soils, Soils Bull., FAO, 1975, 25:107-13.

Physical characteristics such as infiltration rate, water retention and effective porosity differ widely from one type of sand to another and have a great influence and impact on the design drainage rate. In general designed sub-drainage systems will tend to be deep and relatively narrowly spaced on fine sands that have a high silt and clay content, whilst they may be shallower and more widely spaced as the coarseness of the sand increases and silt and clay decrease. Surface drainage may be important on soils that run together easily on wetting. Vertical drainage possibilities may be limited by low soil permeability. Surface drainage systems should promote the lateral flow of perched water through the more pervious layer to the drains. When pipe drains are used, care should be taken to use backfill material that remains permeable. Cover or envelope materials should be well selected in relation to the type of sand. Thin sheets of fibre glass appear unsuitable as cover material. More voluminous materials would seem preferable. Open ditch drains in unstable sands, especially in salty areas, are very difficult to maintain. Using plastic pipes that have been prewrapped in the plant, or that are simultaneously provided with cover material by the trenching machine will in many instances facilitate the installation and reduce the risk of failure.

303

DIELEMAN, P.J. ; TRAFFORD, D. Drainage testing. Irrig. Drain. Paper, FAO, 1976, 28, 172p.

Importance of testing of a drainage system after its installation is discussed and methods to test either a single line or a complete drainage system are provided. Use of envelope, filters and surrounds to achieve different objectives are given. Equations and their applications in the system testing of a drainage project are included.

304

DIERICKX, W. The influence of envelope materials in preventing silting-up of drain pipes. Proc. Int. Drain. Workshop, Wageningen, Netherlands (1978), 1979, 25: 306-30.

A theoretical analysis has been given about the problem of silting up of drain tubes through the underside due to flow pressure. The flow pressure depends on the hydraulic gradient, which is determined by the ratio discharge-soil permeability. A coarse and thick envelope material or a larger drain tube can decrease the danger for silting-up at design norms but cannot prevent it under more severe conditions. To be on the safe side,an envelope material with a higher sand passing resistance be recommended so that the hydraulic gradient may reach at least 2 or 3 times the hydraulic gradient of the design norm.

305

DIERICKX, W. Structural stability of soil and the need for drainage envelopes.4th Natn. Drain. Symp. Proc. ASAE, 1982, Pub. 12-82:79-86.

A weakly aggregated silt loam soil with particle size distribution:16:5% clay,61.4% silt, 22.1% sand and an organic carbon content of 0.4 per cent was used to study the need and performance of different textured envelope materials in the laboratory. The performance of envelope materials has been investigated taking into account the effect of initial moisture content on aggregate stability and of hydraulic gradient on aggregate deterioration for a particular soil. The higher initial moisture content of 15 per cent resulted in less soil deposits in the drain and envelope compared to a lower value of 3 per cent. Synthetic fibre with fine pores was found superior to synthetic fibre with coarse pores and coconut fibre. The last two were at par.

306

DIERICKX, W. Subsurface and surface drainage of grass sports grounds. Rev. Agric. (Bruxelles), 1976, 29:607-23.

Surface water can be removed by a network of trenches filled with, and covered by, river sand and connected if necessary with a subsurface system of drains filled with granular material.

307

DIERICKX, W.; BISHAY, B.G.; MATON, A. Aspects of silting -up of drainage pipes. J. Agric. Engng. Res. 1975, 20:123-37.

A sand tank model has been used to study the silting-up of drainage pipes as influenced by different locations of the perforation lines on the drain perimeter. The water discharge as well as the sand discharge has been studied during 24 hr, until equilibrium has been reached. The theoretical solution of Kirkham has been checked. Further, a comparative study between the sand model tank and a conductive paper model indicate that a real drain filled with sand gave a water discharge value less by 16.8 per cent compared with the discharge values of both the theoretical solution and the conductive paper model. By simulating a real drain, free from sand, difference of only 3.7 per cent was observed. A full agreement was found with a drain filled with sand on the conductive paper model.

308

DIERICKX, W.; MOLEN, W.H. VAN DER . Effect of perforation shape and pattern on the performance of drain pipes.Agric. Wat. Mgmt. 1981, 4: 429-43.

An important assumption of taking the pipe wall as an impermeable surface in arriving at drains flow formulae has been tested using an electrolytic model and it was shown that the calculated values were approximately one half of the measured entrance resistance values. By adding a central source of strength which is half the total strength of the sinks that simulate the perforations, exact results are obtained for several cases of interest. Based on these experiments, modified formulae have been derived for continuous longitudinal slits, circular perforations and discontinuous longitudinal slits.

309

DIERICKX, W.; YUNCUOGLUE, H. Factors affecting the performance of drainage envelope materials in structurally unstable soils. Agric. Wat. Mgmt.,1982, 15. 215-25.

To test the performance of envelope materials in structurally unstable soils, flow tests were carried out in vertical soil columns filled with an unstable aggregated silt loam soil. To discover the most unfavourable situation in evaluating envelope materials, experiments with both downward and upward flow were performed. Also tests were carried out to find out the effect of soil sample height. Further factors affecting soil aggregate stability, such as the initial moisture content and the hydraulic gradient, were considered. Two soils, aggregate conditions, sensitive to degradation, were used in the investigation of the performance of both coarse and fine textured envelopes.

310

DINC, G.; MERVA, E. G.; KIDDER, E.H. Hydraulic roughness of corrugated plastic drain tubing. Proc. Natn. Drain. Symp. Chicago., Illinois, 1971, 11-4.

Hydraulic roughness of corrugated plastic drain tubing has been examined through a laboratory study. The roughness coefficient is found to vary because of corrugations configurations and is also function of corrugation pitch. It has been suggested that the results are not conclusive for want of more data and more elaborate studies are required.

311

DIXON, R.M.; BAY, C.E. PETERSON, A.E. Drainage of a peat soil overlying an artesian aquifer. I. Field determination of some drainage characteristics. <u>Res. Rep</u>. Coll. Agric. Univ. Wis. 1966, 21:1-7.

A field procedure is described by which some physical properties of a deep peat soil were determined for use in drainage system design. The system, which consisted of 0.9 cm galvanized pipes at various depths in the profile, permitted the delineation of soil layers; the measurement of hydraulic heads and conductivities; and the estimation of hydraulic gradient, flow rates and flow direction. Vertical artesian flow from a continuous underlying aquifer contributed about 0.25 cm per day to the excess water in the peat. Determination of drain spacings required for steady-state drainage of rainfall and artesian water indicated that the effect of artesian water on tile spacing would be small and that use of artesian wells to relieve hydrostatic pressure would be impractical. A tile spacing of about 12 m would be required for efficient drainage of this peat.

312

DOERING, E.J.; BENZ, L.C. Pumping an artesian source for water table control. <u>J. Irrig. Drain. Div. ASCE</u>, 1972, 98:275-87.

Pump drainage to control water table in the Grand Forks county (North Dakota) is illustrated. The aquifer is 19.2m thick and the overburden 33.8 m. Aquifer and overburden properties had been determined and explained in relation to water table and salinity control. A 11-well system is described that will reduce by 6.1 m or more the hydrostatic pressure beneath the saline area. Each well (.0315 m^3 /sec) will benefit 6475 ha. Approximately 0.347 m^3/sec of water pumped will be disposed off into the Red River which will increase the salt concentration by 170 ppm and the water will remain suitable for human consumption. The analysis indicates that the scheme is economically feasible.

313

DOERING,, E.J.; BENZ,, L.C.; REICHMAN, G.A. Shallowwater-table concept for drainage design in semiarid and subhumid regions. 4th Natn. Drain. Symp. Proc. ASAE. 1982, Pub. 12-82:34-41.

Water management experiments in field plots and lysimeters have shown that the drainage requirement is about 1m for corn and sugarbeets grown on sandy soils in North Dakota and that irrigation requirements for comparable yields increased significantly as the water-table depth increased. Deep drainage correlates to over-drainage for sandy soils in semiarid, subhumid, and humid regions. Calculations using the Donnan (steady state) equation for drain spacing and 2 combinations of drain depth(2.4 and 1.5 m) and midpoint water-table depth(1.2 and 1.0m) show that the drain spacing ratio for particular drain geometries is a function of barrier depth only. Based on USDI-BR data for drain construction cost , the cost per ha for shallow (1.5m) and deep (2.4m) drains described herein would be equal when the barrier depth was about 4 m. For shallower barrier depths, the shallow drains would cost less to construct than the deep drains. One-time construction cost savings that may result because deep drains can be spaced further apart than shallow drains may be more than offset by recurring annual losses because of reduced yields or increased operating costs.

314

DOERING, E.J.; REICHMAN, G.A.; BENZ, L.C. ; FOLLETT, R.G.Drainage requirement for corn grown on sandy soil. 3rd Natn. Drain. Symp. ASAE, Michigan, 1976, 144-7.

Field plot and lysimeter studies were conducted

during 4 years to determine the optimum water-table depth for corn grown on irrigated sandy soils in the northern Great Plains. Three-year average watertable depths in the field plots (June 1 to August 31) were 140, 195, and 223 cm. Water-table depth increased as the growing season progressed. Constant water-table depths ranging from 46 to 210 cm were maintainedd in the lysimeters. Sprinkler irrigation was applied to provide from 0.5 to 1.5(field) and from 0.3 to 1.3 (lysimeters) times the calculated plant water requirement. A zero irrigation treatment was also included in the field-plot study. Largest yields of both corn grain and dry matter were produced over the shallowest water table in the field plots, regardless of irrigation level. Largest yields were produced in the lysimeters when the water table was 101 cm below the soil surface and the irrigation levels were 0.3 and 0.8 of the plant water requirement. The yields were significantly reduced when the water table was 46 cm below the soil surface in the lysimeters, but they were unaffected by irrigation. Irrigation requirement increased significantly as the water table depth in the lysimeters increased beyond 103 cm.

315

DOERING, E.J.; SANDOVAL, F.M. Chemistry of seep drainage in southwestern North Dakota. Soil Sci., 1982, 132:142-9.

Effluents from subsurface drains from two saline seeps were analysed. EC ranged from 7 to 13 dS/m. Magnesium sulphate and sodium sulphate were dominant ions. Calcium concentrations were consistently low. Drainage is not environmentally acceptable and the best approach is to utilize the soil water while it is nonsaline resource in the rootzone of recharge areas.

316

DONNAN, W.W. Field aplication of the ellipse equation. Proc. Spec. Conf. ASCE Irrig. Drain. Div. ASCE. New

York, 1977, 1:1-13.

Paper traces the development of the ellipse equation and highlights its use under field conditions in evaluating drain spacing. Nomographic solution to the equation are provided. Contractors drain spacing and cost estimator is reported in the form of a slide rule.

317

DONNAN, W.W.; HOUSTON. C.E. Drainage related to irrigation management. In Irrigation of Agricultural Lands. (Hagan. R.M.; Haise, H.R.; Edminster T.W. Eds.) American Soc. Agro. Wissconsin USA, Agronomy Monograph No 11, 1967, 974-87.

Drainage design and systems of irrigated lands are discussed. Use of wells and pumps in drainage and irrigation is also given.

318

DORAI, M. Design considerations for the drainage of padi polders; In Polders of the world. Volume 1. Polder projects. Land and Water Management Aspects. Int. Inst. Land. Reclam.Improve., Wageningen, Netherlands, 1982, 489-98.

Over the last 50 years many lowlands and swamps located along the coasts of Malaysia have been reclaimed and converted to highly productive padi polders, by the construction of embankments, sluices and drains. This paper discusses the principal hydrologic and hydraulic factors that should be considered in the design of drainage channels and sluices within the polders.

319

DORT, J.A.; VAN BOS, M.G. Main drainage systems. In Drainage Principles and Applications. Vol.IV. Design and Management of Drainage Systems. Intern. Inst. Land Reclam. Improve. Wageningen, Netherlands, 1974, 16:123-62.

The design of a main drainage system and the design and construction of canals are described. The limits of application of various types of pumps are discussed.

320

DOS SANTOS, A.G.; YOUNGS, E.G. A study of the specific yield in land drainage situations. J. Hydrol. 1969, 8:59-81.

Results of measurements of all the specific yields obtained in nonsteady state drainage experiments using a completely permeable drain installation and a plastic drain installation in a hydraulic model sand tank, are presented. The variations of the specific yield with time are compared and discussed in relation to the varying moisture profile. With the use of a virtual specific yield given by the air content at the surface in the mid-plane between drain lines, calculated assuming steady state conditions with a stationary water table and also with the use of the experimentally obtained relationship between water table height and drain discharge for a given drain installation, rise and decay curves for water table are calculated and compared with fair agreement.

321

DOTY, C.W. Pump drainage in Caroline bays, J. Irrig. Drain. Div. ASCE, 1973, 99:465-75.

Caroline bays located in the southern coastal plains of United States can be drained by pump assisted surface and tile-drainage system which controlled the water table in the bays at depths that did not adversely affect plant growth. In the bays 89 per cent of the excess water occurred during February and March and June and July. Tile drainage removed 89 mm or 88 per cent and surface drainage 18mm or 1.7 per cent of the rainfall. The tile and surface-drainage systems using small submersible pumps for removing excess water from the bays, can restore the productive efficiency and a gross return above pumping cost of $127.5/ha to $ 172.5/ha per year can be realized on soybeans. The choice between pump assisted surface drainage or tile drainage should be based on comparative cost analysis of the two systems.

322

DOTY, C.W.; CURRIN, S.T.; MCLIN, R.E. Controlled subsurface drainage for Southern Coastal Plains soils. J. Soil. Wat. Conserv. 1975, 30:82-4.

A field study of water table control through subsurface conduits on a Southern Coastal Plains soils showed that in these sandy soils the water table must be kept at 1.06m or less from the surface. Corn silage yields from a field under controlled drainage were greater than those from a nondrained field. For each additional day, between 25 and 55 days, that the water table was less than 1.06 m from the soil surface, silage yields increased by 0.74 to 1.48 ton per hectare.

323

DOTY, C.W.; PARSONS, J.E. Water requirements and water table variations for a controlled and reversible drainage system. Trans. ASAE, 1979. 22:532-9.

A water mound was formed above the estimated natural water table, but could not be maintained at the desired elevation because of flow restrictions in the soil or conduit system. A positive head is required to produce the water mound. Drainage is necessary to produce maxm. yields with the controlled and reversible drainage system (CaRDS). About the same amount of water is required by the CaRDS as estimated for a normal surface applied irrigation system. Water losses from CaRDS to deep seepae and/ or subsurface flow are not excessive. Data indicated that CaRDS can be accomplished with drain lines spaced 32m apart for sandy clay loam soil, if drainage is provided during excessive rainfall; in sandy loam or more permeable soils an even wider spacing can be used, 32m producing the higher yields. Although the results are favourable CaRDS is not recommended for irrigation alone. If, however, a drainage system is needed for sandy clay loam soils, a control structure may be added to the system.

324

DOUGRAMEJI, J.S. Drainage studies at Beni-Tagleb Experimental Station, Euphrates Basin, Syria. <u>Soil Sci. Papers</u>, Arab Centre for the studies of Arid Zones and Dry Lands, 1977, 5, 39p.

A research programme on the reclamation of salt affected soils was started in 1973 at Beni-Tagleb Explerimental Station, 10 km east of Dier Zor. The main objective was to establish criteria for drainage and reclamation. The present study describes some of the drainage studies, including hydrophysical properties of soil, and the parameters selected for the design of the drainage system . The soils studied are alluvial (first terrace of Euphrates river) deposited mainly by flood water and to a lesser extent by run off from the hills. The soils have a weak to moderate structure and are of medium texture with clear stratification. The infiltration rate and permeability are moderate. The soil and groundwater are highly saline possessing the same ionic distribution. For calculating the drainage spacing, the depth of the dewatering zone was fixed at 0.9 m below the surface. Using Hooghoudts steady state and Ernst's non-steady state equations, assuming the hydraulic head at 0.7m and a drainage discharge of 3.5 mm/day, a drain spacing of 92.0 and 90.0 m was calculated.

325

DRABLOS, J.W.C.; WALKER, P.N.; SCARBOROUGH J. L. Field evaluation of corrugated plastic drain tubing. <u>3rd Natn.Drain. Symp. ASAE,</u> Michigan, 1976, 69-74.

Field measurements were made on the deflection of

corrugated plastic tubing 18 or more months after installation. Sizes tested varied from 127 to 203 mm in diameter. The tubing was found acceptable for agricultural drains at depths up to 1.5 m. The minimum stiffness values recommended by ASTM F 405-74 are adequate when recommended installation techniques are followed.

326

DRABLOS, J.W.C.; SCHWAB, O.G. Field laboratory evaluation of 4-inch corrugated plastic drain tubing. Proc. Natn. Drain. Symp. Chicago, Illinois, 1971, 20-4, 27.

Corrugated plastic drain tubing is considered an acceptable material for agricultural subsurface drains at depths upto 1.25 m. Field studies indicate that the tube deflection is affected by the length of time the tubing has been installed, angle of trench bottom groove and strength of tubing. Method of blinding and backfilling and soil type also influence the same. Elongation was found to increase as the parallel plate strength decreases. It also increases but at a decreasing rate with duration of stretching though most of stretch occurs within first minute. The temprature greatly affect the strength and is reduced by 60-70° C and 1.7^{o} C respectively compared to 21.8^{o} C.

327

DUBEY, D.P.; TIRPATHI, R.S. Effects of flooding for different durations at various growth stages of wheat varieties. Harvester, 1979, 21:24-6.

Effect of flooding at crown root initiation and booting stage of wheat on 4 different wheat varieties was examined. The booting stage was more susceptible to flooding than crown root initiation stage. The duration of flooding has a significant bearing on the reduction in yield. The yield for 5 day flooding at booting stage is only 8g/pot compared to 21.5g/pot under control with normal irrigations at critical periods.

328

DUC, T.M. Control of saline water tables on the irrigated experimental area of 120 ha at Richard-Toll (Senegal). Agron. Trop., 1973, 28:894-900.

In an area of the irrigated rice-growing project at Richard-Toll (average annual rainfall and calculated potential evapo-transpiration 250 and 2012 mm, respectively), the effects of various types of drains under different irrigation methods on the control of the rise of a saline water table were studied. It was concluded that satisfactory control was obtained with open drains 250-400 m apart.

329

DUKE, H.R. Capillary flow in agricultural drainage. Stn. Bull., Agricultural Exp. Station, Oregon State Univ., 1977, 629, 99 p.

The process of drainage of soil water is customarily treated as a fully saturated flow phenomenon. Recent work illustrates that the partially saturated zone can significantly influence the overall performance of agricultural drains, particularly with respect to maintenance of a desirable root environment and when the soils encountered are either quite shallow or very fine textured. Capillary properties of soils pertinent to drainage are described. Both laboratory and field measurement techniques are proposed. Necessary considerations for modelling capillary flow systems and models successfully used to predict vertical drainage of the soil profile and horizontal flow associated with a sloping water table are presented. Results of numerical analysis of both sloping aquifers and parallel drain systems are given. The capillary region has considerable influence on the predicted water table depth and depth of soil which is adequtely aerated to allow root development.

330

DUKHOVNY, V.A. Reclamation of salt-affected irrigated lands in Central Asia - Principles and Practices. Int. Symp. Salt Affected soils, Karnal, 1980, 513-24.

In the desert and semi-desert irrigated conditions of Central Asia, only a small portion has adequate natural drainage. The ground water has been observed to be responsible for soil salinization, with a groundwater rise from 0.7 to 3 m per year. The prevalent salinity is of the chloride-sulphate type. It is therefore, necessary to control soil water and salt regime by means of irrigation, drainage and water conveyance techniques to attain the designed moisture level, the uniform leaching regime and maintain a definite groundwater table through drainage. The methodology adopted in context of the objectives is discussed in this paper.

331

DUMM, L.D. Subsurface drainage by transient-flow theory. J Irrig. Drain. Div. ASCE, 1968, 94:505-19.

Uses the transient flow theory as a basis for determining spacing relationships of subsurface drains, as steady-state flow theory does not adequately describe the changing conditions of subsurface drainage; incorporating, research work from Australia and Canada and data from a Bureau of Reclamation project for checking the validity of the mathematical developments of the transient-flow theory under many varying field conditions; cites an actual field problem to show how field investigational data and other available data are used in developing drain-spacing requirements under various specified conditions; includes discharge formulae and verifies their validity by the Australian and Canadian research work.

332

DUMM, L.D. Transient-flow concept in subsurface drainage - its validity and use. Trans. ASAE, 1964

7:142-6, 151.

A new drain spacing formula has been proposed and tested for the case when drain tiles are not resting on the impermeable barrier. The initial water table profile is assumed as a fourth-degree parabola. The theoretical formula is tested with four examples and found applicable.

333

DUMM, L.D., WINGER, R.J. Subsurface drainage system design for irrigated areas using transient-flow concept. Trans. ASAE, 1964, 7:147-51.

The use of transient-flow concept in the drainage of irrigated areas is illustrated. The design spacing and expected discharge over an area of 1500 ha is illustrated month wise. The crops selected are alfalfa, safflower, vegetables, cotton, barley and bermuda grass.

334

DURMFORD. D.S.; PODMORE, T.H.; RICHARDSON, E.V. Drainage: An incremental net benefit approach for optimal design. 4th Natn. Drain. Symp. Proc. ASAE, 1982, 12-82:42-9.

A methodology to determine an economically optimum subsurface drainage system in irrigated areas is presented. The method combines a cost minimization procedure with incremental net benefit search. The optimization routine eliminates the necessity of evaluating a large number of drain spacing and depth combinations and makes evaluation of the effects of particular variables on the results more traceable. The benefit analysis used attempts to estimate benefits in terms of those variables which are often the primary reasons for drain installation in arid regions.

335

DUTHION, C. Effect of excess water on spring beans.

C.R. Acad. Agric. France, 1977, 63:416-21.

In a pot trial with bean (*Vicia faba*) cv. *Strube*, excess water during early growth stages increased the growth period by 15-20 days and increased plant ht. Excess water for 10 days decreased the rate of weight increase with a decrease of 53 per cent with excess water for 10 days beginning 17 days after germination, which also decreased numbers of flower heads and pods, delayed harvest by about 10 days and decreased yields to 104.9 g/pot compared with 122.9 g in the control given optimum rates of water. With excess water for 10 days beginning 17 days after germination or at pod appearance 1000 -seed weight were 572 and 615 g resp., compared with 683 g in the control. Excess water also slowed down or prevented N fixation and affected soluble N and sugar contents.

336

DVORAK, P. Determination of depth and distances in a pipe drainage. *Vod. Hospod.*, 1963, 13:227-8.

Presents a new table of depths and distances of pipe drainage in mineral soils in Czechoslovakia, the table being the result of research of many years; suggest that this table enables more suitable and correct design of a drainage system than that used upto now.

337

DVORAK, P. Dimensioning of subsurface drainage system. *3rd Natn. Drain. Symp. ASAE*, Michigan, 1976, 55-7, 68.

Equations for evaluating maximum length of drain considering steady state and non-steady state are presented. For the steady-state three cases are considered, i.e., (a) the impermeable layer is at a depth less than 1/4 of spacing, (b) the impermeable layer is at a depth more than 1/4 of spacing, (theoretically infinite) and (c) the drainage pipe is placed on an impermeable layer or in a negligible distance above it. Nomographic solutions for different cases are presented.

338

DVORAK, P. Two-layered drainage of heavy soils. 10th Congr. Irrig. Drain. Athens., 1978, 91-9.

Four physical models of two-layered drainage system for tile-drained heavy soils were designed, appropriate to the conditions of heavy soils ($k < 0.1$ m/day and the specific yield $p_d < 2\%$) and the requirements of wide practical applications. In all models the primary cause of waterlogging is poor infiltration of precipitation. The secondary causes related to the groundwater regime were also considered.

339

DYLLA, A.S. Simplifed procedure for transient-flow drainage design. Trans.ASAE, 1966, 9:61-3.

An integrated Donnan's equation has been presented and a procedure to design a drainage system for irrigated arid areas illustrated. Two examples are presented to illustrate the same. It is found that the integrated Donnan's equation is similar to the integrated Hooghoudt's equation with constant, C=1, and after taking care of the convergence correction.

340

EASTON, B.E.; TAYLOR, R.; SWEET, G.; CHENEVEY, G. Polyvinyl chloride corrugated tubing. 3rd Natn. Drain. Symp. ASAE, Michigan, 1976, 80-3.

Experiences in the manufacture and installation of polyvinyl chloride (PVC) corrugated tubing for drainage are described along with a brief historical description of corrugated tubing. Several field trials of PVC tubing for agricultural drainage are described. A deflection analysis for PVC corrugated tubing 1 yr after installation showed a weighted average deflection of about 4 percent.

341

EDE, A.N. Measures for the control of soil water in agriculture. Proc. N.Engl. Soils Discuss. Grp., 1964, 1:30-5.

Methods of controlling groundwater table and the design of drainage system are discussed.

342

EDE, A.N. New methods of field drainage. Agriculture, London, 1963, 70:82-6.

Describes zipper drainage with plastic pipes, an interesting development which the Ministry of Agriculture, Fisheries and Food (Great Britain) have under test; the problems discused are drainage plan, depth of installation, size of tube and the necessity for permeable fill.

343

EDE, A.N. Scheme for the assessment and design of a field drainage system on a hydrology basis. Trans. 7th Congr. Soil Sci. Madison, 1960, 493-505.

Two prime difficulties of drainage design are considered. It is suggested that hydraulic conductivity can be obtained through inadequately drained fields under known water table conditions, and the changes in water storage accompanying fluctuations in water table height. It will provide a more realistic value of the parameter. In the subsequent drainage design a definite buffer zone in the soil is suggested in the depth of the drain calculations to accommodate likely peak storms.

344

EDE, A.N.; SMITHAM, G.M. New materials and machines for drainage. Proc. Drain. for Efficient Crop Prodn. Conf.

Chicago, Illinois, 1965, 58-61.

Use of PVC, polythene and polypropylene for field drains lining and channel stabilization is reported. The principles of operations of various drainage installation machinery such as trenchless drain-installation and labour requirement in the use of these machines are illustrated. The principles of gradient control in relation to gradient tolerance limit in any installation suggest that the technique in which datum line is provided by the sight line of an optical level, which is set up by a ground observer behind the drain layer, is cheap and achieves the accuracy. In this case the observer can sight on a marker fixed to the drainage blade and can maintain the marker on the sight line directly by having control of the tractor hydraulic system.

345

EDMINSTER, T.W. Control of surface water by land forming. 7th Int. Congr. Soil Sci. Madison. 1960, 1:411-7.

Land smoothening, grading and levelling in relation to surface drainage or more so in terms of optimum water management are discussed and costs reported.

346

EDMINSTER, T.W. Developments in equipments and materials for subsurface drainage, Rapp. VI[e] Congr. C.I.G.R., 1964, 2:142-9.

Reports some of the new products for subsurface drainage which are in the developmental stage, e.g. porcelainised steel, spiral supported plastic, interlocking spiral plastic and foamed plastic.

347

EGGELSMANN, R. Acute drainage problems. Wasser Boden., 1969, 21:1-8.

A review of drainage problems arising from the mechanisation of drainage operations and changes in piping materials, especially in peaty soils.

348

EGGELSMANN, R. Comparisons between trenchless and trenching sub-drainage. Proc. Int. Drain. Workshop, Wageningen, Netherlands,(1978), 1979, 25:536-42.

First pedological and hydrological field researches have shown, that in a marsh soil (heavy silty clay) with groundwater influence, the trenchless sub-drainage method had somewhat better effects than the trenching sub-drainage machine with cutter chain, especially if they work under wet conditions.

349

EGGELSMANN, R. Drainage Manual. Paul Parey. Hamburg, German Federal Republic, 1981, 268 p.

The main divisions of the subject matter of this manual following an introductory general section, are: water and soils; field investigations; drainage methods; effect of drainage; hydraulic calculations; drain design; drain material; drain construction; drain maintenance. In this second edition of the compendium, new material has been added with regard to the ecological aspects of the drainage.

350

EGGLESMANN, R. Experiences with various drain materials. 10 th Congr. Irrrig. Drain. Athens, 1978, 4:401-34.

The characteristics and quality parameters of clay and PVC pipes and of drainage filters are described. The hydraulic influence of the impact junctions and slot widths is indicated. Some drainage filters which reduce ochre clogging are discussed.

351

EGGLESMANN, R. Physical effects of drainage in peat soils of the temperate zone and their forecasting. IAHS Publ., 1975, 105:69-76.

The utilization of peat soils is impossible without drainage, but macropore space and permeability decrease after drainage, and micropore space, bulk density and consolidation increase. Data are given to estimate drain depths with regard to subsidence. A diagram is presented showing drain spacing in the fen and high bog in the north-west of the Federal Republic of Germany relative to the moisture content duration and intensity of pre-draining.

352

EGGLESMANN, R. Subsurface Drainage Instructions. Bull. National Commn. Federal Rep. of Germany and ICID. Verlag Paul Parey Berlin, 1978, 6,283 p.

This bulletin contains 10 chapters. Apart from a general chapter on related subjects and another on water and soil, drainage investigations, subsurface drainage methods, subsurface drainage efficiency, hydraulic calculations, drainage project planning principles, drainage materials, construction of subsurface drainage and maintenance of drainage are subjects of discussion. A bibliography, German standards and some photographs have been included at the end.

353

EGGLESMANN, R. Two decades of experience with drainage filters in the Federal Republic of Germany. 4th Natn. Drain. Symp. Proc. ASAE, 1982, Pub. 12-82:115-20.

The paper reviews the work on drainage filters in Federal Republic of Germany. Physical properties of 12 filter materials, oganic and synthetic are tabulated. The studies of new and cheap materials have resulted in decreased cost and improved efficiency of drainage systems.

354

EGGLESMANN, R.; MAKELA, T. The effect of drainage and agricultural utilization on the permeability of moor soil. Maataloust. Aikak., 1964, 36: 77-84.

Subsidence and physical changes (especially specific weight) following drainage resulted in a progressive decrease with time in the permeability of a Sphagnum moor. The possibility of improving drainage conditions by means of mole drains is discussed.

355

EHLERS, M.; WOHLRAB, B. Effects of groundwater lowering on the agricultural utilization of low moors and possible methods of cultivation in the brown coal region of the Rhine. Part I. Effects of the site and on agricultural utilization. Z. Kult. Flurb., 1979, 20:161-7.

The topics considered are: the effects of lowering the ground-water on the properties of low-moor peat; the realization that agricultural utilization of drained low-moors is uneconomic; the deterioration, in the brown coal region, of land that has been excessively drained so that there is no capillary connection between the groundwater and the soil water.

356

EI-MOWELHI, N.M.; HERMSMEIER, L.F. Tile drainage performance compared to theory. Trans. ASAE, 1962, 25:961-3.

The integrated Hooghoudt and Van Schilfgaarde equations were used to predict falling water table levels under subsurface drain conditions and computed results were compared with field measurements. The time of water drop predicted by the Van Schilfgaarde equation was close to the actual times measured in the field during the first year after intallation of the drains and from 1.5 to 2.5 times greater during the second year after installation. The time of water drop predicted by the integrated Hooghoudt eqation was twice the times measured in the field during the first year after installation and four times greater during the second year after installation. Prediction of performance of various drains with and without envelopes based on theories of Deeter in comparison to actual measurements was satisfactory, especially in describing the effect of using polyvinyl chloride (PVC) plastic drains with gravel filters on lowering the mid-space water table.

357

EI-MOWELHI, N.M.; SCHILFGAARDE, J. VAN. Computation of soil hydrological constants from field drainage experiments in some soils of Egypt. Trans. ASAE, 1982, 25:984-6.

On a 45 ha area of the Sakha Agricultural Experimental Station, an experiment was installed to evaluate the effectiveness of a range of subsurface drain materials and filters. Required drain discharge rates (q) were estimated to be 2.4 mm/day on the average, while actual field measurements yielded rates of 2.5 to 2.9 mm/day for 100 mm dia drains and 1.9 mm/day for drains 75 mm in dia. Hydraulic conductivity (K) from actual field measurements of q and h (height of water table above drain axis) resulted in somewhat higher values than those obtained with the auger hole method. The average from the q/h measurements was K = 0.089 mm/day. Transmissivity (T) for the test site was found to be $0.14 \leq T \leq 0.27$ m^2/day on the average. Drainage porosity (f) was determined for all drainage treatments from drain flow measurements. The best average value for the field test site was f =4.1 percent. Drainage intensity factor (a) as computed from actual field measurements ranged between 0.15 and 0.17/day. Consequently the usable period of observations begins 2.5 days after completion of irrigation or after cessation of rain.

358

ELEMA, H.M. Mechanical maintenance of ditches. In Drainage Principles and Applications. Vol. IV. Design and Management of Drainage Systems. Int. Inst. Land Recl. Imp. Wageningen, 1974, 16:237-59.

A brief illustrated review of machines and hand tools used in the maintenance of drainage ditches.

359

EL-DUJAILI, A.M. Effects of a single misalingment and of back pressure on the flow into a uniform tile-drainage system. 7th Congr. Irrig. Drain., Moscow, 1969. 1:C 117-C 156.

The flow of ground water with a free water table, in an isotropic and homogeneous soil overlying a horizontal impermeable layer, from a surface applied rainfall or irrigation supply to a system of horizontal tube drains, was studied by viscous-flow, electrical analogues and numerical (Relaxation) methods. The effect of horizontal, vertical and combined misalignment of a single drain on the free surface, the flow rates and flow pattern to each tube drain was investigated. In theory, the extent of the effect of any misalingnment is infinite. However, experimentally it was found that the vertical misalignment had a significant effect on the free surface, discharge and flownet over the whole range of the experiment while the horizontal one had a noticeable effect only extending over one spacing on either side of the misaligned drain.

360

ELKHOLY, F.H. Field tile drainage project for part of the Nile delta. ICID Bull, 1970., 19:48-51.

The project comprising the providing of an area of about 950,000 feddans (one feddan = 0.42 ha) in the Nile delta with adequate drainage system consisting of pumping stations, remodelling and deepening of upper drainage systems and construction of field tile drains in a period of 7 years is illustrated. The project is likely to result in increased production of annual gross value which is 17 per cent higher than the present levels.

361

EL-LEIOUDB, A.; SHAABAN, K.; EL-DAMATY, A.H.; IBRAHIM, M.M. Desalinization of soils of Memofeya Governorate put under tile drainage. Egypt. J. Soil Sci., 1974, 14:167-76.

An investigation was carried out in an attempt to study the effect of drainage establishment on salinity and behaviour of various ions within certain soil profiles. The tiles were placed at 120 cm and 60 m apart. Results showed that tile drainage, had a depressive effect on the salinity of the tested profiles, the effect being more pronounced at the surface layers. Sodium and chloride ions were more easily leached than ions of calcium and magnesium, sulphate being in between. Results were explained on the basis of solubility product of different salts.

362

EL-MADANY, M. The role of the tile drainage in land development in the United Arab Republic. 5th Irrig. Pract. Sem. NESA, New Delhi, 305-15.

Drainage need and as a development factor is discussed in relation to UAR. Progress made and the procedure for evaluating tile spacing are illustrated.

363

EL-SHAFEI, Y.Z. A study of bearing and frictional

resistance of the mole drainer. Egypt. J. Soil Sci., 1980, 2:19-28.

The action of bearing and frictional resistances on the blade and torpedo of the mole drainer, and the effect of different shapes of mole drainers at high speed on these resistances were investigated. Frictional resistance on blade or torpedo was not significantly affected by speed. Bearing resistances increased with speed on the blade and decreased with speed on the torpedo.

364

EL-SHAFEI, Y.Z. Theoretical and laboratory model for studying mole drainage. Z. Kult. Flurb. 1980, 27:75-86.

A laboratory technique (mole drainage model) for studying the factors influencing mole drain design and efficiency, such as soil miosture content, depth of mole, shape (diameter and front angle of torpedo,length and width of blade) and speed of the mole plow is described. The model consists of a plexiglass soil box, an overhead carriage, a dynamometer, an electric motor with speed reducer used with different diameters of pulleys to obtain various speeds, and different shaped mole plows. The model also provides a constant hydraulic gradient for studying the efficiency of the mole channel. The laboratory model was tested on clay soil and found to give good effectiveness. A theoretical model was also developed to evaluate the power required to pull a mole plow through the soil taking into account the major factors encountered.

365

EL-SHAFEI, Y.Z.; AL-HERR, M.H. Effect of some major factors on the construction and durability of the mole drain. Z. Kult. Flurb., 1980, 21:154-66.

The factors influencing the draft requirements of a mole plow in the soil are discussed. El-Shafei's mole drain model was tested and supported by laboratory experiments. An equation for computing the draft requirements of this plow was developed. The effect of soil moisture content at the time of mole construction on the efficiency of the mole channel was also studied. The optimum soil moisture for clay soils should lie at the plastic limit and in the neighbourhood of 0.33 bar. A linear multiple relationship (R = 0.96) existed between the factors studied and the resistance of the soil to mole plowing.

366

ELTOLA, J. The need of draining in agriculture and forestry. Vesitalous, 1964, 5:17-21.

About 350,000 ha of Finland's cultivated land are in need of basic draining. In recent years, about 25,000 ha of cultivated land have been provided with drainage. Only about 10 per cent of the cultivated land are underdrained. Part of it can be cultivated without draining, and the remainder of it would not be economical to underdrain. The cultivated land suitable for underdraining is estimated at about 1.5 million ha. During the past few years the consumption of wood has exceeded the growth of the forests, and silvicultural measures. Draining of forest lands are therefore, urgently needed to increase the yield of wood.

367

ERICKSON, A.E.; ELLIS, B.G The nutrient content of drainage water from agricultural land. Bull. Agric. Exp.Stn. Michigan State University, 1971, 31, 16p.

Water samples were analysed for N, P, K, Ca and Mg. The greatest loss was of Ca and Mg. NO_3 accounted for more than half the N loss; its concentration in the water was 0.2-11.1 ppm and the loss was 9.1-19.1 kg/ha with maximum loss in the spring. Equivalent figures for P were 0.01-0.3 ppm and 0.9-14.5 kg/ha and for K 0.3-6.8 ppm and 1.91-26.2 kg/ha. From the results fertilizer recommendations are made.

368

ERICKSON, A.E. Short-term oxygen deficiencies and plant responses. Proc. Drain. Efficient Crop Production Conf. Chicago, Illinois, 1965, 11-2.

Response of plants grown at various soil-oxygen diffusion rates show that for most plants critical oxygen diffusion rate seems to be around 5.8-6.7 x 10^{-7} gm cm^{-2} Sec^{-1}. Short term deficiencies do affect the plant growth and the extent of damages can be related to the duration of deficiency, completeness or intensity of the deficiency, kind of crop, stage of development of crop and photosynthetic and transpirative demands on the crop during the deficiency.

369

ERIKSSON, J. Soil functions and drainage. Proc. Int. Drain. Workshop, Wageningen, Netherlands, (1978), 1979, 25:180-212

General facts about climate, soils and drainage in Sweden are given. Especially the soil profile formation, the strucural pattern and drainage properties are discussed. The experience gained in field test program gives the base for the choice of drain intensity e.g. drain space and depth in separate agricultural regions. In a second subsection the root growth and its dependence on the root environment are discussed. The third subsection on changes of functions of the soils as a result of compaction show, using experimental data, how a number of properties of importance for the root environment is influenced, namely pore space distribution, hydraulic conductivity, air permeability, penetration resistance and root growth. The vulnerability of a soil to compaction is primarly related to the water tension acting on the soil skeleton. The primary way of regulating the moisture content and thereby to a certain extent the trafficability, is by drainage, as could be shown by the results of drainge experiments. One has, however, to realize that drainage can influence only a small part of the wide variation in soil water tension. The drainable pore space will be gradually destructed when raising the compaction forces in the soil to 200 kPa. This value will be stated as an allowable limit over which severe deterioration in the root environment and drainage properties will occur.

370

ERNST, L.F. Drainage of undulating sandy soils with high ground-water tables. 1. A Drainage formula based on a constant hydraulic head ratio. J. Hydrol., 1978, 39:1-30.

In the undulating areas in the eastern Netherlands (elevation 5-30 m above sea level) only the largest drainage channels have a perennial discharge, while the smaller channels have an ephemeral discharge which is considerable in wet periods, but absent in dry periods. The usual drainage formulae cannot be satisfactorily applied in such situations, but a considerable improvement can be obtained by taking into account the differences in channel size, the differences in open-water level and the drain density (length per unit area) of all drains exceeding a certain bottom depth. The intensity of groundwater flow and open-water flow depends on the differences in height between phreatic surface, open-water level and bottom depth. Although not strictly necessary, a constant ratio has been assumed for these hydraulic head values relative to the channel bottom in order to obtain a simple and still fairly accurate formula for the relation between drain discharge from the area considered and depth of the phreatic surface below ground surface.

371

ERNST, L.F. Drainage of undulating sandy soils with high ground-water tables. 2. The variable hydraulic head ratio. J.Hydrol., 1978, 39:31-50.

A new drainage formula derived based mainly on a analysis of the extent to which groundwater flow depends on the permeability of the underlying ground and on the often rather complex geometric properties of the channel network in the drainage basin has been considered for improvement. The hydraulic head ratio between groundwater flow and openwater flow strongly depends on the bottom depth of the drainage channel and on the depth of the phreatic surface. By substitution of an adequate expression for this hydraulic head ratio into the basic drainage formula the final expression could be improved, especially regarding the distribution of the groundwater discharge to the large drainage channels. For wet conditions this is not of major significance on the relation between total discharge and groundwater depth.

372

ERNST, L.F. Second and third degree equations for the determination of the spacings between parallel drainage channels. Proc. Int. Drain.Workshop, Wageningen, Netherlands,(1978), 1979, 25:85-107.

The drainage formula proposed by Hooghoudt can be combined with one proposed by the author. This results in a third degree polynomial equation for the drain spacing. The resulting formula can be applied to the steady state ground-water flow to parallel drains in homogeneous aquifers and in two-layer aquifers with the interface at the same level as the open water surface in the drainage channels. For some three-layer aquifers the same equation can be used in combination with a nomograph for the radial flow resistance in a two-layer aquifer. An attempt has been made to obtain simpler formula by adding an empirical coefficient to the radial flow resistance in a two layer aquifer, the result is a slightly less accurate second degree equation. A comparison of these formulae with the results of Kirkham and Toksoz did show only small differences. The formulae presented in this paper can therefore safely be used for practical applications. However, with the conditions that the drain spacing must be atleast equal to four times the total thickness of the aquifer.

373

ESTRADA, L.A.L.; CASTRO, A.P.; MORENO, J. Methods and economic aspects in the operation of the drainage systems in agricultural zones of irrigation district No.38, Navojoa, Sonora-Comparison between annual operating and maintenance costs and the economic benefits accruing to the farmer plus the direct and indirect benefits to the zone, region, state or Nation. _7th Cong. Irrig. Drain. Mexico_., 1969, Q 25 R 6:57-72.

The neglect of drainage maintenance in the Mexican Republic has been highlighted. The cost analysis indicated that there was severe loss of production during the 1959-60 cycle. It was estimated that during the said agricultural cycle the supporting drainage system could have been saved at a cost of $35.10 per ha, which is 1.25 per cent of the per hectare value of the crop.

374

ETDUJAILI, A.M. Drainage and salinity in Iraq with special reference to Dujailah experimental study. _5th Irrig. Pract. Sem. NESA_, New Delhi, 1964, 239-53.

The paper describes the general phenomena of drainage and salinity of irrigated areas in Iraq. The Dujailah soil profile consisted of silty clay loam which at 60 cm changes into silty loam. The results indicated that better drainage can be obtained from narrow spaced drain pipes than from deep and wide spaced open ditches. The theoretical computations for the relation between the hydraulic head and drain discharge showed close agreement with field data.

375

EVANS, D.D.; ASHCROFT, G. Tile drainage for layered soil. Soil Sci. Soc. Amer. Proc., 1961, 25:142-5.

The effect of a restricting layer 30 cm thick on the flow of water to drain tiles was evaluated. Depth to restricting layer, relative conductivity of the restricting layer to the surrounding soil and depth of the drain were variable factors. There was an almost linear relationship between drain flux and depth of drain when the relative conductivity of the restricting layer to the surrounding soil was 1:10. For 1:1000 conductivity ratios, drain flux was almost independent of depth. The surface-intake-rate distribution is presented, for various cases; sizeable tension extending over a large portion of the soil profile would be expected for several of them.

376

EVDOKIMOV, V.N. Wood increment in relation to drainage intensity in the Piceetum equisetoso-sphagnosum in the Komi ASSR. Lesn. Z. 1979, 3:30-2.

In 1932-33 an area of spruce forest 100 yr old, site class V, was drained by a ditch 1 m deep. Increment was determined at various distances from the ditch, and the results are shown in graphs. The effect of the drainage was negligible at distances beyond 35-40 m from the ditch, though some improvement was observed 10-20 yr after the drainage operation. Closer to the ditch, the improvement in increment was considerable (up to 200% or more) in all first 10 yr after drainage. The amounts of absorbed bases, K, and P in the soil increased and the exchange acidity decreased as a result of drainage, but drainage had little or no effect on the amount of hydrolysable N.

377

EVERARD, J. E.; FOURT, D.F. Forest drainage in Wales. Report, Welsh Soils Disc. Group. Forestry Commission

Farnham, UK, 1971, 12:61-7.

The problem of excess soil water on afforestation areas is examined. Estimates of areas of relevant soil types are presented- at least 35 per cent of the total of 120 000 ha in Wales comprises wet soils. Methods of digging, spacing and layout of open ditches are examined and costed, and the research approach to drainage is discussed.

378

EZAKI, K.; YASHIMA, S.; THAVARAJ, S.H. A need for the drainage of paddy fields to enable farm mechanisation for double-cropping of rice in the Muda Irrigation Project. Japan Agric. Res. 1976, 10:119-24.

The load bearing capacity and suitability for machinery of the surface layer(plough sole) of rice soils in the Muda area decreased after the introduction of double cropping. The cone penetration index of the surface layer decreased rapidly as the ground water level rose. The probable benefits of drainage include the maintenance of the subsoil hardpan which is required for support to heavy machinery, improved soil conditions for mechanization, and rapid drying of the soil to allow farm operations during spells of dry weather in July/August.

379

FALCIAI,M.; BRUNO, V. Drainage trials for salinity control (Johar-Central Shebeli) Riv. Agric. Subtrop. Trop., 1982, 76: 261-85.

Four drainage treatments were tested in a sugarcane experimental field at the Johar Sugar Estate (Somali Democratic Republic). No drainage, or drains spaced at 25m , 50 m and 50m without drain filter were the treatments. Effects on water table level, soil salinity and cane yield are reported.

380

FAUSEY, N.R. Flow to shallow drain tubes in layerd soils. 3rd Natn. Drain. Symp. ASAE, Michigan, 1976, 58-60.

Numerical solutions for the equations governing two-dimensional saturated-unsaturated flow are used to evaluate small diameter drain tubes placed at shallow depths in layered soil. Effect of drain spacing and the relative hydraulic conductivity of the soil layers on saturated flow rate and rate of water table drawdown are presented. Drain spacing has no effect on saturated flow rate but has strong effect on drawdown rate. Hydraulic conductivity ratio affects both saturated flow rate and drawdown rate.

381

FAUSEY, N.R. Shallow subsurface drain performance in Clermont soil. Trans. ASAE, 1983, 26:782-4.

Subsurface drains were installed at 45 to 50 cm depth in Clermont silt loam soil with good surface drainage. Water table elevations, drain flow rates, drain water quality and corn yields were measured over a three year period. Water table drawdown was detected laterally to about 3m but crop yields were enhanced to around 6m. The drains showed minimal sediment accumulation, no structural deformation and outflow water was of good quality.

382

FAUSEY, N.R.; HUNDAL, S.S. Role of trench backfill in subsurface drainages-A review. Trans. ASAE, 1980, 23:197-200.

The paper summarizes the available information on the role of the trench backfill in subsurface drainage. General conclusions include; (1)Theoretical analysis shows increase in drainflow rate and water table drawdown rate with trench backfill, (2) Permeable backfill enhances the removal of surface water, (3) Permeable backfill has little effect on midpoint water table, (4) Replacement of natural soil with trench backfill is economically questionable, (5) Permeable backfill is beneficial where shallow permeable layers or mole drains need a permeable hydraulic connection to a deeper drain and (6) Information comparing trencher installed drains with plow installed drains is insufficient to determine which is best. More research on the last two aspects is suggested.

383

FAUSEY, N.R.; NOLTE, B.H.; BEATHARD, B.S. Subsurface drainage maintenance in Ohio. Trans. ASAE, 1982, 25:997-1001.

The kind and frequency of maintenance performed on subsurface drainage systems was determined on the basis of a mail survey of land owners and farm operators. The survey was conducted in 1980 and provided information on 517 systems that were 5 to 15 yr old. Drainage system performance was rated satifactory for 95 percent of these systems. Maintenence had been performed on 20 percent of the satisfactory systems. In those systems maintained, outlet and breakage plugging were the most frequently occurring problems at 24 percent and 23 percent, respectively. Maintenance could corrrect about 57 per cent of the reported problems, while another 25 percent were soil or design related problems. Unspecified problems and routine maintenance account for the remaining 18 percent. Frequency of occurrence of all problems was 38/million m at 5 yr of age and increased to 114/million m at 15 yr of age.

384

FAUSEY, N.R.; SCHWAB, G.O. Soil moisture content, tilth, and soybean (Glycine max.) response with surface and subsurface drainage. Agron. J., 1969, 61: 554-57.

Average soil moisture content at four depths in the upper 30 cm during the seedbed preparation period (April and May), soil tilth immediately preceding seedbed preparation as expressed by crust properties and resistance to surface penetration, and stand establishment, root and top growth, and soybean yield, were compared for plots having different levels of drainage. On plots with no drainage, crop stand was sparse, ponding occurred, and soil structure was poor. Plots having subsurface or combined surface-subsurface drainage had lower moisture content, less resistance to surface penetration, smaller crust units, taller plants, better developed root systems, and higher yields than plots with surface drainage alone.

385

FEDYUKOV. V.I.; RUBTSOV. V.G. The direction of drainage ditches.Lesn. Khoz.,1979, 3:27-9.

The effect of the wind on the condition of drained spruce forests (mortality, debris and fluctuation of radial increment) was studied in relation to the direction of the drainage ditches relative to the prevailing wind. The study was made in picea abies forest (in the Udmurt ASSR) 80-90 yr old, growing on fen-peat soils drained in 1968-72. The stand was severely damaged by a gale in July 1972, and further windthrow occurred in following years. Graphs are given showing the radial increment of the trees in places where the drainage ditches were (a) parallel or (b) perpendicular to the direction of the prevailing wind. Data are also tabulated on the amounts of thrown and dead standing trees. The amount of mortality in (a) was about 5 times that in (b). Accordingly, in planning new drainage systems the drainage ditches should wherever possible be laid perpendicular to the prevailing wind.

386

FEHER, F. Mathematical models of subsurface drainage

in heavy soils. IAHS Publ. 1975, 105:333-41.

Subsurface drainage is one of the potential methods of improving heavy clay soils with poor hydrodynamic characteristics in the hilly regions of Hungary. After exploring hydrological conditions, a mathematical model was devised to assist in the study of a complex soil drainage system. The model consists of a mathematical system comprising a ploughed top layer, subsurface mole drains and a filtered drainage ditch, each of which is described by hydraulic parameters. Mathematical models were used to study the individual elements, their conveying capacity and the flow they have to carry. In this way, it was possible to determine the capacity of the complete model and to check the spacing of drain pipes. From the equations describing the individual elements of the system, various parameters can be calculated. The determination of the permeability coefficients for the loose ploughed layer and the filter in the drainage ditch will be of special interest.

387

FENEMOR, A.D.; BEVIER, B.R.; SCHWAB, G.O. Prediction of deflection for corrugated plastics tubing. Trans. ASAE, 1979, 22:1338-42.

Measured deflections of tubing buried for 5 yr in a silty clay showed that the soil provides minimal side support against deflection. The Iowa formula was the best of 4 methods for predicting deflection but all methods generally underestimated it. Maximum installation depths were calculated for tubing up to 381 mm diam.

388

FERGUSON, D.S. Methods and economics of operation and maintenance of drainage systems in agricultural areas. 7th Congr. Irrig. Drain. Mexico, 1969, Q 25 R 15, 193-204.

The paper deals with the surface drainage of

reclaimed land in the tidal reaches of rivers in an alluvial terrain. The paper describes the planning principles that have controlled the reclamation and development of swamp land in South East Asia. Generally, gravity drainage is possible, by providing a network of open surface drains, discharging through outlet gates into tidal rivers and the sea. Generally, also, the reclaimed land has to be protected from inundation by up-country flood water and salt or brackish water at high tides by perimeter embankments. The basic cropping patterns considered for this paper are coconut growing for copra, and rice. Irrigation of these crops is not considered, but the drainage layouts described can be integrated with irrigation networks and would not be fundamentally changed by provision of irrigation facilities. The cost of operation and maintenance, amortisation and capital depreciation are covered.

389

FINKEL, H.J. Drainage of irrigated fields. In CRC Handbook of Irrigation Technology. Volume 2, (Finkle,H.J. Ed) Boca Raton, Florida, USA; CRC Press. Inc. 1983:13-33.

Drainage-system design, subsurface drainage and surface drainage are discussed.

390

FISKELL, J.G.A.; STEWART, E.H.; CALVERT, D.V. Control of mobility of soil organic matter after trenching. Proc. Soil Crop Sci. Soc. Florida 1970. 30:122-30.

Soil samples from a limed or an unlimed soil taken from the fill of trenches after the installation of drains were leached in the laboratory with deionised water or $CaCl_2$ solution. The deionised water leachates were consistently opaque, due to dissolved mobile organic matter; whereas leachates were clear if the solution contained <50 ppm$CaCl_2$ for unlimed soils (pH 4.0-4.5) or $<$ 200 ppm $CaCl_2$ for limed soil(pH6.5). Application of $CaSO_4$ or fertilizer prevented dissolution of organic matter.

391

FITZ, J.C.; HORNER, G.L.; SNYDER, J.H. The economic feasibility of installing subsurface tile drainage in the Panoche Water District, San Joaquin Valley, California. <u>Bull. Div. Agric. Sci. Univ. California,</u> 1980, 1897, 59p.

On-farm drainage relief systems can be used to reduce water table depths which in turn affect crop yields. A production function for drainage was developed using theoretical relationship between drainage design and water table depth used in connection with crop yield response to water table depths. The production function was then used to estimate the annual returns from varying the lateral spacing and depth of installation. The optimal design was determined by the use of marginal analysis. The profitability of the drainage was evaluated using the annual value method. The optimal solution assumes that the cropping pattern yields and net returns will remain relatively constant over time. It was estimated that a 20 per cent decrease in price in the assumed cropping pattern will result in a zero annual value for the optimal drainage spacing depth combinations.

392

FLORESCU, G.; DUMITRESCU, N.; MAGDALINA, I. Variations in soil moisture content with mole drainage and its influence on oat yield. <u>Lucrari Stiintifice,</u> Instit. Agrono. "Nicolae Balcescu", 1977, 17:106-10.

In field trials in 1970-2, oats were grown over drained shallow soil in the damned area of Lunca Dunarill which had been mole drained using drains 1-2m apart at depths of 0.4-0.6m and diam. of 0.08 m to 0.12m. Fresh weight yields in 1972 ranged from 14.47 t/ha in the undrained control to 36.66 t/ha with drains 1 m apart at a depth of 0.6 m and with a diam of 0.08m.

393

FOKEEV,P.I. Determining the distances between horizontal drains on irrigated soils. Dokl.S-KH. AKAD. TIMIR., 1960, 56:21-6.

Three formulae are discussed for calculating the distances between horizontal closed drains on irrigated land where the groundwater level is deep. The distance should be $<$200m.

394

FOOD AND AGRICULTURE ORGANISATION. Salinity seminar Baghdad. Irrig. Drain. Paper, FAO, 1971, 7,254p.

Report of a regional seminar on methods of amelioration of saline and waterlogged soils held in Baghdad in 1970. The subjects covered include present and potential salt affected and waterlogged areas of the near East, problems of sampling, analysing and mapping salt affected soils, their reclamation and management, dynamics of salts in the soil-water system, improving water use efficiency, irrigation and drainage in relation to salinity and waterlogging, quality of the irrigation water and the limits to its use, salt tolerance and crop management. Short reports for Iraq, Cyprus, Iran, Jordan, Lebanon, Pakistan, Sudan, Syria, Egypt and the Yemen are presented.

395

FOOD AND AGRICULTURE ORGANIZATION. Drainage design factors: 28 questions and answers. Irrig. Drain. Paper, FAO, 1980, 38, 52p.

A number of questions concerning the design of subsurface drainage systems for irrigated land in dry regions are discussed, under the following headings: drainage investigations, design criteria and parameters, engineering maintenance and cost factors. Tentative guidelines are given.

396

FOOD AND AGRICULTURE ORGANIZATION. Drainage of heavy soils, Irrig. Drain. Paper , FAO, 1971, 6, 109p.

The report contains 12 papers presented by various authors at European Commission on Agriculture at Tel Aviv in 1970.

397

FOOD AND AGRICULTURE ORGANIZATION. Drainage machinery. Irrig. Drain. paper. FAO. 1973, 15, 122p.

A report of the working party on water resource and irrigation of the European Commission on Agriculture, Bucharest Romania, 1972. Papers on mechanized drainage operations in the UK, Belgium, Netherlands, France, Poland, German Federal Republic are presented.

398

FOOD AND AGRICULTURE ORGANIZATION. Drainage materials. Irrig. Drain. paper. FAO. 1971, 9, 122p.

Papers (7) by various authors presented at a working party on water resources and agriculture, European commission on Agriculture, Tel Aviv, 1970.

399

FOOD AND AGRICULTURE ORGANISATION. Drainage of salty soils. Irrig. Drain. Paper FAO, 1973, 16, 84p.

Paper presented by the working party on water resources and irrigation of the European Commission on Agriculture, Bucharest, Romania, 1972.

400

FORD, H.W. Characteristics of slime and ochre in drainage and irrigation system. Trans, ASAE, 1979, 22:1093-6.

Certain characteristics of ochre and Fe complexation as influenced by both slime formation and filamentous bacteria are described. Experiments were carried out which monitored bacterial culture chambers resembling drain lines and a field method is described for measuring Fe in discoloured groundwater samples. Under controlled conditions, ochre formed within a few hours and adhered to glass, polythylene (PE), polyvinyl chloride (PVC).neoprene, stainless steel, and soil organic matter. Chemically precipitated Fe^{+3} free of bacteria did not stick under water and the precipitate was porous. Citric acid was found to be an excellent agent for Fe^{+2} in discoloured groundwater samples.

401

FORD, H.W. The complex nature of ochre. Kult. Flurb., 1979, 20:226-32.

Bacteria are necessary for the copious formation of ochre that sticks to drain lines. Colloidal organically complexed Fe in water free of bacteria may contribute to ochre formation. Ferric-Fe free of bacteria and Fe complexes are porous and may not stick readily to the PE and PVC tubing that is being used in the USA.

402

FORD, H.W. Estimating the potential for ochre clogging before installing drains. Trans. ASAE, 1982, 25:1597-600.

Bacteria are responsible for iron reduction in flooded soils. The test procedures described can be used to estimate ochre potential in agricultural drains, drains in wastewater management sites, highways and landfills. The procedures estimate levels of Fe^{+2} that may occur in the soil solution(groundwater) either with soil samples incubated in the laboratory or from groundwater tested directly in the field. The field portion of the procedure consists of separating Fe^{+2} from the groundwater by pressure filtration into sulfamic acid where it can remain for several days without a supplemental energy source, followed by filtration. The methods automatically correct for soil type and pH and are useful in estimating whether the soil zone is deficient in natural energy, reducing potential by bacteria, or amorphous iron. The tests are also helpful in estimating whether Fe^{+2} may be flowing in from surrounding areas or escarpments.

403

FORD, H.W. Low pressure jet cleaning of plastic drains in sandy soil. Trans. ASAE, 1974, 17:895-7.

A low pressure jetting method for the removal of sludges (Ochre and FeS) from plastic drains and the surrounding thin gravel envelope in sandy soils is described.

404

FORD, H.W. Performance of underdrains in three sites utilizing sprinkler irrigaion of secondary sewage and processing effluent. Proc. Soil Crop Sci. Soc. Florida, 1977, 36:41-65.

On soils of the Florida acid flatwoods irrigated by sprinkling with effluent from secondary treated municipal waste water an citrus processing waste drains(polyethlene tubes) 18m apart and at a depth of 120 cm handled 5.6 cm of treated sewage/effluent/day, a steady-state water-table being maintained at a depth of 75 cm midway between drains. The drains were designed to remove 5 cm water/day. Low pressure water jetting of the drains was necessary to remove accumulated sludges and maintain drain flow. Drains 18 to 42 m apart are satisfactory for citrus waste spread on to certain poorly drained flatwood soils provided the rates of effluent application do not pond the surface for periods sufficient to damage the established grass cover. The time required to lower the watertable was decreased by spacing the drains 21 m instead of 42 m apart, and the length of time from which the profile remained aerated between irrigations was consequently increased. Sulphide concentrations of 1-2 ppm in the outflow water from the drains minimized ochre formation in the presence of Fe^{+2} and did not contribute to significant deposition of FeS during irrigations. Jetting the drains three weeks after their installation decreased entry resistance by up to 50 per cent but subsequent jetting was less effective.

405

FORD, H.W.; BEVILLE, B.C. Chemical changes in tile drain filters and ditch banks caused by anaerobiosis. Trans. ASAE, 1968, 11:41-2.

Two filter materials viz.(i) phosphate furnace slag and (2) sawdust filters which had been installed in tile drainage systems for 6 months to 11 years back have been examined. The rapid destruction of slag gravel filters under continuous flow was initiated by H_2S, a bacterial metabolite formed under suitable conditions in an anaerobic flooded environment. Bisulfide ions apparently reacted with clacium in the slag destroying the buffering capacity as well as physical structure. FeS accumulated as deposits around slag and sawdust filters and may be the principal factor reducing infiltration rates. Heavy deposits in the soil were present only on topsoil organic matter used to "blind" the tile. FeS observed in the surface of seepage zone of ditch banks may have an adverse effect on hydraulic conductivity.

406

FORD, H.W.; CALVERT, D.Y. The formation and deposition of hydrogen sulfide on a sawdust drain tile filter. Proc. Soil Crop Soc. Florida ,1970, 29:1-7.

Under conditions of continuous water saturation, sawdust filters serve as a medium for the genration of high H_2S levels which maintain anaerobic conditions in the outflow water and minimize FeS sludge formation.

407

FORD, H.W.; CALVERT, D.Y. ; BEVILLE, B.C. Hydraulic conductivity of soils and Filter materials in florida wetland citrus. Trans. ASAE, 1968, 11:566-7.

The reduction in hydraulic conductivity of soils and drain filters had been studied using soil cores and the results analysed based on accumulation of FeS and $Fe(OH)_3$. Reductions in hydraulic conductivity have been found in sawdust and slag draintile filters, in backfill abutting filters, in the surface of seepage zones of ditch banks, and under furrows of bedded groves. One factor associated with the gradual reduction in conductivcity has been the sorption of FeS on humus with the appearance of a black, greasy, gelatinous sludge. At certain locations, the long-term accumulated deposits plugged sawdust, slag, and fiberglass filters as well as portions of ditches installed in filled-in muck ponds. Tests in the laboratory and observations in the field indicate that FeS does not accumulate on the nonorganic soil fraction. Small particles of humus trapped in and between crevices of sand grains can serve as the focal point for the accumulation of $FeS.Fe(OH)_3$ formed from the oxidation of FeS was found to be as impermeable as FeS.

408

FORGAC,K. Amelioration of the water regime of heavy soils in the East Slovakian lowland by Drainage. Ved. Prac. Vyzk.Ustavu. Melior. Zbrasl. 1973, 13:71-88.

A combination of tile and mole drainage was more effective in the heavy soils of the East Slovakian Lowland than either tile or mole drainage alone.

409

FOROUD, N.; BROUGHTON, R.S. Effect of different

drainage rates on duration of high water tables in South-Western Quebec. Can. Agric. Engng., 1978, 20: 71-5.

Precipitation and evapotranspiration data for 27 years from three stations in Quebec have been used together with a simulation model developed by Broughton and Foroud(1978) to obtain the duration of particular water table depths to be expected. An IBM 360/75 digital computer was used to make computations of the water table depth to be expected each day for 60 different combinations of soil characteristics and drainage rates. The results from these combinations have been prepared in tabular and graphical form to show the effect of drainage rates on water table depth. By using a criterion for the number of days the water table may be at or less than, a particular depth, one can select the minimum drainage rate that will meet that criterion as the drainage coefficient for the design of sub-surface drainage system.

410

FORSYTHE, W.M.; VICTOR, A; GOMEZ, M. Flooding tolerance and surface drainage requirement of Phaseolus vulgaris L. In Papers Intern. Conf. Role of Soil Physical Properties. Intern. Inst.Trop. Agric., Ibadan, 1977, 18p.

In greenhouse trials a few hours of flooding substantially reduced yields of Phaseolus vulgaris var 27-R. In field trials on mixed bean and maize cropping on a typical distripept, bean fields were upto 40 per cent higher on a cambered beds than on normal fields. Use of the cambered bed system lowered the effective water table level and improved surface drainage.

411

FOUND, W.C.; HILL, A.R.; SPENCE, E.S. Economic and environmental impacts of Agricultural land drainage in Ontario. J. Soil Wat. Conserv., 1976, 31:20-4.

Investigation of the impact of government-assisted agricultural land drainage in Ontario was centred on seven townships that represent the full range of drainage conditions in the province. 37 drainage projects completed during the period from 1965 to 1970 were analysed. On the whole, the drainage projects have generated benefits for agriculture at the local level without creating undue environmental damage. A minority of projects have not generated sufficient agricultural benefits to justify their construction. Considerable variation in the benefit-cost balance is evident among drains, and there is a tendency for the traditional centre for farming activity, southwestern Ontario, to benefit most.

412

FOURNIER, C. Critical study of guiding systems for drain-laying operations. <u>10th Cong. Irrig. Drain.</u> Athens, 1978, 4:17-34.

Techniques for achieving accurate drain-laying, including laser systems, are described and the advantages and disadvantages of each are discussed.

413

FOUSS, J.L. Corrugated plastic drains plowed-in automatically. <u>Trans. ASAE</u>, 1968, 11:804-8.

From a research and development standpoint, structural stability of the corrugated plastic drains is of the prime concern. For the small-diameter plastic drains, grade control is probably more important than hydraulic capacity. The drain tube plow installation of corrugated plastic tubing looks promising as a means for more effective and more economical field drainage. The drain tube plow es'timates the need for ditching and backfilling; the corrugated plastic drain tubing is light weight, low cost, and can be easilily handled in coils; and the laser-beam automatic grade-control system makes high-speed installation of subsurface drainage a reality. With the drain-tube plow equipment 600 m or more of corrugated plastic drain can be installed per hour under many field conditions. This is about the amount of draintile laid one-half day with present trenching machines. The low installation charge with drain-tube plow could result in a cost reduction of one-third or more for the installed farm drainage system in comparison to the cost for conventional tilling.

414

FOUSS, J.L. Drain tube materials and installation. In Drainage for Agriculture . (Agronomy No.17: Schilfgaarde, J. Van, Editor). Amer. Soc. Agron. Wisconsin, 1974, 147-77.

Current drainage tube materials, the effect of soil type on the flexibility of newer plastic materials, specification tests and installation and handling equipment are reviewed.

415

FOUSS, J.L. Plastic drains and their installation. Proc. Drain. for Efficient Crop Prod. Conf. Chicago, Illinois, 1965, 55-7.

Different types of plastic drains in use for tile materials are reported. The new inovations in the installation of these tiles, and problems of automaic grade control in these installation methods have been commented upon.

416

FOUSS, J.L. Status of specifications for corrugated plastic drain tubing. Proc. Natn.Drain. Symp. ASAE, Chicago, Illinois, 1971, 32-3, 37.

Excessive deflection, excessive ring-compression in the conduit wall, buckling of the tube wall or a combination of these may cause failure of flexible type conduits. Present status on the specification of these items and methods of testing have been discussed in the paper.

417

FOUSS, J.L.; FAUSEY, N.R.; REEVE, R.C. Drain tube plows: their operation and laser grade control. Proc. Natn. Drain. Symp ASAE, 1971, 39-42.

Different types of drainage plows and their functioning has been described. The flow diagram for adoption of laser automatic depth and grade control has been reported which is a advance technique in this field and has a wide scope for future.

418

FOUSS, J.L.; HOLMES, R.G.; SCHWAB, G.O. Automatic grade control for subsurface drainage equipment. Trans. ASAE, 1964, 7:111-3.

The fluid-dampened pendulum as a vertical reference for automatic grade control on a tile trencher is recommended; it is economical and can be easily installed on most trenching machines at a cost of $300. Because of high ground speed of the floating-beam mole plow, the accuracy of this system is reduced compared to a tile trencher. Efforts to improve this system and/or to install other system for mole plow are listed.

419

FOUSS, J.L.; SCHWAB, G.O. Research on plastic-lined mole drains in the United States. Rapp. VI. Congr. C.I.G.R., 1964, 2:150-8

The paper deals with the development of non-rigid plastic drain tubes and related equipment necessary for installing them in a mole drain channel.

420

FOUSS, J.L.; WINGER, R.W. Status of specifications and standards. 3rd. Natn. Drain. Symp. ASAE, Michigan,

179 Global Research on Drainage in Agriculture

1976, 14-7, 21p.

Specifications, already approved by the American Society of Testing Materials, that pertain to the manufacture and testing requirements and installation practices for corrugated plastic drain tubing are summarized. The specifications cover requirements and methods of test for materials, marking dimensions, workmanship, elongation, environmental stress cracking, pipe stiffness, and perforations for tubing sizes from 75 to 200 mm in dia. Also described are some of the latest efforts to develop additional or improved tests for these specifications.

421

FRAMJI, K.K. Review of watelogging and drainage problems and their solutions in India. ICID Bull., 1974, 23:36-41.

The paper reviews the waterlogging that occurred almost everywhere in India on the introduction of canal irrigation and which grew to menacing proportions in the absence of any in-built anti-waterlogging measures in the irrigation projects. The causes, effects and extent (12 mha) of waterlogging have been outlined and the important lessons learnt from experience and from the historical studies on waterlogging and drainage have been touched upon with a cautionary note against expectations from transference of knowledge and experience to other problem areas. Research needs for obtaining scientifically precise anti-waterlogging measures and their design criteria have been indicated.

422

FRASER, A.I. Studies on drainage and soil moisture for tree growth. Res. Dev. Pap. For. Commn., 1967, 50, 8p.

Many afforested catchment areas in Britain possess ill-drained soils where tree roots are confined to the surface layers. The tree species most commonly used in these plantations have been shown to root very deeply on well-drained soils. Artificial drainage of heavy textured soils lowers the winter water table, affects, and increases the amount of water stored in the profile. This is expected to result in greater depth of rooting for the trees and a better distribution of run-off.

423

FROGGE, R.R.; SANDERS, G.D. USBR subsurface drainage procedure. In Water Management For Irrigation and Drainage. Proc. Spec. Conf. ASCE Irrig. Drain. Div. ASCE, New York., 1977, 1:30-45.

The paper outlines the U.S. Bureau of Reclamation procedure to design and install a subsurface drainage system. Drain spacing computations by transient flow and steady procedure have been compared and effect of angle of incidence between the drainline and direction of ground water approach on inflow rate into the pipe has been tabulated. Design efficiency is tested on the basis of benefit/cost index. Challenges for improved design procedure are listed.

424

FRY, R.K.; SPOOR, G. Electrical analogue studies of the influence of soil compaction and smear in the vicinity of a pipe drain on drain performance. J. Agri.Engng. Res., 1983, 28:207-16.

An electrical analogue has been constructed to examine the influence on water table behaviour of local impermeabilities such as beneath the pipe and smear alongside the drain. The two increase the water table height, the latter being more detrimental. The combined effect of the two result in rapid and unacceptable water level reduction, smear adjacent to the pipe must be removed immediately.

425

FUJIOKA, Y.; KITAMURA, T. Studies on tile drainage (1) On the discharge and ground-water table after the disappearance of ponded water. J.Agric. Engng. Soc. Japan, 1964, 32:317-21.

The mechanism of ground water movement concerning the case when ground water is rapidly removed through the drains has been discussed. As a result of tests, it has been found out that the relation between ground water table and discharge from drain could be expressed, in practical sense, by the following linear equation: $q = FkH+G$ (q, discharge from drains; k, hydraulic conductivity; H, height of ground water table above the drains; F and G, shape factors depending upon drain spacing, drain diameter and the depth to the impermeable layer), Further asssuming the decrease of the ground water table as quasi steady state, derive the equations, which explains the mechanism of such phenomena.

426

FUJIOKA, Y.; TAKAHASHI, T. Fundamental studies on the tile drainage in layerd soils (1)- on the resistance network analogue. Trans. Agric. Engng. Soc. Japan, 1967, 21:13-8.

An equation has been derived for the resistance at the drain perimeter in layered soils by assuming that the water in the vicinity of the drains is flowing radially towards the drain and that the equipotentials form concentric circles. It confirms results on homogeneous soil with Kirkham theory.

427

FUJIOKA, Y.; TAKAHASHI, T. Fundamental studies on the tile drainage in layered soils (3) - Estimation of drain discharge using the theory of conformal mapping. Trans. Agric. Engng. Soc. Japan, 1968, 23:19-24.

The paper estimates with the theory of conformal mapping the discharge from a drain in paddy fields, where a large permeable middle layer existed duly; accounting for the effects of a backfilled trench. It compares the estimated values with the experimental values derived from resistance network analogue which agree well with the former.

428

FUJIOKA, Y.; TEJIMA, S. Study to rationalisation of drainage for low-flat paddy field zone. ICID Bull., 1966, 15:40-50.

Rationalisation of drainage for Oguraika Land Improvement district and its binder basin of 5250 ha has been reported. It is suggested that runoff water from the back basin should be separated by the rice field zone and runoff water due to the rainfall in the rice fields themselves should be stored in the rice fields as much as possible till the safer drainage time of the pump. Drainage method using the controlling section at many stages is most suitable for this area.

429

FUKUDA, H. Sub drainage in heavy soils. Theoretical considerations. Soil Sci., 1964, 97:281-5.

Some theoretical considerations are given on drainage discharge from tile drains and the determination of the dia. of tiles. Diagrams show the movement of drainage water in clay soils to tile drains, and an electrical device for measuring drainage discharge. Graphs show the similarity of theoretical and experimental relationships between Q/hk and w/h where q = drainage discharge in cm^2/sec, k = coefficient of permeability in cm/sec, h= water head in cm and w = width of the drain trench. By theroretical calculation the quantity of water to be drained through a trench to a tile can be estimated and this estimate can be used as a guide to determine the dia of the tile.

430

FUKUNAKA, H. Subsurface drainage by plastic drains. 2. Working efficiency and drainage effect of different types of subsurface drains in peat bog. Res. Bull. Hokkaido Natn. Agric. Exp.Stn., 1980, 127:123-34.

The working efficiencies of tile drains, PVC pipes (4.5 cm diameter x 4 m length), drain-hose (continuous iron coil covered with nylon fibre matting), continuous corrugated PVC piping, used with three different covering materials and mole drains were compared. With regard to the time required for construction, these drains showed a ratio of 1:0.5:0.6:0.01 respectively. The drainage capacity was greater for plastic drains than for tile or mole drains because of their continuous nature, or firm connection between the segments, which resisted stresses caused by differential settlement. As a result of the increased efficiency of the plastic drains groundwater levels and hence surface water levels dropped more markedly than with tile or mole drains.

431

FUKUNAKA, H. Subsurface drainage by plastic drains. 1. Construction methods and drainage effects. Res. Bull. Hokkaido. Natn. Agric. Exp. Stn., 1980, 127:109-22.

The working efficiency of a bio-drain (continuous polyethylene pipe) was 20-30 times that of a tile drain and about 2.3 times that of a mole drain. The flux was also much greater than with a tile drain, particularly when laid in a weak foundation such as peat bog. Differences between the fluxes increased with time. Mole drains showed least flux.

432

FUKUSHIMA, T. Underdrain machine made in trial for heavy clay soil and its performance. 10th Congr. Irrig.Drain, Athens, 1978, 5:161-73.

Heavy clay soils on Ariake polder remain in a state of super-saturation unless effective subsurface drainage can be installed. The characteristics of these soils and the development of a high-speed drain-laying machine for use under such conditions are described. The results of performance trials in Ariake polder are promising.

433

FUXA, A. Maintenance of drainage systems in the Czechoslovak Socialist Republic. 10th Congr. Irrig. Drain. Athens, 1978, 6:125-30.

The principles and standards laid down for the maintenance of state-owned drainage installations are outlined. Some machines used for this work are briefly described.

434

GALVIN, L.F. Land drainage survey II. Ir. J. Agric. Res., 1969, 8:1-18.

Statistical information for 16336 schemes on 48865 ha is presented on a country-wide and on a county basis. Drainage problems encountered were mainly seepage and springs (37.8%) impervious subsoil (31%) and water tables (23.8%). Tile drains were used on 71 percent of the total area, open drains on 15 percent and stone drains on 11.7 percent.

435

GALVIN, L.F. Reclamation of peats and impermeable soils. Proc. Int. Drain. Workshop, Wageningen, Netherlands, (1978), 1979, 25:243-52.

The paper deals with the designs and development of drainage techniques and equipment for the reclamation of peats and impermeable soils. Blanket peats upto a depth of 0.5 m are usually ploughed to get a satisfactory peat/subsoil mix and an increase in surface strength. Drainage may also be required, depending on the permeability of the subsoil. Where the depth of peat lies between 0.5 m and 1 m, drainage is usually limited, by economic considerations, to surface grading to open collector drains. Where the peat depth exceeds 1 m, the "tunnel plough" or the "gravel tunnel" machine may be used to install an intensive drainage system at an economic cost. Impermeable soils (K-values 0-0.1 m/day) require some form of subsoil disruption. Mole drains can be successfully installed on cohesive soils whose plasticity and shear strength characteristics are suitable. However, there are no reliable criteria available for the selection of suitable mole drainage soils and investigations to establish an adequate classification system are urgently required. Soils unsuitable for mole drainage are sub-soiled, ripped or have a system of "gravel filled moles" installed. In the reclamation of all impermeable soils, particular attention is paid to surface grading to eliminate ponding and reduce depression storage to a minimum.

436

GAMBRELL, R.P.; GILLIAM, J.W.; WEED, S.B. Denitrification in subsoils of the North Carolina Coastal plain as affected by soil drainage. J.Environ. Qual., 1975, 4:311-6.

In a tiled moderately well-drained soil, over 200 kg NO_3^- N/ha was found distributed throughout the top 3 m. However, in a poorly drained soil, relatively low levels of NO_3, were found in the top 1m and very little persisted in the saturated zone beneath 1 m. Oxidation-reduction potential (eh) measurements in the tiled soil indicated well-oxidized condition (500-700 millivolts) to 3 m and in the poorly drained soil indicated favourable conditions for denitrification. A marked decrease in the NO_3-N/C ratio with depth in the poorly drained soil supported the contention that denitrification was occurring. The water soluble C content of the subsurface soil water was 2.5 mg/L in the moderately well-drained soil and 10-15 mg/L in the poorly drained soil. Due to limited denitrification in the moderately well drained soil, much of the fertilizer N not utilized by crops moved into shallow aquifers and surface water. However, sufficient organic C was present in the poorly drained soil to bring about a reducing soil environment and subsequent denitrification. Thus, little NO_3 moved from this soil into surface waters.

437

GAMEDA, S.; JUTRAS, P.J.; BROUGHTON, R.S. Ochre in subsurface drains in a Quebec fine sandy soil. Can. Agric. Engng., 1983, 25:209-13.

The deposition of iron ochre in drains and around outlets has been associated with continual reduction in the efficacy of an experimental drainage system. Ochre formation was primarily due to the action of the micro-organism *Gallionella* and *Sphaerotilus*. The highest amount of ochre deposition took place at ground water temperatures between 10 and 14° C. Draining rates and head losses around drains were significantly affected due to the deposition of iron ochre associated with fine soil particles.

438

GARCIA, S.F.; SEVILLA, G.J.C. Drainage system design for the reclamation of a highly saline soil in the cooperative Cartavio Ltda No. 39, Bo Tec ICIA, 1978, 7:53-69.

Causes of salinity and of bad drainage (lateral infiltration, impermeable layers, and high water table) in the area studied are described, and reclamation measures (desalinization and drainage with a 40 m spacing between the laterals) are discussed.

439

GARDNER, W.R. Approximate solution of a non-steady state drainage problem. Soil Sci. Soc. Am. Proc. 1962, 26:129-32.

Flow in the unsaturated zone during one-dimensional drainage is taken into account approximately by assuming the water content to be a linear function of the hydraulic head. The problem is then treated as one in unsaturated flow. After a brief initial period, the rate of outflow should decrease exponentially with time, with the outflow rate proportional to the soil water diffusivity. Agreement with experimentally determined out flow rates is reasonable. The treatment is extended to two dimensions by the equivalent of the Dupuit-Forchheimer assumptions. The total amount in the profile at any horizontal distance from a drain is calculated rather than the position of the water table. The theoretical outflow is in good agreement with results of sand tank and field measurements.

440

GARDNER, W.R. Present knowledge of the inter-relationships between soil moisture, irrigation, drainage, and water use efficiency. In Soil-Moisture and Irrigation Studies. Proc. Panel FAO/IAEA, Vienna, 1967, 77-82.

A few selected problems of evaporation and transpiration, infiltration, field capacity, drainage and irrigation efficiency, and salt movement (NO_3 and Cl) are discussed.

441

GARLAND, K.R.; LONES, G.O. Reclamation of saline soils at Kerang, Victoria, by ponding and pumping from an aquifer. Aust. J. Exp. Agric. Anim. Husb. 3, 1963, 10:177-9.

Unproductive saline soils cover 12,000-24,000 ha

of the Torrumbarry irrigation area in North-Western Victoria, Australia. Trials were conducted in which one test area was flooded while at the same time water was pumped from an aquifer; the other test area was flooded only. In both instances it was found that the salt content of the soil was rapidly reduced and was maintained at a tolerable level for at least 14 months, and that the pH rose after reclamation. Application of gypsum did not increase the efficiency of the treatment. The trials did not indicate whether pumping is justified or not.

442

GASANOV, T.G. Laying a mole drain with heat strengthened walls. Mekhan. Elektri. Sotsial. S-Kh., 1978, 5:24-6.

In the system under examination, fuel and a compressed oxygen/ air mixtures are fed to the drain being formed by the mole plug to strengthen the walls of the drain by heat. This can be done only in mineral soils, laying drains of 50-70 mm diam. The consumption of fuel is 0.6-0.9 kg/m and that of oxygen 0.047m^3 (flask pressure 15 MPa). The rate of work, cycle time, thickness of strengthened layer in the drain, specific fuel consumption etc. are analysed. The discharge from such drains is 0.010-0.035 L/s and is virtually the same as from a plastic pipe.

443

GASTLER, A. Technological and technical aspects of ditchless construction of drainage and ground water lowering: installations in the GDR. 10th Congr. Irrig.Drain. Athens, 1978, 5:75-85.

The development in the GDR of a drainage machine (the Meliomat) for the trenchless laying of PVC and clay pipes is described. Equipment for pipe transport and for maintaining the correct grading of the drainage channel, and technical aspects of the use of porous backfill in trenchless drainage, are also discussed.

444

GE, X.Y. Reclamation of poorly drained soil by shallow wells combined with tile drainage in polder land of the Lixia River, northeastern Jiangsu province. Acta Pedol. Sin., 1980, 17:79-84.

Groundwater levels were lowered by employing a system of pumping in shallow drainage wells combined with tile drainage. This resulted in decreased salt percentages in the top layers (0-0.4m) and improved available N and available P status of the soils.

445

GEMACHLING, M.C. Drainage of alluvial plains of the lower Rhone protected by the embankments. 7th Congr. Irrig. Drain. Mexico, 1969, Q 25 R 3:23-40.

A short review of the natural, geographical and hydrological conditions of the area and informations concerning the works which permit to master the regime of the water table in agricultural parts located below reservoirs and canals are presented. Large ditches dug at the foot of embankment control the water table by collecting overflow of water table coming from slopes. Usually single ditch or sometime double ditch paral-lel to the first one are sufficient to do the job. Main drainage system of about 140 km covers an area of 10,000 ha. Maintenance consists of mowing twice a year with the help of specially equipped boats and by application of herbicides. More careful cleaning once in 10 years is undertaken.

446

GHOSH, B.C.; MITTRA, B.N.; PAWDE, H.K. Identification of critical stages in semi-dwarf varieties of rice to varying depth of flooding. Harvester, 1975, 17:17-21

Field experiments were conducted with rice varieties Jaya and IIT 60 to find out some measures against flood damage. The crop growth phases seedling establishment to maximum tillering, maximum tillering to flowering and flowering to maturity were subjected to flooding amounting to 25, 50 and 75 per cent of plant height. Among the growth phases seedling establishment to maximum tillering was found to be most critical at 75 per cent level of flooding. Hence drainage of flood water and keeping the submergence to 25 percent of plant height throughout a period of 10-55 days after transplanting, causes minimum damage.

447

GHOSH.R. The problem of waterlogging in the protected area of the eastern Kosi, flood embankment - Its causes, effects and remedies. Proc. Symp. Waterlogging Causes and Measures for Prevention, CBIP Publication No.118,1972, 1:11-6.

It deals with a special type of problem of waterlogging mainly attributable to the construction of flood levee along the eastern bank of Kosi, due to the peculiar geographical conditions and topography of the area. To avoid waterlogging, a combination of surface and subsurface drainage is required. Subsurface drainage can be accomplished either by subsurface drains or tubewells.

448

GIARDINI, L. Subsurface drainage problems and perspectives in Italian agriculture. Irrigazione, 1981, 28:5-24.

The effect of waterlogging on soil properties, agricultural operations and crop growth and productivity are briefly discussed. The installation and the problems and advantages of various subsurface drainage systems are described in detail, including backfilled trenches, mole drains and pipe drains. An analysis is presented of the optimum slope, depth and spacing of pipe drains to remove a defined amount of water per day. Subsurface pipe drainage and surface drainage methods are compared in relation to hydrological factors, land reclamation, irrigation, cultural operations, fertilizations, salinity, weed control, crop production and economics. Results of initial investigations into subsurface pipe drainage system in the Tre Venezie area of Italy indicate that such systems will improve the success of land reclamation in Italy.

449

GIBBS, D.B. Drainage in hill country. N.Z. J. Agric., 1978, 136:62-5.

A trial was carried out to compare graded- bank (terraced)with non-graded-bank area near Kaikohe under a sheep rotational-grazing pattern. There was no significant difference in DM producation between drained and undrained plots, but there was much better stock performance on the graded-bank area. Live-weight grain/ha yr increased 30 percent on graded blocks during the 3 yr trial period. This is considered to be the result of better utilization of the pasture grown rather than an increase in pasture quality or quantity.

450

GILLIAM, J.W.; SKAGGS, R.W.; WEED, S.B. Drainage control to diminish nitrate loss from agricultural fields. J. Environ. Qual., 1979, 8:137-42.

In an attempt to reduce nitrate movement to drainage waters, flashboard riser-type water level control structures were installed in the mains or outlet ditches at two locations in the North Carolina Coastal Plain to raise the water table to increase denitrification during winter. A large reduction in nitrate movement through tile lines occurred in moderately well-drained soils because of reduction in effluent volume. In such soils, there was no indication of increased denitrification. In poorly drained soils, drainage control had no influence upon soil profile oxidation-reduction potentials but resulted in approx. 50 percent reduction in nitrate movement through drainage ditches. This was due to increased water movement into and through deeper soil horizons (below 1 m). Nitrate concentrations and low eh values in all profiles below 1 m indicate that nitrate which moved to this depth underwent denitrification.

451

GILLES, K.P. Improvement and maintenance planning of an irrigation and drainage system. 11th Congr. Irrig. Drain., New Delhi, 1981, 2:643-7.

The planning of the selection and application of machines for the maintenance of drainage net works is discussed. Economic and technical correlations between constructional and maintenance activities are described. Methods and operations applied to carry out the planning are indicated. Particular reference is made to potential improvements in machine application. The procedures used and the different applications are illustrated by practical examples from study of a maintenance program for an irrigation project covering 45,000 ha.

452

GIL SANCHEZ, I.; MARTINEZ BELTRAN, J. Drainagae of peat soils in the polder of Pego-Oliva, Alicante, Spain. In Polders of the world. Int. Inst. Land Reclam. Improve., Wageningen, Netherlands. 1982, 306-19.

In the eastern coastal region of Spain about 6,200 ha of former rice fields remain uncultivated due to severe drainage problems. A plan has been put forward to reclaim these fields for horticultural production. Natural conditions in a pilot polder of about 1000 ha are described. The design of the drainacge system is shown. Subsurface drainage was evaluated in an experimental plot of 27 ha.

453

GIRDWOOD, J.G. Performance of an on-farm evaporation basin for disposal of saline groundwater. In Hydrogeology of the Riverine Plain of South East Australia. Proc. Symp. Griffith, Australia, 1977 (Storrier, R.F., Kelly, I.D., Eds). Aust. Soc. Soil Sci. Inc., 1978, 269-78.

The input and output components of the water balance, except seepage-out, have been measured over the last four years for an unsealed evaporation basin near Cohuna, Victoria. Seepage has been calculated by difference and is approximately equal to evaporation from the basin on an annual basis. The rate of seepage is increased by operating the basin at high levels and by low groundwater pressures outside the basin. About one quarter of the seepage is intercepted by a surface drain around the perimeter of the basin. Most of the remainder must recharge the regional groundwater as there is no evidence of salt accumulation in the soil surrounding the basin.

454

GIZBILI, M.M. Experience gained in Turkey for the irrigation and drainage in deltaic, coastal and low-lying areas. 8th Congr. Irrig. Drain. Varna, Q 27 R 26:403-8.

General aspects of drainage in deltaic, coastal and low-lying lands are presented. A good design should be performed in order to reduce size of the area to be drained by pumps, the amount of pumping discharge and number of pumps. A careful operation and field management is suggested to reduce the amount of water to be discharged by the pumps. For this, irrigation by means of ground water is suggested with precaution to avoid sea water intrusion.

455

GLATHE, H.; OTTOW, J.C.G. Ecological and physiological

aspects of the mechanism of iron oxidation and ochre formation. Z. Naturwiss. Abteilung, 1972, 127:749-69.

It is a review paper dealing with iron oxidation in drained soils leading to ochre formation in the drains. The paper lists 98 references on this subject.

456

GLEBOV, F.Z. (Ed). Integrated evaluations of wet peatlands and waterlogged forests in relation to their improvement. Novosibirsk USSR, Nauka, 1973, 238 p.

A collection of 14 papers dealing with various aspects of the study of wet peatlands (bogs and fens) and the forests on such sites, mainly in the Tomsk region of W.Siberia. The papers relate in the main to the classification of these sites, detailed description of their soils, flora and environmental conditions (especially hydrological conditions), and the effect of drainage in improving conditions for forest growth.

457

GLOVER, R.E. Use of mathematical models in drainage design. Trans.ASAE, 1966, 9:210-2.

When the conditions of continuing for the flow of ground water is expressed in terms of the Dupuit-Forchheimer idealization and a constant aquifer thickness, differential equations of the forms occurring in the theory of heat conduction in solids are obtained and the extensive resource in this older field becomes immediately available for application to the design of drainage systems. The cases for parallel drains, deep drains, drainage by pumping, canal seepage functions and drainage in more than one direction are illustrated and it is concluded that first-approximation solutions are or can be made available to cover almost any conceivable need in the drainage field. In shallow aquifers, where the drainable depth may be significant part of the original saturated thickness, the linear differential equations of the theory must be replaced by differential equations of non-linear form the solution for which is difficult but a great stride is now made to solve the same.

458

GNEDENKOV, V.I.; GORELOV A.G. Influence of individual factors on the degree of return on capital investment in drainage in the Far East region. Sib. Vest. S-kh. Nauki, 1975, 1:95-105.

The results are reported of investigations into the profitability of drainage installations on the Island of Sachalin. Certain close connections are shown between the following dimensions:yield per ha, investment expenditure per ha, degree of utilization of the improved areas, product price. A method is proposed which would make it possible to ascertain the minimum yields for the different crops which are necessary to amortize the invested capital within laid down periods. The relevant data for nine crops and for amortization periods of 2,3,5 and 10 years are presented for ditch drainage and pipe drainage.

459

GOEHRING, L.D; SWADER, F.N.; THOMAS, L.E. Corrugated plastic drain tubing stretch during drain plow installation.Trans. ASAE, 1982, 25:88-92.

Investigations of the amount of stretch occurring in corrugated plastic tubing when installed by a drain plow were conducted with four different brands of tubing installation in each of three locations. At each location, tubes were installed with and without power feed device. Stretch ranged from -0.09 to 9.53 per cent without the power feed, from -3.21 to 2.63 per cent with the power feed. The data indicate that the power feed unit significantly reduced the tubing stretch.

460

GOETTELMAN, R.C.; GRASS, L.B.; MILLARD, J.P.; NIXON, P.R. Comparison of multispectral remote-sensing techniques for monitoring subsurface drain conditions., NASA Technical Memorandum, 1983, No. 84317,16p.

The multispectral remote-sensing techniques were compared to determine the most suitable method for routinely monitoring agricultural subsurface drain conditions; airborne scanning, covering the visible portions of the spectrum through thermal infrared (IR); color-IR photography; and natural-color photography. Color-IR photography was determined to be the best approach, from the standpoint of both cost and information content. Aerial monitoring of drain conditions for early warning of tile mal-function appears practical. With careful selection of season and rain-induced soil-moisture conditions, extensive regional surveys are possible. Certain locations, such as the Imperial Valley, Calif., are precluded from regional monitoring because of year-round crop rotations and soil stratification conditions.

461

GOINS, T.;LUNIN, J.; WORLEY, H.L. Water table effects on growth of tomatoes, snap beans and sweet corn. Trans. ASAE,1966, 9:530-3.

Tomatoes, snap beans and sweet cron were grown in greenhouse in soil columns where the water table was maintained at 15,30, 45, 60 and 80 cm below the surface. The resulting plant growth water use efficiency and plant composition were determined on three textural classes of soils (Runford loamy fine sand, Othello loam and Othello silty clay loams. Tomatoes and sweet corn yield increased with increasing depth to water table except of corn on Othello silty clay loam. Snap beans yields increased with increasing depth to water table only on Runford soil. Yields were influenced depth to water table only on Runford soil. yields were influenced by soil texture as well as water table depth. The total water use generally increased linearly with amount of dry matter produced. The increased amounts of Ca, Mg and N found in the plant tissues on the deeper water available for root extraction..

462

GOLOVKO, D.G. Effect of draining peat soils of the Beshcher lowland on their moisture-air and nutrient regimes, Pochvovedenie, 1970,12:127-41.

Shallow drainage, which permits continuous rise from the groundwater into the rooting zone, gave potato, herbage and grain yields 11-58% higher than deep drainage did. Deep drainage often decreased soil moisture to the wilting point. Aeration was best after deep drainage , but such aeration resulted in excessive mineralization of peat and formation of very large amounts of NO_3 for high yields.

463

GOOR, G.A.W.W. VAN DER. Drainage of rice fields. In Drainage Principles and Applications. Vol IV. Design and Management of Drainage Systems. Intern.Inst. Land Reclam. Improve. Wageningen, 1974, 16:341-82.

The different types of rice culture are described, with particular emphasis on lowland rice. The purpose, time and rate of drainage of lowland rice fields are discussed in their raltionship to existing information about expected rainfall, permissible depth of water on the soil, and rate of percolation of water in the soil.

464

GOOSSENS, F.;DIERICKX, W.; CABRIELS, D.;BOODT, M.DE, Soil conditioners for stabilising trench backfill in an unstable loam soil of low permeability.J.Sci. Fd. Agric., 1979, 30:1139-47.

In the laboratory, permeabililty tests and wet-sieving were carried out to evaluate six different soil conditioners for stabilising a loam soil, to increase the structural stability of the backfill of drain trenches. Loam soil samples treated with polymer emulsions must be air-dried to become stable and highly permeable. At higher moisture contents the emulsions become ineffective because of the lack of coagulation between soil particles. Polfmer solutions are effective with regard to soil stability and water permeability even when the soil remains wet. There is good agreement between the results of the wet-sieving test and those of the permeability test.

465

GOOSSENS, F.; NAUDTS, L. Use of soil conditioners in drainage, Rev. Agric., 1979, 32:1413-21.

Laboratory experiments on the use of conditioners for stabilizing filling materials in drainage systems showed that in sandy-loam soils, treatment with polymer emulsions gave best results when the soil had a water content of approx. 10 percent. Polymer solutions gave satisfactory results at all the soil moisture levels tested. With clay soils, the most stable aggregates were obtained at a soil water content of field capacity, for treatments with both emulsions and solutions.

466

GORA, A. The results of laboratory and model experiments for the subsurface two-way control of water on soils with impeded drainage. Arch. Acker. Pflan. Bodenk., 1976, 20:631-6.

Laboratory and model experiments on the spacing of pipes for the subsurface two way control (by irrigation and drainage) of water in soil with a natural impermeable horizon below the main rooting zone showed that the hydraulic conductivity should not be less than 0.52m/day and that the hydraulic head should not be much greater than the depth at which the pipes are laid. Optimum design conditions are then obtained by the use of corrugated PVC pipes with 0.8 mm slits. The hydraulic conductivity of the impermeable horizon should be less than 0.009m/day, a condition that is satisfied by most impermeable horizons _in situ_

467

GORA,A.; SCHWARZ, K.; WERNER, D. Technological principles and design criteria for the amelioration of waterlogged soils. _Trans. 10th Int. Congr. Soil Sci._ 1974, 10: 251-9.

Nomograms for German soils show that 70-75 percent of the waterlogged layer must be loosened, 15-25 t of burned lime should be applied and the optimum drain spacings should range from 25 to 35 m. Some figures are given characterizing the success of soil reclamation which resulted in an average yield increase of 14.5 grain weight units/ha.

468

GORBACHEV, G.F. The economic basis of forest drainage. _Lesn. Khoz.,_ 1976, 4:35-8.

The paper discusses the problem of assessing the economics of drainage in the context of the Soviet socio-economic system, in which the discounting rate is taken as a standard 8%, although many Soviet forest economists believe that a rate of 3 % is more appropriate for forestry. The economic effect of the stand age at which drainage is carried out is shown,, using as an example 1 ha of Pinetum herbose-sphagnosum in the Leningrad region. The effect on wood production is greater, the earlier the drainage is carried out, but from the economic standpoint, drainage should be carried out where the effect will be achieved in a comparatively short time, i.e. 20-30 years before final felling. Some of the practical and economic implications of this conclusion are discussed.

469

GORDIICHUK, A.S.; GORDIICHUK,, L.O. Effect of water regime of peat soils on yield of potatoes. Kartopl. Res. Mizhvid Temat. Nauk., Zb. 1975, 6:92-7.

In pot experiments, tuber yields of potatoes grown in peat soil with water table depth at 0.8-0.9 m were decreased by 5.2 t/ha compared with water table depth at 1-2.2m. Results were confirmed in field trials.

470

GOSLING, W. Accuraccy of land drainage installation by mole-plough. 10th Congr. Irrig. Drain. Athens, 1978, 5:183-94.

A simple theory of motion of the long beam mole-plough is used to describe the grade accuracy obtained when using the mole-plough for drain installation. The principles of the laser control system, and technical means for employing the error information for corrective action are presented. It is concluded that the high speed potential of mole-ploughing will only be fully achieved with the more widespread use of laser control with fast reacting servo hydraulics.

471

GRAAF, A. VAN DER; RISSEEUW, I. The inspection and maintenance of drainage system Bol.Technico, ICIA, 1975, 4: 18-26.

A method for testing the performance of existing drainage systems is described, and some drainage problems are discussed in relation to soil water movement. Methods and equipment for the maintenance of tile drains and open ditches are described.

472

GRANDE, R.; LOSADA, A. The drainage systems of the

Guadalquivir River delta: its operation, maintenance and evolution. 7th Congr. Irrig. Drain, Mexico., 1969, Q 25 R13:169-79.

The operation and maintenance of the drainage network in the Guadalquivir marshes(Spain) meant for draining and leaching is described. The causes of excess water are flooding, lateral infiltration and rainfall or irrigation. Open ditches, tile drains and plastic pipe drains are being used at a spacing of 10-20 m, with depth varying from 70-90 cm and at a slope of 1 percent. The maximum length of these drains is 250 m for case of cleaning. The operation and maintenance of the system are carried out by a committee elected by the community of users.

473

GRASS, L.B. Tile clogging by iron and manganese in Imperial Valley, California. J. Soil Wat. Conserv. 1969, 24:135-8.

Results of a survey of tile drainage system indicate malfunctioning of system because of deposition of red(iron oxide compounds) and black(Manganese oxide compounds) precipitates which fill the joints. Silt or roots have also affected some systems. Formation of red or black deposits is not related to location, soil type, kind of tile material, or length of time drains had been in service. The affected systems can be effectively reclaimed by treatment with sulphur-dioxide gas and water.

474

GRASS, L.B.; MACKENZIE, A.J. Restoring subsurface drain performance. J. Irrig. Drain. Div. ASCE, 1972,98:97-106.

Clogging of tile drains openings have been discussed and techniques to restore the drainage system reported. Chemical treatment with 2 percent SO_2 by weight gas and water has been found to be successful in case of clogging by iron and manganese precipitates. Gypsum deposits can be removed by high pressure water jetting equipment. Travertine ($CaCO_3$) would be susceptible to acid treatment. High pressure jetting equipment is good for soil sediments and small roots. For larger root systems long flexible rotating rods with cutting blades are best.

475

GRASS, L.B.; MACKENZIE, A.J.; MEEK, B.D., SPENCER, W.F. Manganese and iron solubility changes as a factor in tile drainage clogging. II. Observations during the growth of cotton. Soil Sci. Soc. Am. Proc., 1973, 37:17-21.

Manganese and iron solubility can be reduced by controlling or regulating the oxidation-reduction status of the soil profile by reducing the waterlogging period after irrigation to a minimum. It will reduce the chances of tile drain clogging.

476

GRASS,L.B.; WILLARDSON, L.S. High- pressure water jet cleaning of subsurface drains. Trans. ASAE, 1974, 17:886-8.

Equipment for cleaning subsurface drains with high pressure water jets is described.

477

GRASS, L.B.; WILLARDSON, L.S.; LEMERT, R. A. Soil sediment deposits in subsurface drains. Trans. ASAE, 1979, 22:1054-7.

Field data on some of the physical and chemical characteristics of soil materials found in drain pipes are reported, and compared to the characteristics of the adjacent soils in order to establish possible causes of drain failure. Deposits varied considerably in texture and included sand, loamy sand, sandy loam, loam, silt loam, clayloam and silty clay loam, but particle sizes were predominantly in the medium sandy loam range in most drain deposits. Most of the silting problems were in clay or bituminous pipes. Salinity was not an important factor in silt accumulation. Particle size distribution on soil deposits in drains, compared with that of adjacent soil, showed the possible mode of sediment entry into the drain.

478

GRAUL, W.; WERNER, D.; SCHWARZ, K.; HOFMANN, A.; STRACKE, W; The influence of ameliorative measures on soil physical properties, water balance and yields on a Pleistocene loam-stagnogley in northern GDR. Arch. Acker. Pflan. Bodenk., 1979,23:105-14.

Deep loosening plus pipe drainage, with or without liming, was compared with conventional drainage only as ameliorative treatments for a mottled stagnogley in a three-year trial in the Rostock district. Pore space and air capacity were improved by deep loosening while the moisture equivalent was little affected. Apart from plots with conventional drainage only, the ameliorative measures led to improved water regimes even following heavy rainfall. Water availability in dry periods was also improved due to the increase in the storage zone suitable for root growth. Yields from the rotation, rape-winter barley- rape, were increased in all years. Continued liming was necessary to maintain the increases.

479

GREEN, F.H.W.Ridge and furrow,mole and tile. Geograph. J., 1975, 141:88-93.

In Berkshire, Buckinghamshire and Oxfordshire, the distribution of mole and tile drainage in 1973 coincided to a high degree with the distribution of mediaeval ridge and furrow. The distribution of mole and tile drainage is generally correlated with topography, solid geology and soil series.

480

GREINERT, H.; LYDUCH, L.; NIEDZWIECKI, W.;ZABLOCKI, A. Effects of lowering of the ground water level on soil properties and plant communities near a power station building site. Probl. Postepow Nauk. Roln., 1975, 169:135-42.

A reduction in the soil water table of 15 cm at the building site produced a fall in the water table of the surrounding land to a distance of 2 km, to below 2 m in the profile. Cracking of the soil surface occurred, and the rate of decomposition of organic matter increased, particularly in peat and muck soils used for arable crops or grassland. Filling the reclamation ditches with pumped water had little effect on the reduction in water table level, due to low soil permeabililty. Grassland communities showed changes in species composition.

481

GRICE, S. L. Developments in field drainage design. J. Sci. Fd. Agric., 1980, 31:415-6.

For soils with a high clay content subsoil, the necessary drain spacings derived from equations developed for groundwater control would be uneconomic for plastic or clayware pipes. Field experiments have consistently shown, however, that subsoiling or moling over drains, with permeable backfill achieves better control of water, produces higher crop yields, and gives greater improvements in trafficability than pipes alone. Hydrologic data from these sites indicate that classic drain spacing equations are also valid for surface water control.

482

GRICE, S.L. The work of the Field Drainage Experimental Unit. In Land Drainage (Gardiner,M.J.Ed.) A.A. Balkema, Rotterdam, Netherlands, 1982,, 297-302.

A brief description of field and laboratory experiments at the FDEU.

483

GROOT, J.M. The performance of surface and subsurface drainage on heavy clay soils in Yugoslavia. Neth. J.Agric. Sci., 1974, 22:160-74.

The soils of the Sava River basin(Yugoslavia) have a permeable topsoil of about 0.3 m depth, which overlies an impermeable subsoil. During rainfall, horizontal flow through the topsoil occurs. Appropriate formulae for steady and non-steady flow have been derived, which, together with field observations on field accessibility, have permitted the determination of drainage requirments. In selecting the best drainage system the projected mechanization of agriculture in the area has been considered. Partly for this reason, partly the first year of cropping, and partly for long-term economic reasons (initial costs are higher), subsurface drainage (1.0-1.2m depth and 15 m drain distance) is preferred to surface drainage. Factors determining peak discharges of field drains are discussed.

484

GROOT OBBINK, J.; SEEKAMP, J.V. Plastic pipes for subsurface drainage installations. J. Aust. Inst. Agric. Sci., 1964, 30:29-32.

Modern advances in drainage methods due to the use of plastics, fibreglass and installing machines are outlined.

485

GROVER, B.L.; KIRKHAM, D. A glassbead-glycerol model for non-steady state tile drainage. Soil Sci. Soc. Amer. Proc., 1961, 25: 91-4.

A reduced- scale model for studying a falling water table is described and illustrated. Data on discharge rates can be obtained as well as photographs of shape and position of the surface of saturation. The model can be used to obtain information valuable in drainage theory and design.

486

GROVER, B.L.; KIRKHAM, D. Solving the drainage problems by using model data. Iowa St. Univ. Agric. Home Econ. Exp. Stn. Res. Bull., 1964, 523, 48p.

Over 800 "water tables" in a glassbead-glycerol model were photographed. Data, tabulated and in part graphed, show how water tables fall with time in simulated tile-drained soil. The "soil" is either homogeneous or stratified. The tabulated basic data are depths to water table versus time for positions (a) midway between pairs of drain tile, (b) above the tiles, and (c) at two intermediate points, so that the whole water-table arch is defined. The tile spacings are 50,100,200 and 400cm, depths below the tile axes to an impermeable barrier are 12 cm and 0.5 cm. One cm of model may be conveniently taken as 30 cm in the field but other scales are also suitable.

487

GULATI, O.P.; AGARWAL, M.C. Investigation of waterlogging problems and drainage project planning. Cent. Bd. Irrig. Pwr. Pub. New Delhi, 1972, 118, 2:145-51.

The causes of waterlogging in local areas of Hansi and Rohtak of Harayana state and overall increase in waterlogged area after rainy season have been examined. For reclamation of waterlogged and salt affected soils, the techniques of drainage investigation have been presented for guidance. The effect of various factors on design of drainage system has been critically discussed. The critical water-table depth has been found to be approximately 25-50 cm and 300 cm in sandy clay loam soils of Haryana for designing drainage systems in waterlogged, and salt affected soils, respectively.

488

GULLICH, P.; WERNER, D.; SCHWARZ, K. Amelioration problems of Chernozem and Rendzina soils overlying Keuper and Upper Shell Limestone in the Thuringian Basin. <u>Arch. Acker. Pflan. Bodenk.,</u> 1980, 24:777-90.

Waterlogging occurred in soils containing >30 percent clay during important periods in the agricultural cycle and was not remedied by drainage. For Keuper-Rendzine soils containing <45% clay (byweight) deep loosening was recommended. Loosening was not necessary for the A-h horizon of Keuper-Chernozem soils. Due to swelling, Muschelkalk (Shell Limestone)- Rendzine soils in general, and Keuper-Rendzine soils containing >45percent clay could not be improved with any degree of permanance by loosening. Waterlogging over subsoil of low permeability could be remedied by means of seepage slot drainage over inflow positions, the envelope material being installed up to the lower-boundary of the top soil.

489

GUPTA, O.P. Waterlogging, causes and remedial measures therefore. <u>ICID Bull.,</u>1962, 11:55-63.

The paper deals with the problem of waterlogging in general with particular reference to the conditions of Uttar Pradesh in the Ganga-Yamuna Doab. The various causes contributing to waterlogging conditions are analysed and remedial measures suggested mainly guided by the experiences of Uttar Pradesh. The pumping by tubewells is suggested as one of the measures for drainage. In the four Doabs namely Sengar-Yamuna,, Sengar-Arind, Isan-Pandu and Karon-Yamuna, respectively, 19,81,21 and 21 tubewells of 15 cm diam strainer are suggested so as to maintain the water balance over the year and 230,560, 176 and 254 tubewells to pump extra seasonal ground water potential in these doabs.

490

GUPTA, S.K. Probability analysis of dry days for drainage planning in the catchment. Indian J. Soil Conserv., 1982, 10:12-6.

For proper planning of an area dry spell analysis alongwith the rainfall characteristics is essential as this provides an estimate of the water requirement in the area during this period. It was observed that in the study area the requirement of rice cultivation in normal lands is 30-45 cm, during continuous dry spells at 25 percent probability level. The requirement for alkali lands for the same period is only 40 percent of this value. From the analysis it is concluded that runoff in the area could be augmented by planting normal lands with crops like maize and bajra for subsequent utilization in lowland alkali lands planted with rice.

491

GUPTA, S.K. Sub-surface drainage for waterlogging and salinity control-field experiences. Proc. 51st An. R & D Session, CBIP New Delhi, Pub. No. 169, 113-24.

Extent of waterlogging problem in India has been discussed. The work on critical water levels has been reviewed and reported which is helpful in deciding the depth of tile drainage and also to workout alternative cropping pattern for waterlogged lands. The work on sub-surface drainage in India shows that provision of drainage helps in maintaining aeration status of the soil during critical periods and thus helps in establishing better crops compared to undrained lands. The work done at Central Soil Salinity Research Institute, Karnal shows that drainage is essential for waterlogged saline lands as it helps in creating favourable salt and water balance in the area. Irrigation wells can serve as a measure of vertical drainage in fresh water regions whereas disposal of saline water is the major constraint in the installation of vertical drainage in saline regions.

For bio-drainage, Eucalyptus seems to be good plant for drainage in fresh water regions; for high water table in saline areas, identification of plant species which can tolerate high soil and water salinities is essential. The paper lists major areas of research activity requiring attention of the research scientists.

492

GUPTA, S.K. Variations of water table in Yamuna drainage basin of Harayana-implications and management strategies. Proc. Natn. Symp. "Strategies for Irrigation Development", 1983, 4:38-49.

Ground water development in Haryana has taken place with a tremendous speed but has been limited to fresh water regions. It has resulted in imbalance so much so that sub-soil water table continues to rise in the areas with saline ground water underneath whereas water table in fresh water regions is declining. This trend is giving rise to development of salinity/alkalinity in areas of high water table. It is feared that if this trend continues, the fresh water aquifers may also get contaminated in the long run. It is, therefore, essential to develop saline ground water for irrigation which constitute 80 percent ground water presently available for development in the state. The use of saline water is the major problem in any such development for which one or combination of more than one out of three management practices suggested in this paper can be utilized.

493

GUPTA, S.K.; DHRUVANARAYANA, V.V. Surface drainage methods for saline-alkali soils. Proc. Symp. Waterlogging Causes Measures for its Prevention. CBIP New Delhi, Pub.1972,118:129-34.

The problem of drainage in India is due to the water stagnation on flat lands and the rise in water-table. There seems to be no precise data on the actual extent of the problem area and rough estimates put it at 6 mha. For permanent reclamation of such lands and to conserve good quality rain water for subsequent usage as irrigation water, the Central Soil Salinity Research Institute, Karnal has started a study to develop a three tier system of surface drainage. In this system , dug-out ponds of sufficient capacity have been constructed to store the excess rainfall, through gravity flow and pumping. The characteristics of the three catchments differ in magnitude, slope and the land use. Preliminary data, from the three storms in early July, indicate that the storage efficiency varies from 40 to 60 percent. The water collected in the two ponds with relatively salt free soils in the catchment had a pH in the range of 7.6 to 8.6 and EC from 0.14 to 0.25 dS/m. The water in the third pond had a pH of 9.5 and EC 0.39 dS/m.

494

GUPTA. S.K.; DHRUVANARAYANA, V.V. Design considerations for dual purpose ponds. Soil Conserv. Digest, 1974, 2:1-7.

A three tier system of surface drainage has been adopted at the Central Soil Salinity Research Institute, Karnal, in which excess rainfall during the monsoon is not allowed to remain standing in the fields, thereby causing crop damage. The runoff is led into dugout ponds of sufficient storage capacity in low-lying areas and either used for subsequent irrigation or else allowed to go into the drains when they are no longer full. Expected water yields during the monsoon at Karnal farm and construction of the ponds are discussed.

495

GUPTA, S.K.; PANDEY, R.N. Designing irrigation-cum-drainage ponds for alkali land. Indian J. Agric. Sci. 1980, 50: 428-30.

A simple procedure based on water-balance analysis

to predict the runoff from agricultural catchments planted with rice has been proposed. The procedure was tested by conducting an experiment wherein water-balance components were measured in a 6 ha rice catchment in an alkali land. The predicted values of runoff were in close agreement with the observed data. Runoff values from a storm for 2 days with a 10 years return period could be chosen for the design of surface drainage systems and storage reservoirs in the region.

496

GUPTA, S.K.; TEJWANI, K.G.; BABU, R. Drainage coefficients for surface drainage of agricultural land for different parts of the country. Irrig. Pwr., 1971, 28:53-60.

Tentative recommendations for drainage coefficients and drainage runoff formulae have been made for various parts of India based on the soil, rainfall and crops grown.

497

GUPTA, Y.C. Efficient irrigation and drainage for better crop production. Symp. Water Mgmt. Udaipur, (1966), 1968:208-12.

Management of alkali soils through application of amendments and drainage at National Dairy Research Institute Karnal farm show that gypsum # 5 ton/ha coupled with leaching has given encouraging results for rice-berseem rotation. Open channels around the experimental plots served as a measure of drainage. Pumping of ground water is suggested to control ground water table.

498

GUREGHIAN, A.B.; YOUNGS, E.G. The calculation of steady state water-table heights in drained soils by means of the finite-element method. J. Hydrol., 1975, 27:15-32.

The use of a Galerkin-type finite element method for the solution of steady-state drainage problem in homogeneous and heterogeneous soils is described. Comparison of the computed results with those obtained by known analytical solutions showed that the numerical procedure is satisfactory for particular types of ditch drainage in uniform and layered soils with given incident rainfall. The use of the method for a drained layered soil in which the backfill over a pipe drain creates vertical band of soil of different conductivity is also described.

499

GUREGHIAN, A.B.; YOUNGS, E.G. The use of the finite element method in obtaining soil-water content distributions in drained land with a uniform steady rainfall incident on the surface. In Mathematics of Finite Elements and Applications. Proc. IMA Conf. Brunel University(1975), Academic Press, 1976,289-97.

A finite element solution is presented for the two-dimensional problem of a system of parallel equally spaced drain lines with a uniform vertical flux incident on the soil surface. This is illustrated by the computation of the water-table position and streamline pattern in an effectively infinite depth of soil for an installation of drainage pipes of equivalent radius 2.5 mm, spaced 20 m apart at a depth of 0.7 m below the surface when the flux q/k_{sat}=0.01, and the associated water pressure and content distributions are above the water-table.

500

GUSTAFSON, J. Forest drainage-methods, techniques and planning.Redog. Forsk. Skogs., 1979, 1, 42p.

An account is given of developments in Sweden and Finland.

501

GUSTAFSSON, J.Forest drainage-prerequisites and planning. Skog. forut.Plan. Ekon. Forsk. Skogs., 1979, 3,4p.

A brief discussion of biological, technical, organizational, legislative, environmental and financial aspects of forest drainage in Sweden, where the annual drainage area in the 1970, s has been 30000-40000 ha.

502

GUYMON, G.L. Digital computers and drainage problem analysis. Part III-Finite element method. In Drainage for Agriculture (Agronomy no 17; Ed.Schilfgaarde,J.Van). Am. Soc. Agron. Wisconsin, 1974, 587-602.

A variational approach to the finite element method for numerical solution of seepage problems is described and solved by the Ritz method(i.e. selection of a trial sequence of functions that are substituted into the variational principle). The technique is then applied to a ponded drainage problem.

503

GUYON, G. A drainage formula obtained from analog model. ICID Bull., 1978, 27:55-75.

Drainage design formulae for various situations are presented based on studies on a analog model. Tabulated values and graphical figures are presented showing the flownets and the design procedure.

504

GUYON,G; WOLSACK,J. Subsurface hydrology and agricultural drainage.Acad. Agric. Fr., 1979, 65: 1123-33.

Theoretical and experimental investigations leading to the determination of drain spacing in subsurface drainage networks are briefly reviewed. The application of Darcy's law to drainage problems is considered, and recent formulae are presented involving the equivalent horizontal hydraulic conductivity and drainage porosity. The experimental determination of these two parameters in the field is discussed.

505

GUYOT, J.; CONCARET, J.; JACQUIN, M., Ochre deposits in drains. The role of iron in processes by which they are formed. C. R. Acad. Agric. Fr., 1979,65:101-9.

The analysis of deposits of ochre in plastic drains in the field and of the Fe concentration in water that had percolated through soil in plastic cylinders in the laboratory suggests that clogging of the drains begin with the deposition of fine particles within the drains and on their orifices and that these obstructions then become coated with hydrated Fe compounds.

506

GYUK, I.; SORIANO, A.; KARADI, G.M. Flow towards periodic tile drains. J. Hydrol., 1973, 19:113-29.

Method of conformal mapping is applied to the analysis of transient flow towards parallel periodic drains in a semi-infinite aquifer taking into consideration the non-linear boundary conditions on the free surface. The mapping function is expressed as a power series in time and the seepage domain is mapped on to a domain of an auxiliary complex variable. Mapping is performed in such a manner that the free surface will always remain the real axis. Calculations are carried out for different ratios of drain depth to drain spacing using various drain diameter to depth ratios.

507

HACK, H. The interactions of pre-sowing watering,

early watering and surface drainage on the growth of Acala cotton at Kenana Research Station. In Cotton Growth in the Zezira environment (Siddig, M.A.;Hughes, L.C. Eds.) Agric.Res. Corp. Sudan, 1970, 188-32.

In a 3 year experiment, cotton on an impermeable clay soil with and without surface drainage was subjected to various treatments of pre-sowing and early (17-19days after sowing) irrigation. Early irrigation had a harmful effect on cotton growth. Plants on drained plots which received no pre-sowing irrigation recovered from the effects of early irrigation but plants on undrained plots which received pre-sowing irrigation did not. Where irrigations were delayed for nearly a month after sowing, pre-sowing irrigation proved useful and drainage harmful.

508

HACK, H.R.B. Components of error in field experiments with cotton, groundnuts, kenaf and sesame in the central Sudan rainlands. II Variation in plant measurements through the season and a discussion of surface drainage as source of variation. Exptl. Agric., 1976, 12:225-40.

Dry weight, stem, and branch length and node number, and flower and fruit number where appropriate, were recorded through the season on cotton, groundnuts, kenaf and sesame grown with and without a short or long pre-sowing flooding in the central rainlands of the Sudan in 1965. For each sampling date, variation within plots and between replicate plots within blocks as well as treatments X blocks. The main components were usually within plot and between replicate plots within block errors. Uncontrolled variation was unacceptably great, and was thought to be in large part the result of irregular surface drainage conditions. It seems that including surface drainage contrasts in field experiments subject to high intensity rainfall on this impermeable clay soil is more likely to be useful than attempting to increase precision alone.

509

HAGTING,A.; ZUIDEMA, F.C. Application of operational research during the execution of drainage systems. <u>10th Congr. Irrig. Drain. Athens,</u> 1978,7:707-22.

A detailed description, including time-motion and cost -effectiveness studies of the operations involved in laying pipes for field drainage systems, has been given.

510

HAKANSSON, A. Drainage experiments with different drain distances. <u>Grundforbattring</u> (special No. 4), 1961, 98p.

A relationship between crop yield and drain spacing was obtained by laying plots in long narrow strips between the drains. Plots equidistant from the tiles are treated as replicas of the same drainage intensity. A relationship between drain spacing and crop yield is calculated by successively excluding larger and larger portions of the area between the drains and calculating the average yields on the portion remaining. The values obtained represent yields to be expected from a spacing corresponding to the remaining portions. A relationship is thus indirectly obtained for a spacing smaller than or equal to the tested spacing. Besides giving more precise results the method requires much less land than the older method.

511

HAKANSSON, A. Water inflow into subsurface drains of different design. <u>Grundforbattring,</u> 1968, 21:157-164.

Filter material on or around the drains increased water inflow. For smooth plastic drains the increasing order of inflow was, no filter $<$ foam plastic filter= fibre -glass binding $<$ glass-fibre strip over pipe= glass-fibre strip above and below pipe. Inflow into a gravelled tile pipe was about equal to that into smooth plastic pipes with fibre glass binding. There was a linear relationship between groundwater level and the drain flow. A low water inflow resulted in a higher groundwater level and a lower total drain flow.

512

HAKEN, D. Sequence of soil formation and increase in soil fertility of drained peat bogs in the Krusne mountains. Ved. Pr. Vyzk. Ustavu. Melior. Zb., 1972, 12:71-84.

Soil formation after drainage was affected by the poor permeability of the peat overlying the mineral soil. Fluctuations in the groundwater level affected the airwater relations of the rooting zone. In soil under arable crops and grassland, organic matter decreased by 5.3 and 2.1 per cent and total N by 16.2 and 42 per cent respectively, and the rates of ammonification and nitrification were higher under arable crops than grass.

513

HAKEN, D. Study of optimum groundwater level on marsh soil. Sb. UVTI-Melior., 1975, 11:25-33.

Data from pot experiments showed that for most crops optimum groundwater levels were between 60 and 100 cm, but that for beans and carrots , 40-60 cm levels were better.

514

HALKIAS, N. Basic data in drainage design. Deltion, 1969, 39-40:19-20.

The paper deals with the Greek's practice of drainage design where 350,000 ha needs drainage. It also deals with the type of drainage required, capacity of ditches, drainage outlets, soil and subsoil conditions, design of open ditches, rainfall, and agro-technical data.

515

HALIGREN, G.; WIKLANDER, L. Precipitation of ferric hydroxide in drain tiles. Grundforbattring, 1966, 19: 345-51.

The reactions involved in the formation of gley, and precipitations of $Fe(OH)_3$ leading to the clogging of drain tiles are briefly discussed. These processes mostly occur in loamy soil rich in FeS, which are very acid and show oxidation of the FeS afer drainage; Fe precipitation may also occur in drainage tubes in calcareous sandy loam overlain by fen soil.

516

HAMID, A.; WARKENTIN, B.P. Lateral water movement in ste. Rosalie clay subsoil. Can. J. Soil Sci., 1967, 47:139.

Measurements with I^{131} showed lateral drainage below the water table at two sites with soils of two textural groups. The tracer moved only downslope from the point of application; when placed at the surface of water table,no horizontal water movement was detected. An estimate of water balance indicated significant water loss by drainage. Low measured values of hydraulic conductivity imply doubtful benefits of artifical drainage, but responses to tile drainage occur.

517

HAMMAD, H.Y. Depth and spacing of tile drain system. J. Irrig. Drain, Div. ASCE, 1962, 88:15-34.

The paper disusses saturation of agricultural soil by over-irrigation or rain; studies downward seepage flow through porous soil to system of parallel tile drains; develop theory of hydrodynamics for providing scientific basis for design; obtain new spacing formula based on time variation of water table; deals with effect of evaporation from soil surface on spacing formula.

518

HAMMAD, H.Y. Design of tile drainage for arid regions. J. Irrig. Drain. Div. ASCE, 1964, 9:1-15.

The paper investigates design of tile drainage systems for irrigated zones; studies the distribution of moisture in the capillary zone, the drainage and irrigation requirements, the water-table fluctuations, and the depth and spacing of tiles; gives the solution consisting of a set of equations sufficient for the determination of the differet parameters necessary for the design of tile drainage; bases the treatment on the soil tolerance of the quality of irrigation water, and the salt tolerance of plants.

519

HAMMAD, H.Y. Natural drainage of river valleys. J. Irrig. Drain. Div. ASCE, 1965, 91:39-49.

The paper discusses ground-water movement in the geological formation prevailing in the Nile valley and the similar valleys with the object of studying the drainage problem in the valleys; concludes that no artificial field drainage system will need to be installed for most zones of the Egypt after regulation on the High Aswan Dam except for some limited zones opposite the Nile barrages, and the Mediterranean coast in the north of the Nile delta.

520

HAMMAD, H.Y.; HATHOOT, H.M. Effect of tile drain height above an impervious layer on drain discharge. ICID Bull., 1980, 29:47-51.

An analysis to show the effect of soil thickness below a system of equidistant drains on drain discharge is presented. The cause of clay layer underlain by an impervious stratum is considered and analysed by theories of Hammad, Kirkham and Hathoot. The effect of changing the clay layer thickness below drains on the discharge, including the special case of partial embedment of drains in the impervious layer is investigated in two different cases. A system of curves are plotted according to different theories and the results are discussed and analysed.

521

HAMMAD, H.Y.; HATHOOT, H.M. Tile drainage assisted by natural drainage. 11th Congr. Irrig. Drain. New Delhi, 1981, 2:31-41.

The downward seepage flow of water through porous soil to a system of parellel tile drains under excessive irrigation was studied. The soil layer was underlain by a sand aquifer characterised by relatively low piezometric pressure. Thus, some of the excess irrigation water will flow downward to the sand aquifer and the remainder will be absorbed by the tiles.

522

HANSEN, B.; KJELLERUP, C.M. Drainage, liming and tillage depth on marsh soils. Meddel. Statens Planteavl., 1977, 75, 4p.

Data are given on yields of wheat, oats, barley and sugarbeet grown on various marsh soils (high Na contents) in Denmark, drained using different drain depths and between drain-spacings. Drainage reduced year to year crop yield variation markedly and also improved conditions for vehicle passage and autumn sowing. On higher marsh soils, it was concluded that drains should be at least 110 cm deep and upto 10 - 25 m apart. On heavy marsh soils, measures to remove superficial water were also necessary and a between-drain spacing of 81-12 m was advisable. Results of trials using lime and gypsum as soil amendments for various crops indicated the crucial importance of lime on marsh and, in extreme cases, of gypsum, for crop yields. On heavy soils, it was concluded that a soil pH of 7-7.5 should be aimed at, with application of additional gypsum to replace soil Na and Mg by Ca. Use of gypsum improved the yield dependability of spring cereals and sugarbeet.

523

HARDJOAMIDJOJO, S.; SKAGGS, R.W. Predicting the effects of drainage systems on corn yields. Agric. Wat. Mgnt., 1982, 5, 127-44.

Stress day index (SDl) models were incorporated in the water management simulation model, DRAINMOD, to quantify the effect of soil water stresses on corn yields. The effects of a combination of excessive and deficient soil water conditions were approximated by a smiple first-order crop response model, $YR=YR_w \times YR_d$, wherre YR is the overall relative yield, and YR_w and YR_d are relative yields due to excessive and deficient soil water conditions, respectively. The accuracy of the modified water management model was evaluated by comparing predicted and measured corn yields for 16 plot years of experimental data on the Tide -water Research Station near Plymouth, NC. The predicted and measured results were in good agreement with the model describing 63 per cent of the variation in yields for the 12 year period. Use of the modified water management model was demonstrated by simulating the performance of several drainge system designs for a Portsmouth sandy loam soil. The results of the simulation show that a maximum long-term relative yield of 80 per cent of the potential corn yield can be obtained with a drain spacing of 40 m or less with good surface drainage. Higher yields could not be obtained without irrigation to reduce deficient soil water conditions. The response of long-term average corn yields to surface drainage varies inversely with the intensity of subsurface drainage. The 25-year average yield for 100 m spacing was only 47 per cent of the potential yield when the surface drainage was poor as compared to 61 per cent of potential yield for good surface drainage.

524

HARDJOAMIDJOJO, S.; SKAGGS, R.W.; SCHWAB, G.O. Corn yield response to excessive soil water conditions. Trans. ASAE, 1982, 25:922-7.

The data from a long term drainage experiment in Ohio were analysed to develop a relationship between relative yield and stress day index (SDI). The water management model, DRAINMOD, was modified to compute SDI for four treatments for each of the 12 yr of the experiment. Relative yields were obtained for each case by dividing the measured yield by the potential yield. It is observed that a strong relationship exists between relative yield and stress day index for corn. A linear regression was fitted to the data and the regression model was tested against three other sets of data from the literature-two from India and one from Iowa. The model was in good agreement with results from the independent experiments even though there were wide differences in soils, climate and experimental conditions.

525

HARET,C.;SIMIONESCU,I.; SALAJAN, G.; VELEA, C.; LAPUSAN, A. Researches concerning the effect of mole drainage and drainage through open channels upon the moistening excess of the slopping soils in Huedin. In Studies on Land Improvements. Instt. of Hydrotechnical Res., Bucharest, Rumania, 1965, II:203-34.

The paper reports that the study has shown that mole drainage as well as drainage with open channels have modified in a 3 year period the soil hydrological regime such that the natural vegetation has changed, proceeding from the hydrophile species of a low economical value to mesophile precious species; further in the Huedin conditions, mole drainage is advantageous on arable fields and drainage by open channels improves moistening excess in orchards.

526

HARPER, L.A.; HAMMOND, L.C. Drainage characteristics of Leon and Felda soils simulated with an electrical resistance network. Proc. Soil Crop Sci. Soc. Fl., 1966, 26:155-63.

An electrical simulation of water flow in two Florida soils gave hydraulic potential distributions and water tables which compared favourably with those found in previous field studies. The relative effectiveness of various drainage systems on layered soils could be determined from laboratory and field test data programmed into electrical resistance analogues.

527

HARRIS, M.D. Effect of initial flooding on forest vegetation at two Oklahoma lakes. J.Soil Wat. Conserv., 1975, 30:294-5.

Observations and infra-red aerial photographs were made at Keystone and Oclogah Lakes after impoundment for flood control. Trees and shrubs which were completely submerged were nearly all killed. Trees with a small portion of the crown above water, and trees submerged for only a few days, subsequently developed stress symptoms, including small, malformed, yellowing leaves and discoloured cambium. Small trees exhibiting these symptoms usually died, but trees larger than 2.5 cm, 7.5 m in height recovered. Hardwood species showing greatest resistance to flooding are reported.

528

HARRIS, W.S. Minimum requirements for field drains in row crops in the Delta. Spec. Rep. Ark. Agric. Exp.Stn., 1967, 15:3-24.

Temporary row drains were used extensively on clay and clay loam soils, less extensively on silty clays and silt loams, and seldom on fine sandy loams. Cultivated field drains were mainly on clays and clay loams; uncultivated field drains on silty and sandy loams. One-half the field drains were used mainly on silty and sandy loams. Minimum field-drain depths were half the row bed height and had to be increased when the beds were higher.

529

HARRIS, W.S. Row-crop drainage in the Arkansas delta. Trans. ASAE., 1964, 7:64-6.

The row drainage studied in the delta area of Arkansas which includes broad alluvial areas and loessial plains had been investigated to determine the requirement of adequate drainage. Depressions that pond or pocket surface water after runoff has ceased, were found to be a major cause of inadequate drainage. Four signs of inadequate drainage were observed: (a) reduced stand and of vigor, (b) reduced yield, (c) excessive weed and/or grass growth and (d) interference with or delay in mechanical farming operations. Unrestricted outlets and a minimum slope of 0.05 per cent and length of slope not exceeding 30 m if it is less than 0.05 per cent were some of the recommendations to improve drainage.

530

HARRISON, P.L.; SPOOR,G.; GODWIN, R.J. Water flow to mole drains. J. Agric. Engng. Res., 1982, 27:81-91.

A study has been made of the mechanism by which water flows to a mole drain. Drain discharge hydrographs and fast response tensiometers have been used to compare the differences in performance between mole drains having large soil fissures above the drains, with drains having no fissures. The results show that in a well structured clay soil, mole drains with large fissures have a significant faster drainage response to rainfall than drains without them. Mechanically induced fissures become increasingly important for rapid water flow to mole drains as the soil structural conditions deteriorate.

531

HART, M.G.R; CARPENTIER, A.J.; JEFFER, J.W.O. Problems in reclaiming soils in Sierre Leone. Agron. Trop. Paris, 1963, 18:800-2.

Mangrove soils become very acid when subjected to heavy drainage. The acid formation is the result of oxidation of elemental sulpuhur by the bacterium Thiobacillus thio-oxidans. Field trials conducted in Sierra Leone have shown that the following reclamation scheme may be recommended: (1) impoldering, (2) drying as thoroughly as possible, (3) reintroducing tidal wash, (4) permanent impoldering, (5) liming, and (6) leaching with rain water until the salt content is sufficiently low.

532

HASSOUNA, N.G.; BARAKAT, M.A.; IBRAHIM, I.A; SALAM, M.A. Studies on red leaf diseases of cotton. 1. Effect of water table and alkalinity on cotton growth. Egypt J. Phytopath., 1975, 7:5-12.

Giza 67, cotton plants (Gossypium barbadense L.) were grown in heavy clay soils, in deep glazed earthenware cylinders. Water tables were kept at 25, 45, 70 and more than 95 cm. Alkalinity levels were 0.0 and 3.0 me/L for the first year, and 0.0 and 6.0 me/L for the second. Red leaf always occurred with water table depths down to 70 cm, while the water table kept at more than 95 cm did not cause such reddening. Red leaf was accompanied by an increase in the linear growth of the shoots and roots, number of falling leaves, number of open bolls, and seed cotton yield per plant, but a decrease in seed per boll. Sodium carbonate concentrations upto 6.0 me/L had no effect on reddening or on cotton growth, except with the water table kept at 25 cm when some damages were apparent.

533

HATHOOT, H.M. Artificial protection of soil against an upward potential gradient. ICID Bull., 1981, 30:44-6.

A new solution for the problem of an agricultural soil subjected to an upward potential gradient is presented. The problem is hydrodynamically treated using the complex func-tions. The basic formulae, the velocity potential and the stream functions are first established. A new formula for discharge through the tile drains and a new spacing formula are presented. The spacing is designed such that waterlogging of the clay layer is avoided. The numerical example using actual field data, has been given.

534

HATHOOT, H.M. New formulas for determining discharge and spacing of subsurface drains. ICID Bull., 1979, 28:82-6.

The problem of draining a top clay cap underlain by an impervious layer with a system of tile drains is mathematically solved. The mathematical treatment is based on hydrodnamical lines and takes into account the actual flownets and proper boundary conditions. A new discharge formula is established and approximated to a simpler one. The unsteady behaviour of the subsoil water surface is studied and a new formula expressing the time variation of the water surface is presented. Finally a new design formula is derived and an actual field problem is numerically solved.

535

HAUTALA, E.; MENEFEE, E. Artificial gravel produced by sintering soil. J. Sci. Fd. Agric., 1981, 32:508-12.

A montmorillonite-type clay soil was sintered at 800° C to form a permeable gravel substitute suitable for use as a drainage envelope material. The sintered soil was water-stable with deformation characteristics that resembled sand more than the original soil. Hydraulic conductivity of the soil was increased more than ten-fold by sintering. The economics of sintering may be favourable depending on the costs of mining and hauling gravel from local deposits.

536

HAYNES, H.D. Machinery and methods for constructing and maintaining surface drainage on farm lands in humid areas. Trans. ASAE, 1966, 9:185-9, 193.

Equipments for land forming including that for site preparation, rough grading, finished grading and for maintaining farm land are reported. Drainage ditch diggers and ditch maintaining equipments are discussed. The cost of land forming with equipments and for various slopes and length of slopes are presented.

537

HAZAN, R. Evacuation of surface runoff and drainage waters by means of injection into the aquifer. 5th Congr. Irrig. Drain. Tokyo, 1963, Q 18 R 12:161-8.

The paper deals with protection of Doukkala plain in Morocco against surface runoff and drainage water by injection into the aquifer. Method of injection, preliminary research tests; geology, geophysics, reconnaissance drillings and determination of the hydrological characteristics of soils likely to be absorbent; full size tests; installation of a permanent system in a pilot zone are described. Extension probability and cost analysis are discussed.

538

HEDSTROM, W.E.; COREY, A.T.; DUKE, H.R. Models for subsurface drainage. Hydrol. Paper Colorado State Univ. Fort Collins, 1971, 48:1-56.

The drainage theories are presented. The effects

of drainable water above the water table on drainage behaviour were analysed to determine their magnitude and the extent to which they are influenced by soil parameters. The Brooks-Corey scaling theory was first shown to be valid for two-dimensional transient flow drainage and was then applied in the analysis of the problem. Results from the experiments and the numerical solution showed that drainage was affected by pore size distribution as measured by the index. A practical implication of the results is that a design method which accounts for the water above the water table should be developed. A number of transient flow drainage design methods, presently being used, were shown to yield results which are appreciably in error.

539

HEIKURAINEN, L. Drainage conditions and tree stand on peatlands drained 20 years previously. Acta Forest. Fennica, 1980, 167, 39p.

Tree-growth and groundwater measurements were made, and the condition of the ditches assessed, in 1978 at Virrat and Alavus, central Finland, on 35 sample plots laid out on nutrient-poor peatlands drained in 1957-62. The results indicated that water-table had the greatest effect on stand volume, volume increment and radial growth; the lower the water table the better the growth. The water table was dependent, in the first place, on the depth and condition of the ditches. Ditch spacing (which in this study varied between 35 and 70 m) had no significant effect on the tree stand. It is concluded that in order to keep the water table sufficiently low, ditches should be kept deeper than 70 cm.

540

HEIKURAINEN, L. Hydrological changes caused by forest drainage. IAHS Pub., 1975, 105:493-9.

The drainage of peatland for timber production is

being performed on an extremely large scale in Finland; by 1972 a total area of about 4 m ha had been drained, and the annual area drained in the late 1960s was between 200,00 and 300,00 ha. This paper deals with possible post-drainage changes in the magnitude and variations of runoff, in the humus content of the drained soil, in the moisture characteristics of mineral soils surrounding the drainage area, and in groundwater resources. It is concluded that the hydrological changes caused by drainage are a function of the time that has elapsed since draining occurred.

541

HEIKURAINEN,L.; LAINE, J. Effect of fertilization, drainage and temperature conditions on the development of planted and natural seedlings on Pine swamps. Acta Forest. Fennica, 1976, 150, 38p.

Twenty-four experimental areas were laid out at various latitudes from the south of Finland to southern Lapland, in 1965 and 1966, to compare climatic, drainage and fertilization effects on natural and planted Pinus sylvestris. NPK was applied broadcast at 500, 1000 and 1500 kg/ha. Ditch spacings were 10, 20 and 30 m. Interim results are expressed in terms of height growth in 1973. In all cases best growth was obtained with the most intensive drainage and with fertilization, but the 1500 kg application gave little or no-improvement as compared with 1000 kg. It is noted that broadcast fertilizing (as compared with the spot fertilizing now in general use) gives a long lasting effect but increases weed growth. The variable with the greatest effect was accumulatd temperature. Few planted seedlings survived below 850 day degrees Celsius (thereshold 5° C). Differences due to drainage and fertilization were greater at higher accumulated temperatures. The results support the use of natural regeneration; after an initial check caused by the draining and fertilizing operations the natural seedlings have grown faster than the planted ones.

542

HEINSOHN. W. Methods of drainage used in the bottom land between Elbe and Weser. 8th Congr. Irrig. Drain. Varna, 1972, Q 27 R 2: 17-34.

The paper gives a description of the procedure applied in the bottom lands between Elbe and Weser to convert the insufficient drainage systems used so far to optimal field drainage. Successful procedures are highlighted to focus attentions for inclusion in planning. Procedures discussed are: (a) natural drainage via open outlet ditches (b) artificial drainage by pumping with piped outlets or open outlets. Alternatives for pumped water discharge are discussed. Drainage polders with piped outlet channels widely used in the bottom areas are described in detail.

543

HELLINGA, F. Soils and water control programmes. Proc. U.N. Sci. Conf. Conserv. Util. Resour., 1961, 4:361-6.

Reclamation of polders starts with intensive draining and in 2-3 years the soil is fit for normal cultivation. Soils inundated during the winter were restored by drainage and application of 3000 to 8000 kg/ha of gypsum. Tabulated data show the influence of depth of water-table on the yield of different crops.

544

HERBERHOLD, R.; KRAHMER, U. Effectiveness of different subsurface drainage methods in pseudogleyed clay soil. Z. Kult. Flurb. 1983, 24:228-37.

Several types of subsurface drainage were tested in a weathered Cretaceous pseudogleyed soil, using different drain spacing and transverse mole drains with or without a filling of polystyrene. By considering meteorologial data, the discharge of individual drain plots, and the distribution of soil moisture tension between drain it was possible to establish an accurate water balance picture to assess soil water movement and to describe the physical properties of the soils. After a 10 year test period, it was possible to reach conclusions on the life and effectiveness of the drains and to make recommendations for the agricultural use of these soils.

545

HERGERT, G.W.; BOULDIN, D.R.; KLAUSNER, S.D.; ZWERMAN, P.J. Phosporus concentration-water flow interactions in tile effluent from manured land. J. Environ. Qual., 1981, 10:338-44.

Phosphorus concentrations in tile effluent during drainage events increased with increasing flow and decreased as flow decreased. Nitrate and chloride concentrations were inversely related to flow rate. Nitrate/chloride ratios of tile effluent during the course of drainage events, together with NO_3/Cl ratios in soil samples from the plow layer and subsoil, suggested that the flow of water to the tile was dominated by downslope movement in the plow layer and by downward movement to the tile in the relatively permeable backfill material above the tile. Calcium or salt concentration in soil solution and tile effluent during the course of a drainage event influenced the P concentration. Increasing Ca concen-trations decreased P whereas decreasing Ca concentrations increased P concentrations.

546

HERGERT, G.W.; BOULDIN, D.R.; KLAUSNER, S.D. ZWERMAN, P.J. Effects of dairy manure on phosporus concentrations and losses in tile effluent. J. Environ. Qual., 1981, 20:345-9.

Dissolved P concentrations and losses in effluent from tile-drained silt loam were compared before and after manuring. More than 95 per cent of the effluent samples had dissolved molybdate reactive P (DMRP) concentrations $<$ 30 ppb before manuring,

During 3 years when the manure was applied, 53 and 92 per cent of the effluent samples contained > 30 ppb P for the 200- and 35- metric-ton applications, respectively. The flow-weighted DMRP concentrations in tile effluent were 116, 441, and 97 ppb from plots receiving 200 metric tons of manure/ha per year, during the 3 years manure was applied. The tile-flow accounted for about 40 per cent of the drainage water lost from the plots, but the P contents of tile flow were 10 per cent or 19 per cent of total dissolved P losses in surface runoff plus tile flow, from 35 or 200 ton plots, respectively. P additions from 30 to 40 wet metric tons/ha of manure would be adequate to meet crop needs but should not produce excessively high P concentrations in the tile flow.

547

HERMSMEIER, L.F. Drainage characteristics of Fargo Bearden soils. Trans. ASAE, 1965, 8:371-3, 376.

The soils of Fargo Bearden having 50-80 percent clay and 2-20 percent silt at depths to about 1.5 m are slow in drainage because of poor hydraulic conductivity. Natural water table drawdown occurs at the rate of 1.9-6.3 cm per day from 30 cm below the surface to 1.8-3 m below the surface. The tile drainage spaced at 3-27 m did not help much in alleviating the drainage problem. Land forming and shallow ditch drains for surface drainage were found to be beneficial. Use of sod crops to maintain high hydraulic conductivity was found to be helpful in reducing drainage problem and increasing the amount of water available to plants.

548

HERMSMEIER, L.F. Economical envelopes for subsurface drain in irrigated lands. 3rd Natn. Drain. Symp. ASAE, Michigan, 1976, 18-21.

Comparison of laboratory studies on some of the physical and hydraulic properties of a number of synthtic materials and gravel as drain envelopes are reported. Amount of sandy loam soil passed through Mirafi 140 (M), Nylone Sock (K), Typar 3-1/2 (T) and Drain guard (D) was 0.5, 0.4, 1.8 and 3.9 respectively. Tensile strength of Mirafi and Typar was nearly 4.5 times (5.5-5.6 kg/cm) compared to Drain guard (2.3 kg/cm). Flow rates were maximum with gravel compared to drain guard but is similar or more after 2 days upto 9 days.

549

HERMSMEIER, L.S. Shallow drain performance in a heavy soil. Trans. ASAE, 1973, 16:92-4,96.

Shallow drains installed at 122.0 cm depth and 60 m spacing in a clay and loamy clay soil removed only from 12 to 7.2 per cent of the salt added by irrigation water over a 2.5 year period. Salt removed by natural drainage however, helped to maintain the salt balance. But during sugarbeet cropping season there was overall unfavourable salt balance when only 7.2 per cent salt was removed by the drains. A relatively high water table was seldom more than 25.4 cm higher 22.5 m away from the drains than at a distance of 3 m. Effluent flow was very slow. For maintaining favourable salt balance in clay and clay loam soil with irrigation water quality of 1.5 dS/m it requires good management and drainage.

550

HERMSMEIER, L.F. Yield of tile and surface drains and their effect on the watertable in a wet soil. Trans. ASAE, 1968, 11:86-9, 93.

Surface, subsurface and a combination of these drainage systems on glacial lake bed soils which are black to dark grey, silty clay loams and silty loams to depths of 1.5 m or more had been studied. Water entered tile lines installed at 1.05 m depth at a rate equivalent to drainage coefficient upto 10 cm when the water table everywhere in the test area was at the soil surface. However, the rate of inflow decreased rapidly as the water table in the immediate vicinity of the tile approached the tile depth. With tile at 36 m spacing and flow capacity based on 0.95 cm drainage coefficient, the rate of drainage was limited by the capacity until the water table midway between the tile lines reached a depth of 72 cm. After the soil was saturated, one day was required for the water table to be lowered 60 cm in plots with tile drains, two days were required in plots with surface drains, and five days were required in plots with no improved drainage. Surface drains in this experiment had little effect on peak tile flow rates, in as much as the rate of water application was only 8.25 to 11.25 cm per day. With higher intensity rainfall, surface drains in combination with tile drainage systems probably would result in considerable reduction in peak tile flows. Surface drains are very effective in removing excess water from these soils and surface drains should be incorporated in most tile-drainage systems. An exception may be when it will be more economical to carry water through a tile line than making a deep cut for a surface drain. Hydraulic conductivity rates based on tile outflow rates and rate of water lowering, using the van Schilfgaarde equation, corresponded very well with those obtained by the piezometer tube method. Hydraulic conductivity rates based on core samples were about one-third as large as those obtained by the piezometer method for depths below 60 cm. For depths less than 60 cm, hydraulic conductivity rates based on core samples were three to four times those obtained with the piezometer method.

551

HERMSMEIER, L.F.; WILLARDSON, L.S. Friction factors for corrugated plastic tubing. J. Irrig. Drain. Div.ASCE, 1970, 96:265-72.

Hydraulic tests on corrugated plastic drains tubing were carried out at the southwestern irrigated field station, California. Mannings 'n' and Darcy-Weisbach, 'f' were determined for 3.42 cm to 10 cm diam pipes. An average design value of

Manning's 'n' of 0.016 is recommended for 5 cm through 7.5 cm diam corrugated plastic tubing for grades over 0.001. For lower grades the value is 0.017 for these diam pipes. For less than 5 cm diam pipes the recommended value of 'n' is 0.018. Humps and depressional variations upto two diameters from a uniform grade do not increase head losses in full-flowing pipes.

552

HIGASE, S. Studies on the premature flowering phenomenon in tobacco in waterlogging soil conditions. Scientific Papers Central Research Institute, Japan, Tobacco & Salt Public Corporation, 1975, 117:55-108.

In tobacco, cv. Bright Yellow grown in the Takamatu region, leaf yield was negatively correlated with precipitation in the period Apr-June. Leaf yield reduction was due to soil moisture level rather than the mechanical action of precipitation. Acceleration of flowering was observed in seedlings grown in waterlogging conditions for $>$ 15 days at the 12-leaf stage, but flowering was not hastened by treatment at earlier stages. Root zone O_2 concentrations $<$ 15 per cent induced premature flowering. 47 out of 54 cultivars flowered prematurely due to waterlogging; this was also observed in tobacco scions grafted on to tomato root stocks. Premature flowering was not induced by nutrient deficiencies, but was induced by addition to the soil of ethanol or acetaldehyde which accumulated in roots during waterlogging. A high level of soil N was necessary for induction of premature flowering by ethanol at the 12-leaf stage, but not at the 15-leaf stage. No effect of ethanol was found before the 10-leaf stage. The presence of ethanol was necessary for induction of premature flowering at the early stages but not during several days before flower bud differentiation.

553

HILER, E.A. Drainage requirements of crops. 3rd. Natn.

Drain. Symp.ASAE, Michigan, 1976, 127-9.

Various drainage design criteria, such as static and dynamic water tables, salinity control, and rootzone aeration are reviewed and tested for meeting the drainage requirements of grain sorghum. Based on these studies, a method is proposed for incorporating crop drainage requirements and drainage theory into a rational drainage system design.

554

HILER, E.A.,; CLARK, R.N.; GLASS, L.J. Effects of water table height on soil aeration and crop response. Trans. ASAE, 1971, 14:879-82.

Effect of water table levels on soil oxygen diffusion rate and crop response was studied. Grain sorghum was grown in lysimeters with water table at 30, 60, 90 and 120 cm and filled with soil classified as Travis fine sandy loam. Drastic reduction in quantity and quality of grain sorghum were observed at 90 cm. Oxygen diffusion rate in 30 cm water table case never exceeded 3.3×10^{-9} g/sqcm/min in the root zone. For 60 cm the oxygen diffusion rates were $6.7 - 15 \times 10^{-9}$ g/sq./cm/min which shows the oxygen stress limits of the crop sorghum. Leaf temperature and leaf diffusion resistance were not appreciably affected by oxygen deficiencies in the crop root zone.

555

HILF, J.W.; COHAN, H.J. Use of synthetic materials in irrigation and drainage by Bureau of Reclamation. 8th Congr. Irrig. Drain. Varna, 1972, Q 28 R 5: 69-96.

Large number of tests conducted by Bureau of Reclamation (US) on many of the newly developed materials for such use as pipes for water distribution and drainge; canal and reservoir lining for seepage control; improvement of concrete by polymer impregnation for use in pipes and a variety of structures, new joint sealing techniques and more durable repair and maintenance methods are included.

556

HOFFMAN,G.J.; SCHWAB, G.O. Tile spacing prediction based on drain outflow. Trans. ASAE, 1964, 7:444-7.

Field measurements of tile flow and the water table height midway between tile lines were taken following 5 irrigation periods over a 33-year period on stratified, anisotropic soils typical of the lakebed region in Ohio. From the relationship between tile flow and the midplane watertable height the effective hydraulic conductivity was determined. Spacing then was calculated using theoretical formulae developed for homogeneous soils. It is observed that surface drainage increased the permissible tile spacing at least 50 percent.

557

HOLLOWAY, R.S. Drainage requirements in the Markham Valley.Papua New Guinea Agric. J., 1973,24:119-30.

Preliminary estimates are made of the discharge and capacities of suitable systems for disposing of large quantities of water discharged into the valley floor from foothill catchments along the northern margin. Surplus water from direct rainfall and soil chemical conditions also necessitate the installation of a field drainage system. Detailed tests for drain spacings are necessary because of the wide range of hydraulic conductivity of the soils. Appropriate formulae for calculation of field drainage spacings are presented.

558

HOLMSEN, G. Experiences on the damage to soil by land drainage systems. Norg. geol. Unders,1963, 218,205p.

A discussion, mainly of the possibly harmful effects on soil of groundwater levels altered by drainage.

559

HOLSAMBRE, D.G.; RAPTE, S.L.;ACHARYA, H.S. Laboratory experiment to study the subsurface drainage systems.Res. Bull. Marathwada. Agric.Univ., 1977, 1:33-4.

In Iysimeter experiments on a black cotton soil, stone drains gave a higher rate of water discharge than tile drains.

560

HOLSAMBRE, D.G.;VARADE, S.B.;ACHARYA, H.S.;RAPTE, S.L. Drainage characteristics of vertisols. J. Indian Soc. Soil Sci., 1982,30: 116-121.

A subsurface brick drainage system was installed in an ill-drained field of 1.64 ha for finding out the drainage characteristics in vertisols. Soil drainage coefficient, drainable porosity and hydraulic conductivity were determined using daily outflow through drains and water table under rainfed and irrigated conditions is illustrated giving an example. The results reveal that the heavy initial investment can be justified only under irrigated condition.

561

HOOGERKAMP, M.; WOLDRING, J.J. Effect of groundwater level on yields of arable crops and grassland on marsh soils. Z.Acker. Pflan.,1968,127:1-19.

On heavy clay soil, yields of arable crops (especially cereals) increased with increasing depth of drainage. The drought damage due to deep drainage rarely occurs. Yields from permanent grassland were generally high in spring and during wet periods, and low in dry summers, when groundwater levels were low. Decreased yield at high groundwater levels are mostly associated with a low supply of soil N, and would be better remedied by N fertilizing than by increasing the depth of drainage; the latter should maintain the groundwater level at 65 cm depth.

562

HOON, R.C. Reclamation of saline lands under irrigation. 6th Congr. Irrig. Drain., New Delhi, 1966, 19:301-17.

The salient features of reclamation of deteriorated lands in the states of Punjab, Maharashtra and Gujarat and of the saline coastal land of the States of Maharashtra and Gujarat are discussed. It has been observed that reclamation of alkali lands generally takes 4-5 years to bring them to normalcy from the point of view of crop yields. The use of soil surveys and hydrographical data, the latter in respect of water table, surface and sub-soil drainage etc. is stressed for obtaining the best results from reclamation operations.

563

HOORN, J.W.VAN. Drainage of heavy clay soils. In Drainage Principles and Applications. Vol.IV. Design and Management of Drainage Systems. Int. Inst. Land Recl. and Improve., Wageningen, 1974, 16:297-326.

A soil profile of heavy clay is often characterized by layers of different hydraulic conductivity. Principles and applications of draining such layered profiles are discussed.

564

HOORN, J.W.VAN. Drainage of heavy clay soils. In Land drainage (Gardiner, M.J.Ed.) A.A.Balkema, Rotterdam, Netherlands, 1982, 57-72.

The drainage of marine and basin clays in the Netherlands is discussed.

565

HOORN,J.W. VAN. Fundamentals of field drainage. ICID Bull., 1964., 13:58-94.

Drainage design history and the relationships between various soil physical parameters and design depth and spacing of the system are presented mostly as used in the Netherlands conditions. The Nomographs for drainage design are presented.

566

HOORN,J.W.VAN; MOLEN,W.H.VAN DER. Drainage on sloping lands. In Drainage Principles and Applications. Vol.IV. Design and Management of Drainage Systems. Int. Inst. Land Reclam. Improve. Wageningen, 1974, 16: 327-39.

Drainage problems on sloping lands are caused by inflowing groundwater from up-slope or local water from precipitation or irrigation. Principles and applications of commonly used equations for appropriate interceptor and field drains are discussed.

567

HOOVER,J.R.; SCHWAB,G.O. Effect of tile depth, spacing, and cropping practices on drain discharge.Trans. ASAE,1969, 12:150-2.

The long term tile discharge data (1951-1966) of a tile drainage experiment at Ohio had been examined. The tiles in this case were placed at 60 and 90 cm depth and spaced at 9 and 18 m. The data for the 16 years had been presented in tabular form to highlight the effect of percipitation and crop combination. Corn-oats gives the highest discharge at all probability levels compared to oats-corn or corn-meadow. Peak flow and precipitation correlation during March-September was also attempted. Probability analysis of tile discharge data is presented in graphical form and its use in designing a drainage system is highlighted.

568

HORE, F.R.; TIWARI, H.C. Cover materials for tile drains. Can. Agric. Engng.,1962,4:2-3.

Some of the cover materials in use in Canada are discussed and their performance evaluated.

569

HOVE.P. Filter materials for drains-experiences in Norway.Z. Kult. Flurb.,1982, 23:104-9.

Results of laboratory and field experiments in Norway indicate that use of a suitable filter is of great importance for the efficiency and life-span of drains. Filters of large volume (sawdust, fibrous peat and gravel of suitable particle size distribution) in thickness of 10-20 l/m around drains are recommended. Organic materials should not be used where the drain line will be permanently submerged. Sawdust and peat used as drain filters showed about 50 per cent decomposition over a period of 20 years.

570

HUDSON, J.A.; ROBERTS, G. The effect of a tile drain on the soil moisture content of a peat. J. Agric. Engng. Res., 1982,27:495-500.

Several neutron access sites were installed in a line at right angles to, and on either side of, a tile drain across a slope within a grassland improvement area. The effect of the drain on moisture content of peat was determined and found to be negligible at distances 1 m from the drain.

571

HULSBOS, W.C. The determination of the drainage factor as a criterion for the soils of the Indus Plains. Neth. J. Agric. Sci.,1968, 16:25-8.

Drainage factors (DF) were used as design criteria

for comparing relative costs of horizontal and tubewell drainage of mainly medium-texture silty loams, especially in saline groundwater areas. DF calculated from recharge estimates were higher than those calculated from leaching requirements. For tubewell drainage (normally maintaining a water-table lower than is possible with horizontal drainage), DF from recharge estimates should serve as criteria for establishing the capacity of the installation. For calculating the required spacing for horizontal drainage, DF slightly higher than those derived from leaching requirements are desirable to take into account,drainage or seepage from line sources. If wide spacings are calculated with relatively small DF,the reaction of the water-table halfway between the drains to intermittent recharge may be too slow, and a smaller spacing may be required. Because of limited accuracy of the DF additional data on hydraulic conductivity of the soil are desirable.

572

HULSBOS, W.C.; BOUMENS, J.H. Leaching of saline soils in Iraq.Neth.J. Agric. Sci.,1960, 8:1-10.

Salinity data from a few project areas on the Lower Mesopotamian Plain are summarized. An example of the degree of salt movement in the top soil and of overall discharge of salt from an area is cited; it is concluded that even with intensive drainage and leaching, the subsoil and drain water will remain salty for at least 15-20 years. A leaching curve for one site is constructed; it is compared with others and the limits of its use are explained; with the curve and knowledge of the salinity before leaching the amount of water needed to produce a particular salinity level can be estimated. To assist such calculations a leaching graph has been constructed to show the drainwater depth per unit depth of soil as a function of the salinity before leaching and of three different levels after leaching.

573

HUNDAL, S.S.; SCHWAB,, G.O.; TAYLOR, G.S. Drainage system effects on physical properties of lakebed clay soil. Soil Sci. Soc.Am. J.,1976, 40:300-5.

Soil physical conditions were evaluated for four drainage treatments, namely; undrained, surface drainage, subsurface drainage, and combined surface and subsurface drainage. Treatments without subsurface drainage showed more severe surface crusting, greater surface penetration resistance, higher soil strength, and lower hydraulic conductivity than treatments with subsurface drainage.

574

HUNDAL, S.S.; TAYLOR, G.S. Drainage investigations of the cultivated Willard Marsh soils. Res. Circ. Ohio Agric. Res. Dev. Centre, 1979, 249, 13 p.

Drainage impediments in cultivated organic soils were evaluated at four sites by field inspection of drains and laboratory analysis of soil, ground water and drain sediments. The plough layer was highly decomposed, with 75 per cent OM, saturation moisture content (SMC) 0.75 cm^3/cm^3, and bulk density 0.41 g/cm^3. The subsoil was partially decomposed, with 85 per cent OM, SMC, 0.95 cm^3/cm^3, and bulk density 0.16 g/cm^3. Hydraulic conductivity of the subsoil was low (0.03 to 1.1 cm/hr). Chemical analysis of soils were similar at all sites. There was least sediment in drains where soils were least decomposed. Ochreous drain sediments occurred where quantities of reduced iron were high. Jetting with water of sediment-affected drains removed sediments. Measures to improve drainage in these soils are proposed.

575

HUNDAL, S.S.; TAYLOR, G.S.; FAUSEY, N.R. Backfill effects on subsurface drainage of Toledo Soils: 1. With grass-legume cover. Res.Circ. Ohio Agric. Res. Dev. Centre, 1980,257, 14 p.

A highly permeable backfill was created by trenching and back-filling and by subsoiling directly over drains installed 6 years earlier in an Ohio lakebed clay soil. The drains were 5 cm diam plastic tubing installed at 40 cm depth. The site had a cover of grasses and legumes. The full potential of backfill alterations could be evaluated because of the limited capacity of the drains. Trenching and backfilling increased flow rates by a factor of two to four during periods of ponding but had no effect on flow rates after ponding ceased. Trenching to drain depth gave greater flow rates than trenching to within 5 cm of the drain. Trenching and backfill had no effect on water table drawdown, apparently because of excellent surface drainage. After 90 weeks, drain flow rates were essentially unchanged by trenching within 5 cm of the drain but had declined 20 to 25 per cent with trenching to the drain. Subsoiling to 16, 26 and 34 cm depths gave stepwise increase in drain flow rates during ponding. Subsoiling to 34 cm resulted in flow rates that were approximately the same as trenching to drain depth. The flow rates for subsoiled drains have not declined significantly after 82 weeks.

576

HUNTER, J.M.; TRAFFORD, B.D. An economic apparaisal of six years results from the drainage experiment on clayland. Exp. Husb., 1979, 35:1-6.

At the pipe spacing used (15,30 or 60 m), drainage has, on average, increased winter wheat yields by 16 per cent. With pipe drainage plus moling the yield increase was 28 per cent with no large variation for drain spacing. Economic analysis are presented. As mole drainage, over widely spaced pipes, is both the cheapest and provides the best performance, this treatment is to be preferred.

577

HURLEY, P. A. Predicting return flows from irrigation.

J. Irrig. Drain. Div. ASCE, 1968, 94:41-8.

The paper presents an analytical method for computing return flow of drainage from irrigation applications; computations consider aquifer properties, drain spacing, and deep percolation remaining in transient storage. By using the graph and simple digital procedures, realistic estimates of irrigation return flow can be datermined. The method was used to determine monthly quantities for 13 years of return flows in the Mesilla Valley.

578

HUSSAIN, Z. Problems of irrigated agriculture in Al-Hassa, Saudi Arabia.Agric. Wat. Mgmt., 1982, 5: 359-74.

The Al-Hassa region in Saudi Arabia is the largest oases of which nearly 20000 ha of land is irrigated by underground sources through a number of artesian springs. Salinization and deterioration of land drainage caused by shifting sand dunes has reduced the arable land to 50 per cent. The paper gives a description of the problem area together with the results of the reclamation work. For water management, border and sprinkler irrigation have been recommeded. For drainage of sandy loam soils drain spacing of 30-45 m and for silty soils with high carbonate content 20-30 m is found suitable. The drainage water is led to deep depressions where it gets evaporated or lost through seepage. The effect of quality of irrigation water on yields of cauliflower, Rhodes grass, sugarbeet, barley, okra and alfalfa are presented which can be utilized to find out the possibilities of use of drainage water for efficient water management in a oases.

579

HWANG, R.B.; LUTHIN, J.N.; TAYLOR, G.S. Effect of backfill on drain flow in layered soils. J. Irrig. Drain. Div. ASCE,, 1974, 100:267-76.

An implicit finite difference procedure is used to solve the problem of flow to drains in layered soils for the case where soil layers are unchanged by drain installation and the case where a trench is backfilled with surface soil. Assuming that the backfill has the same hydraulic conductivity as the surface soil there will be a significant effect on the rate of water-table drawdown. Backfilling with permeable soil can more than double the flow into the drain if the drains are located below the interface in a two-layered soil where the top layer is more permeable.

580

IBRAHIM, W.R. Drainage effects on soil characteristics and crop production. 3rd Natn. Drain, Symp., ASAE, Michigan, 1976, 117-22.

Drainage evaluation studies in Egypt were made by direct evaluation, indirect evaluation and the combination of the two procedures. The direct method consisted of crop sampling from inside and outside of the project area to find out the increase in crop yields to quantify the net economic returns. The method is approximate as to keep all other factors constant over large areas is difficult. The indirect method consisted of measuring and comparing the soil physical and chemical properties. The relationship between these properties and crop yields then yielded the net effect of drainage. The limited knowledge on these relations limits the use of this method. The integrated method consisted of the combination of the two methods and appeared to work well.

581

ICID, AUSTRIAN NATIONAL COMMITTEE. The drainomat - an Austrian contribution to the development of trenchless drain-laying. 10th Congr. Irrig. Drain. Athens, 1978, 5: 87-9.

The advantage of the Drainomat self-propelled one-man trenchless pipe-layer include a high daily capacity, laser system for depth control, and suitablity for use on soft, stony or swampy soils.

582

ICID, COMMITTEE ON IRRIGATION AND DRAINAGE CONSTRUCTION TECHNIQUES. ICID standard 109- construction of surface drains and ICID standard 10- construction of sub-surface drains. ICID Bull.,1982, 31:47-66.

Detailed specifications as approved by the committee on irrigation and drainage construction techniques are presented for surface and subsurface drainage.

583

IHSIZUKA, Y.; OGATA, S.; SEKIYA, S. The depth of drainage of peat land for agricultural reclamation. 1. The relationship between depth of water table and growth of forage crops. J.Sci. Soil Tokyo, 1963, 33:483-8.

In pot experiments, the optimum level of water table was 20 cm for oats and azuki bean and 10 cm for Italian ryegrass and red clover. With 10 cm water level, the top soil of peat, holds moisture beyond 100 per cent of its water-holding capacity. Nitrification of NH_4-N was as high as in topsoil having a lower contents of peat topsoil as compared with mineral soil were not small when expressed as weight percentage, but volume prcentages were small because of the smaller specific weight of peat soil. After reaching of maximum at the end of June, NO_3 disappeared from the topsoil even where no crop was cultivated, its disappearance being more rapid in topsoil having a higher water table. N deficiency thus appeared with high water table and drought injury with low water than with high water level, suggesting that oxidative decomposition was more rapid.

584

IL'IN, N.I. Water exchange between the units of a drainage system and the ground. Vest. S-kh. Nauki, Mosk., 1969, 11:78-81.

Water exchange is determined by introducing $Na_2{}^{35}SO_4$ into the drain.

585

ILNICKI, P.; BURGHARDT, W. Subsidence in repeatedly drained highmoor on the north-west German Plain, 6th Report: Influence of climatic water balance and discharge on relief form, and on surface and drain subsidence. Z. Kult. Flurb., 1981, 22:112-21.

In a repeatedly drained (1920-25, 1952, 1965-68) highmoor, a surface subsidence ranging from 22 cm (fairly compact peat) to 38.3 cm (loose peat) occurred during 1968-1979. The drain subsidence was 5.0-36.8 cm. Subsidence occurred in all layers measured and caused a peat compaction for the whole peat profile of 1.2-3.1 m. Compaction was greatest(200%) in the top (20 cm)layer; it was 33 per cent in the layers below 60 cm. In the 0-20 cm and 20-60 cm layers the increase in ash content of the peat was twice as great as the increase in volume weight. In drained areas subsidence was greater in the proximity of the drains. Deficiency in the climatic water balance in dry years could produce similar subsidence to drainage. In wet years a partial peat reswell was observed. A correlation was found between discharge and peatland subsidence.

586

INGRAM, B.F. Planning drainage systems in coastal north Carolina. 8th Congr. Irrig. Drain. Varna, 1972, Q 27 R 6:79-96.

The physical charactristics of the area in relation to drainage are discussed. The area is generally flat and is subject to frequent and intense rainfall. The land development scheme, the planning and design of field and outlet drainage systems for the area including special problems relating to wind tides, saline intrusion, lack of basic hydrologic data and project effects on fish, wildlife and timber resources are presented in the paper.

587

INSTITUTION OF ENGINEERS. Preprints Symp. Proc. Papers Hydrology and Water Resources Symp, Barton, Australia, 1980, Institution of Engineers, Australia, 1981, 191p.

These papers contain compilation of 43 papers on various aspects of irrigation, drainage and hydrology. Effect on soil available water because of drainage operations is discussed in papers related to drainage in symposium preprints volume.

588

INTERNATIONAL COMMISSION ON IRRIGATION AND DRAINAGE. 9th Congr. Irrig Drain, Moscow, 1975, 9 Vol. New Delhi, India, 1975.

The main themes were:(1) Introduction of automatic facilities for the needs of collecting and processing of information, planning and managing in the fields of irrigation and drainage; (2)environmental control for irrigation and flood control projects, (3) sediment-controlling irrigation intake structures; (4) development of irrigation by pumping, (5) planning of development of groundwater for irrigation and (6) comparison of advanced methods of irrigation.

589

INTERNATIONAL INSTITUTE FOR LAND RECLAMATION AND IMPROVEMENT. Drainage Principles and Applications. Vol.I Int. Inst. for Land Reclam. Improv., Wageningen, 1972, 16, 241p.

Deals with the introductory subjects on drainage;

which include hydrogeology, soils and soil properties, physics of soil moisture, salty soils and electrical models etc. It provides basic information to any drainage design engineer.

590

INTERNATIONAL INSTITUTE FOR LAND RECLAMATION AND IMPROVEMENT. Drainage Principles and Applications Vol-II. Int. Inst. Land Reclam. Improv., Wageningen, 1972, 16, 374p.

This volume of the four volume set deals with theories of field drainage and watershed runoff. Subsurface flow into drains, salt balance and leaching requirement, effects of irrigation on drainage, field drainage criteria, flow to wells and rainfall-runoff models are some of the topics discussed in this volume.

591

INTERNATIONAL INSTITUTE FOR LAND RECLAMATION AND IMPROVEMENT. Drainage Principles and Applications. Vol.III, Surveys and investigations. Int. Inst. for Land Recl. Improv., Wageningen, 1974, 16,368p.

Based on ten lecture-series presented at a annual International Course on Land Drainage at Wageningen.

592

INTERNATIONAL INSTITUTE FOR LAND RECLAMATION AND IMPROVEMENT. Drainage Principles and Applications. Volume IV. Design and management of drainage systems. Int. Inst. Land Reclam. Improv. Wegeningen, 1974, 16,476 p.

A series of papers based on lectures delivered at the International Course on Land Drainage at Wegeningen. Chapters cover field drainage systems, main drainage systems, control of aquatic weeds, ditch maintenance, drainage of various specified soil types, drainage of rice fields, drainage surveys and economic evaluation and management of drainage schemes.

593

INTERNATIONAL INSTITUTE FOR LAND RECLAMATION AND IMPROVEMENT. Land Reclamation and Water Management. Int. Inst. Land Reclam. Mgmt., Netherlands, 1980, 27, 191 p.

A collection of thirteen articles dealing with problem soils and their management. The two chapters specifically dealing with drainage problems are included i.e. (1) The study of effects of drainage on agriculture and (2) Developments in subsurface drainage techniques. The book can serve as a good reference source on engineering aspects of problem soils viz. hydrology, irrigation and drainage.

594

IQBAL, M.M.; WARKENTIN, B.P. Nitrogen and phosphorus content of flow from drains fertilized with cow manure slurry. Pl. Soil, 1981, 63:517-21.

The rate of flow and the nitrate and phosphorus contents of the water from four drained sandy and clayey plots cropped to continuous corn were determined following two annual applications of different rates (0, 260, 390 and 520 kg N/ha) of cow slurry. The drain flow was directly related to the rainfall and was greater in sandy than in clayey soils. The N losses were greater in 1972 (7.8 to 19.1 kg N/ha) than in 1971 (0.4 to 7.8 kg N/ha) because of more summer rainfall. Nitrogen and phosphorus losses were larger from the sandy plots than from the clayey plots. The manure application rates had no apparent effect on nitrogen losses in the drain water.

595

IRIZARRY, H.; SILVA, S.; VICENTE-CHANDLER, J. Effect of water table level on yield and system of plaintains. J. Agric. Univ. Puerto Rico, 1980, 64:33-6.

Plaintains,cv. Maricongo, were grown in lysimeters in which the water table was maintained at a depth of 12-48 cm. Lowering the water table from 12 to 24 cm greatly increased fruit numbers, the effect of further lowering being slight. Yields were increased from 5650 to 37 800 kg/ha by lowering the water table to 36 cm after which they tended to decline. Root development in the upper 15 cm soil layer was little affected by changes in the water table, but at lower soil depths (15-30; 30-45cm) root development was successively greater with each lowering of the water table.

596

IRWIN, R.W. Drainage problems with organic soils. Trans. ASAE, 1967, 10:427-8.

Drainage problems related to organic soils are discussed under the heads: water control, open ditches, tile drains, installation problems.

597

IRWIN, R.W. Trenchless installation of drain tubing. Proc. Natn. Drain. Symp, Illinois, ASAE, 1971, 37-8.

The Badger system of trenchless installation of a drainage system is described. Effect of Badger system plow on compacting the soil adjacent to drain tube is investigated. The preliminary results are to be compared with other machines and more work on this aspect is suggested for future.

598

IRWIN, R.W.; BROUGHTON, R.S. Experience with drain-laying machines in Ontario and Quebec, Cananda. 10th Congr. Irrig. Drain. Athens, 1978, 5:1-11.

About 47 000 km of subsurface drains are installed each year on farms in Ontario and Quebec. Approx. half of those are installed with wheel or chain trenchers and the other half with trenchless drain laying plows. Backhoes are used to a small exent, mainly in stony ground. Laser grade control systems were used for approx 75% of the drains installed in 1976. Increases in rates of installation with trenchless plows in the future will come primarily from improvements in the organization of related aspects of work and not from increasing the forward speed of the plows. Time studies have shown that the plows are installing drains only 20 to 25% of the field working time. The paper gives production rates, performance problems, advantages and disadvantages with three main types of machines.

599

IRWIN, R.W.; HORE, F.R. Drain envelope materials in Canada. Proc. Int. Drain. Workshop, Wageningen, (1978), 1979, 25:283-96.

Envelope materials have always been used with problem soils in Canada. Originally this was grass sod, straw or tar paper. About 1957 thin synthetic envelope materials became available and field and laboratory tests were conducted periodically to evaluate their use. The results of these tests have been printed. No failures have been reported in the field where envelope materials have been properly applied. Envelopes add 23 per cent of the total cost of drainage and should be used only if necessary under the conditions.

600

IRWIN, R.W.; TSANG, G. Hydraulic roughness of corrugated plastic tubing. Trans.ASAE, 1972, 15:290-295.

Flow calculation theory is presented. The Manning's 'n'and Darcy-Weisbach 'f' values for four types of corrugated plastic tubing and for ceramic drain tile with minimum joint spacing as possible and mean joint spacing of 1.27 cm are tabulated. The value of Manning's 'n' for corrugated tubes vary from .0148 to .0176 whereas for ceramic tiles it is .0108 and .0102 for little space and 1.25 cm spaces joints, respectively.

601

ISHIZUKA, Y.; OGATA, S. Depth of drainage of peat land for agricultural reclamation. 3. Effect of fluctuation of water table on the growth of forage crops and soil properties. <u>J. Sci. Soil Tokyo</u>, 1963, 34:195-6.

The water table was adjusted to a constant level of 10, 20, 30 or 50 cm and was elevated by 10 cm for two days a week. Yields of Italian ryegrass, red clover and oats at the most adequate water levels (10 cm for ryegrass, clover and 20 cm for oats) or at a lower level were reduced markedly by the elevation treatment. With excess drainage, elevating the water table reduced the air phase of the topsoil with 10 cm or 20 cm levels of water table. Even 5 days after elevation of the water table was discontinued, percentages of the air phase did not rise to the previous level. Decreasing the air phase produced anerobic conditions reducing Fe^{3+} to Fe^{2+} and transforming insoluble soil P to soluble form. Denitrification proceeded in the topsoil with 10 cm or 20 cm levels of water table but was as light or absent with 30 cm or 50 cm levels.

602

ISITOTAS, D. The planning of irrigation and drainage projects based on new thoughts. <u>Deltion</u>, 1969, 38:10-3.

The paper recommends two considerations on Greek experience (i) the irrigation of crops is done at a percentage based only on their yearly needs of water so that more crops may be grown from the saving of water, (ii) protection of plants from excessive moisture be had through deep ploughing and, not through expensive drainage works which otherwise waste useful water.

603

ISTOK, J.D.; KLING, G.F. Effect of subsurface drainage on runoff and sediment yield from an agricultural watershed in Western Oregon U.S.A. J.Hydrol., 1983, 65:279-91.

Rainfall, watershed runoff and suspended sediment concentrations for three small waterheads (0.46, 1.4 and 6.0 ha in size)were measured for four winter rainfall seasons. Following two rainfall seasons of data collection, a subsurface drainage system (consisting of a patterned arrangement of 10-cm plastic tubing at a depth of 1.0 m and a spacing of 12 m) was installed on the 1.4 ha watershed no.2. Perched water tables were lowered and seepage was reduced on watershed 2 after the installations. The reductions were quantified with a water table index (cumulative integrated excess). Watershed runoff and sediment yield from watershed 2 were decreased by 65 and 55 per cent respectively. These reductions were estimated from double mass curves and by statistical regression on a set of hydrograph variables. Maximum flow and average flow rates were decreased and the time from the beginning of a storm to the peak flow (lag time) increased. It is concluded that subsurface drainage can be an effective management practice for erosion control in western Oregon.

604

ITURRIBARRIA, J.A. Procedures and maintenance cost of the net-work of drainage canals in irrigation district. No.41, Rio Yaqui Sonora and benefits obtained. 7th Congr. Irrig. Drain. Mexico, 1969, Q 25 R 8:87-103.

The study presents activities of drainage and maintenance cost of drainage canals in an area of 20,000 ha under cultivation. The mean drainage length in the area is 5.71 m/ha. The cost of maintenance was approximately six million peso.

605

IVANOVA, N.A. The effect of drainage on the utilization of nitrogen by barley on podzolic gley soils. <u>Agrokhimiya</u>, 1975, 2:72-6.

The effects of drainage on the redox potential, the contents of available Fe and Al, and the number of micro-organisms in the soils are tabulated. In micro-plot experiments, drainage caused some increase in the utilization of labelled $(NH_4)_2SO_4$ by barley and greatly increased the utilization of soil N.

606

IVANOV, YU.N. Effect of drainage on non-timber products. <u>Lesn. Khoz.</u>, 1976, 4:31-5.

Describes measurements and observations made in 1974 on sample plots on drained and undrained areas of forest in the Vologda region (N. Russia), to determine the yields of non-timber products viz. berries, fungi, hay and resin. The drainage had been carried out 8-12 years previously (mainly in Scots Pine Forests) with drains 0.8-1.2 m deep. Data are given on the yields of cloudberry (<u>Rubus chamaenorus</u>), cranberry (<u>Oxycoccus</u>), whortleberry (<u>Vaccinium uliginosum</u>), cowberry (<u>V.vitis-idaea</u>), edible fungi (<u>Boletus</u> etc.), and <u>oleoresin</u> from resin-tapping. Results in general indicate that drainage does not adversely affet the yields of non-timber products, and may in some cases increase yields.

607

IVANOV, YU.N. Effect of drainage on the yield and merchantability of boggy Scots Pine stands. <u>Lesn. Khoz.</u>, 1974, 9:46-50.

Reviews experience in the Kalinin and Kirov regions of Russia on the effect of drainage measures carried out 50-60 years ago on the growth of three Scots Pine stands (of different forest types), now aged 50, 65 and 95 years. Data are presented showing the decrease in yield with increasing distance from the drainage channel. Regular cleaning and maintenance of the channels are shown to be essential. To ensure more uniform yield of timber over the whole area drained, the drainage channels should be closer together than has been usual hitherto; the following spacings are recommended: 80-100 m in the Eriophorum/ Sphagnum type, 100-130 m in the Carex/ Spagnum type, and 130-180 m in the herboso-uliginosum type.

608

IVANOV,YU.I.; KARLOV, YE.M.; KASATKINA, T.B. Drainage water as an indicator of changes in drained soils.Pochovedenie,1977, 5:101-4.

Changes in the chemical composition of drainage water and agrochemical characteristics are examined on the basis of 3 years of observations of drained land. Drainage water is found to reflect clearly the changes in the soil, whereas the agrochemical characteristics behave conservatively.

609

IVARSON, K.C.; SOJAK M. Microorganisms and ochre deposits in field drains of Ontario. Can. J. Soil Sci., 1978, 58: 1-17.

The effluents and ochreous deposits in several partly blocked field drains in Ontario were studied. The effluents contained several million per milliliter of the iron-oxidizing bacterium Gallionella. Sphaerotilus, a sheath-forming bacterium, was also numerous. Both micro-organisms were generally encrusted with amorphous ferric iron oxides. They were also found in the ochreous deposits inside the tiles and in the deposits adjacent to plugged filters surrounding the tiles. Thiobacillus ferro-oxidans was not found. Filamentous structures resembling the iron-oxidizing bacterium Metallogenium were frequently observed but found to be aggregates of colloidal Fe $(OH)_3$. Calcite, dolomite, quartz, feldspar were common constituents of the effluents and the ochreous deposits. In some cases, gypsum, plagioclase, chlorite and illite were present. The ochreous deposits were generally high in iron and contained elements normally found in soil, with occasionally traces of magnetite. The remaining non-silicate iron was present as amorphous ferric oxyhydroxide. The results and a literature survey indicated that Gallionella plays a major role in the oxidation of Fe^{3+}. A pathway for the formation of ochreous substances from Fe^{3+} is proposed.

610

IVITSKII, A.I. New way of determining the drainage norm of ameliorated soils. Doklt. Akad. Nauk. belorus, 6:795-7.

CO_2 & O_2 contents of soil air in bog soil depend on ground-water level. Optimal air regime and crop yield on peat soil occur only after lowering the groundwater level to 120-130 cm. Drainage norms of bog are obtained with greater accuracy by investigating composition of the soil air and the relationship between CO_2 & O_2 content of the soil air groundwater level.

611

JACKSON, M.B. Rapid injury to peas by soil waterlogging. J. Sci. Fd. Agric., 1979, 330:143-52.

Short-term responses of pea cv. Sprite to waterlogging were examined under glasshouse conditions. The symptoms of injury arising from 1-4 days of waterlogging included extensive desiccation and chlorosis of the foliage and lower rates of transpiration,stem extension, and growth of shoots and fruits. Premature quiescence of the apical bud of waterlogged plants reduced the number of fruiting nodes and extending internodes. Flowering plants at the 9- to 10- leaf stage were found to be damaged more severely than young vegetative plants bearing only 2 or 3 leaves. Waterlogging inhibited the uptake of N from soil sufficiently to reduce its concentration in vegetative and fruit tissues. The uptake of K and P was less severely affected and the overall concentration in shoot tissues was slightly higher in waterlogged plants than in controls. The concentration of Mn and Ca were not changed greatly by waterlogging. The equilibrium concentration of O_2 in the soil water fell to ` 1 per cent after 24 hr of waterlogging while the amounts of CO_2 nitrous oxide and ethylene increased considerably. The injury to plants grown in nutrient solutions supplied with O_2 and CO_2 at concentration present in waterlogged soil was much less than that sustained by plants growing in the waterlogged soil itself.

612

JACOBSON, P. Drainage design in humid areas. Proc. Drain Efficient Crop Production Conf., Chicago, Illinois, 1965, 69-70.

Drainage criteria in humid areas requires imrovements for its efficient functioning. In open-ditch design the development of criteria has probably been adequate to meet the design needs in most cases. One of the major factors which still plagues the engineer is allowable ditch velocities. Considerable study is now being made to determine these factors. Much of the tile drainage for humid areas is still predicated on rule-of-thumb methods. This will persist until drainage engineers universally adopt one of the more sophisticated design procedures. A new field for drainage engineers could be opened up in the design of a disposal system which would change topography in order to store water on land near the place where it falls. In many areas, with our present knowledge, this could provide the most permanent flood-control system design.

613

JAGER, A.W.De. Drainage with plastic pipes. 5th Congr.

Irrig. Drain. Tokyo, 1963, Q 17 R 18:267-78.

In Holland, 10-20 per cent of the total drainage work is with plastic. Technical data concerning the plastic pipes of hard PVC commonly used are: diam. 4 cm, length 5-6 m, wall thickness 0.8 mm, weight 140 g/m, number of saw slits 60/m, length and breadth of saw/slits 2.5 and 0.6 mm, respectively. In sand and sandy loam soils a layer of fibergrass of a thickness of 0.45 mm is recommeded.

614

JAIN, J.K. Hydrologic aspects of the salt problem in agriculture and its control through improved water management in arid and semi-arid region (with particular reference to the problems in India). Proc. Symp. Salinity Probl. Arid Zone, Tehran, UNESCO (1958), 1961, 111-6.

Salinity problem in India has been discussed in relation to drainage, leaching requirement and seepage in canal irrigated areas. Suitable methods to rectify different types of problems are suggested.

615

JAISWAL, S.R.; DHRUVANARAYANA, V.V. Design of subsurface drainage system. Symp. Waterlogging Causes and Measures for its Prevention, CBIP New Delhi Pub. No. 118, 1972, 1:91-100.

The design is based on the field data of hydraulic conductivity, the drainage requirements on the basis of rate of water table fluctuations and infiltration rate, and the depth of impermeable layer. The average hydraulic conductivity of the top 1.5 m of soil profile is 1.2 cm/hr, drainage coefficient is 6.0 mm/day and depth of impermeable layer is assumed to be at 4.5 m below ground. On this basis, the drain spacing has been computed as about 20 m. The drain size is fixed at 15 cm diam. In the experiment, three drain spacings (10, 20 and 30 m) and two drain types (tile and open ditch) were being tested.

616

JAISWAL, S.R.; PANDEY, R.N. Water table responses to subsurface drainage. Proc. Int. Symp. Develop. Ground Wat. Resour., Madras, 1973, II (b): 29-36.

Tile and open drains having 10, 20 and 30 m spacing were installed at a depth of 1.5 m in a severely salt affected area for the control of the rise of water table. Observations in the first monsoon season indicate that water table fluctuations follow the rainfall distribution pattern. Rise of water table could be controlled by installing the subsurface drainage system.

617

JANERT, H. Problems of the reclamation of waterlogged and saline salted irrigated soils of arid regions. Trans.8th Congr. Intern. Soc. Soil Sci., Bucharest-Romania, 1964, 2:909-21.

The failures of efforts on reclamation of waterlogged saline soils are traced to emphasize on the surface drainage through open drains only. Subsurface drainage is recommended and modern developments towards reducing the cost of subsurface drainage have been discussed.

618

JENAB, S.A. New theories in transient drainage and their applications in low-lying areas of Fars. ICID Sp. Sess., Varna, 1972, 33-53.

Ahow-Char area is the lowest portion of Ramjerd-Beiza basin which is periodically subjected to over-flowing from the Kor river, flood water from neighbouring mountains and the return flows from upper-lying areas. For designing the drainage system of this particular area Glover's and Jenab's transient drainage formulae were used which gave good results. A four way drainage plan viz. (a) underground drainage to control water table, (b) surface drainage to catch precipitation and return flow, (c) main drain to take the water to Kor river and finally (d) the pumped drainage system to pump the water in main drain from depressions is suggested for the area.

619

JESSEN, T.; MOLLE, K.G. Experiments with drainage, lime and fertilizers on low-lying, fine-grained sandy soil in Vollerum 1960-69.Tidsskrift Planteavl, 1972, 76:331-63.

In a 9 year experiment on reclaimed sandy soil low in organic matter and P in which clover-grass was grown during 1961-64, followed by oats or barley in 1965-69, and initial applica-tion of 5000 kg/ha $CaCO_3$ and an annual P application of 20 kg/ha P seemed adequate. For well establised clover-grass, 150-200 kg/ha K and 50-100 kg/ha N were sufficient. For oats and barley, 75-100 kg/ha K met the K requirements, and with the exception of crops immediately folllowing clover-grass, 100-130 kg/ha N was most effective. No significant differences in yield response were observed between tile drainage and surface drainage.

620

JOHANSEN, V.P.Methods of land drainage in Denmark. Proc. Int. Drain. Workshop, Wageningen, Netherlands, (1978), 1979, 25:477-80.

Methods of land drainage, covering materials and soil mechanics aspects in land drainage in Denmark are discussed.

621

JOHN, K. Functional testing of pipeless permeable strip drainage in a field model.Arch. Acker. Pfla. Bodenk., 1979, 23:115-20.

Permeable strip drainage without trenches or pipes proved successful in draining a stagnogley provided that appropriate envelope materials (free from fine particles) were used. The results of previous laboratory studies were confirmed in model field experiments. Pipeless permeable strip drainage was compared with strip drainage (with pipes) and the combination of measures given in TGL 28587.

622

JOHN, K.; GORA, A.; SCHWARZ, K. Aspects of pipeless permeable strip drainage on waterlogged mineral sites with deep groundwater table. Arch. Acker-U, Pflan. Bodenk., 1978, 22:635-42.

The principles of introducing materials into soils which are wet because of waterlogging, or because of film-water rentension, and which overlie deep groundwater tables was tested. Of the various envelope materials examined, raw gravel sands containing a low proportion of sand were found to be suitable for this type of drainage, where the projected average discharge was <0.4 1/s per ha. Investigations on the surface contact between soil and the permeable material did not suggest the likelihood of any stability risk under the expected hydrological conditions. Drainage based on this principle should be confined to sites which have, in addition to the above mentioned conditions, a percentage of clay (<0.002 mm) of 15 per cent in the subsoil.

623

JOHNSON, D.B.; KELSO, W.I. Micro-organisms and field drainage failure. Rep. Welsh Soils Disc. Gr., 1980, 21:80-90.

The failure of field drains attributable to the oxidation of ferrous iron by soil micro-organisms producing hydrous ferric oxides and hydroxides giving deposits of 'ochre' is discussed. Filamentous ochre is derived from the oxidation of ferrous iron in an aerobic environment such as a field drain. Pyritic ochre, associated with more acid conditions, accumulates in drainage systems under soils containing quantities of pyritic minerals. The nature of pyrite oxidation is examined and the detailed research at U.C.N.W. into pyritic environments and associated micro-organisms is summarized. Reference is also made to other types of microbial growth which cause field drain problems.

624

JOHNSON, E.A.G. Land drainage in England and Wales. In River Egineering and Water Conservation Works, Butter-Worths, 1966, 29-46.

The paper traces the administrative background and describes the objectives, drainage problems, and types of works; states effects of drainage works on the river flows and underground water, field drainage, and channel improvements; describes up-to-date progress made and the future requirements.

625

JOHNSON, H.P. Hydraulic roughness in drain tile. Proc. Natn. Symp. Chicago, Illinois, 1971, 15-9, 27.

The paper analyses the data on Mannings' roughness coefficients of drain tiles since 1920 which indicates that it varies from 0.011 to 0.017. For excellent placement the variation was from 0.011 to 0.013 only. The friction coefficient can be expressed as a function of the friction coefficient for smooth pipe, tile length and displacement of the longitudinal axis.

626

JOHNSTON, W.R. Drainage installation problems in the San Joaquin Valley, California, USA. Proc. Int. Drain. Workshop, Wageningen, Netherlands, (1978), 1979, 25:603-19.

The area of the San Joaquin Valley, California, U.S.A., which has a subsurface agricultural drainage problem is described, as are the proposed solutions to the problem. The rate at which drainage problems have developed over the past ten years is tabulated for the 240,000 ha Westlands Water District. The costs and benefits of subsurface agricultural drainage in the San Joaquin Valley are described and compared with expected crop yield decrements due to not providing of farm drainage after the ground water rises within 1.5 m of the ground surface. Estimates indicate that the average benefit from drainage is about $ 375 per ha which is approximately five times greater than the annual cost of a drainage system. Specific problems pertaining to drain envelopes and pipe materials; trenching and backfilling during the installation of subsurface drainage systems in unstable soils with an extremely high ground water table are also discussed.

627

JOHNSTON, W.R. Drainage problems and the proposed solution for a large irrigated area in the San Joaquin Valley of California (U.S.A.). 7th Congr. Irrig. Drain. Mexico, 1969: 26.233-26.246.

The 240,000 ha Westlands Water District developed perched water table over approximately 1/6th of the District land, causing severe drainage problems and high salinity conditions in the crop rootzone. The district drainage system proposed to be constructed in stages will consists of deep open-joint collector drains, shallow tight-joint carrier lines, drainage sumps and on-farm drainage systems. Under full operation drainage disposal service, including operation and maintenance costs and repayment for construction of the drainage system will cost the district $ 7.90/ha of the district land.

628

JOHNSTON, W.R. Local solutions to drains problems.

Proc. Natn. Conf. Irrig. Return Flow Quality Mgmt., (Law, J.P.; sko-gerboe, G.V. Eds.) Colorado State University, Fort Collins, USA, 293-5.

Tile drainage systems in San-Joaquin Valley are being installed to alleviate the problem of high water table. Local solutions to dispose of saline subsurface drainage waters have been reported for three areas. Whereas discharging in the San Joaquin River is followed by northern part of the Delta-Mendota Service, Southern part of it disposes of its water into the grassland water district for irrigating native pasture and for wildlife habitat. Tulare lake Drainage district uses evaporation ponds for disposal whereas Kern Country Water Agency is contemplating the use of this water for cooling the reactors in a nuclear power plant.

629

JOHNSTON, W.R.; LETEY, J.; PILLSBURY, A.F. Tile drainage performance compared to theoretical predictions. Trans.ASAE, 1965, 8:548-9, 552.

Observed and predicted drainage rates were compared. It was found that of the four theoretical equations compared in this paper, the Van Schilfgarde equation most accurately predicted a water-table recession rate, especially after the initial ten-day drainage period which started with a saturated soil profile. This equation overestimated the time required for the water table to recede during the first ten days of drainage, as did the integrated Hooghoudt and the integrated Toksoz and Kirkham equations for the entire drainage period. The Luthin and Worstell equation did a good job of estimating the water-table recession rate during the first ten days of drainage, but underestimated the recession rate for the remainder of the drainage period. Difficulty arises in determining hydraulic conductivity and drainable porosity values for use in the equations.

630

JOHNSTON, W.R.; STEINERT, B.C.; STROH, C.M. Benefits from the drainage of heavy irrigated soils. 4th Natn. Drain. Symp. Proc. ASAE, 1982, Pub. 12, 82:171-7.

On-farm subsurface drainage can hold the level of the water table below the crop rootzone so that in San Joaquin Valley productivity and/or changes in cropping patterns can be prevented or reversed. The annual net on-farm income maintained by drainage in Westlands Water District is estimated at $370 per hectare with the grower's costs for a complete system from on-farm facilities to ultimate disposal of the saline effluent-estimated at some $ 99 per hectare. Clearly, investment in drainage-particularly under circumstances where drainage effluent is too salty to be recirculated in the irrigation supply or discharged to a nearby watercourse is profitable. The beneficial economic impacts of drainage are felt beyond the confines of the farm and the grower's profits. Avoiding Westlands' $ 17,000,000 drainage related production loss for 1981 would have meant about $ 37,000,000 more income in the entire San Joaquin Valley and an aditional $ 13,000,000 more income in the rest of the State. All citizens would benefit, directly or indirectly, from investment in agricultural drainage.

631

JONES, L.H. The effect of soil moisture gradients on the growth and development of broad beans (Vicia faba.).Hort. Res., 1963, 3:13-26.

Water tables were situated at intervals between 5 and 55 cm below the soil surface. With the highest water table,seeds rotted and growth was restricted, and the lowest water tables resulted in poor emergence and death of some plants from wilting. Early growth rates were slow in these treatments, but by flowering time growth rates were similar for all treatments. The yields were closely related to haulm weight which was determined by the preflowering growth rate. Limited available soil volume in the higher water-table treatments may have restricted root growth and may have resulted in exhaustion of nutrients, but good plants were produced where the water-table was 15 cm down. Maximum root development always occurred where soil-water content was at or above field capacity. Beans appeared to be very tolerant of waterlogging but not of early drought.

632

JONG, DE, M.H. Drainage of structured clay soils. Proc. Int.Drain. Workshop, Wageningen, Netherlands, (1978), 1979, 25:268-80.

Soil properties in relation to subsurface drainage of a structured clay soil in Netherlands is discussed. Swelling and shrinkage of these soils and their accounting in a model of testing drainage criteria is yet unresolved and requires considerable research efforts in this regard.

633

JURCZUK, S. Grassland drainage on peat. Wiad.Melior., 1967, 8:175-7.

The paper attempts, on the Stepnica-Brylanty peatland (Szczecin Province), to determine experimentally the drainage effect in sowing conditions, water conditions and the yield.

634

JURCZUK, S. Effects of supplementary drainage in different organic soil types. 11th Congr. Irrig. Drain., New Delhi, 1981, 1:151-5.

On the experimental plots, a widely spaced network of ditches, representing the original drainage, was suplemented with tile drains of 5 cm diam. installed at spacings of 10, 20 and 30 m and at depth 0.8-1.0 m. For comparison, a part of the area was kept without drains, with only ditches present (spacings 120-160 m). Supplementary drainage was found to have beneficial effects on soil moisture capacity and yields of meadow vegetation in the case of moist soils or soils only periodically semi-arid.

635

JURCZUK, S. Peat soil permeability changes after drainage. Wiad. Melior., 1970, 7-8:233-4.

It has been observed that in peat soils drained and utilized for several decades, with low or medium mucking degree, a reduction of filtration coefficient takes place in conditions of more dense reclamation network, and this reduction was almost twofold within 3-4 years in the layer of 20-100 cm, at which it was the highest in the layer situated immediately under sod and decreased along with the profile depth; recommends a twofold reduction of filtration coefficient value of the layer situated above drain for the design of reclamation.

636

JURY, W.A. Solute travel-time estimates for tile-drained fields: II Application to experimental studies. Soil Sci. Soc. Am. Proc., 1975, 39:1024-8.

Model calculations are applied to published reports of tile-drain effluent concentrations and flux and are shown in all cases to be in good agreement with observed data. Measures proposed for improving water quality by required standards for effluent discharge are discussed and criticized by use of model simulations of projected policies.

637

JURY, W.A. Solute travel-time estimates for tile-drained fields: I.Theory. Soil. Sci. Soc. Am. Proc., 1975, 39:1020-4.

A model based on earlier work by D. Kirkahm is proposed for calculating solute travel times and effluent concentration in tile-drained soil profiles. Field systems are represented by a characteristic travel-time parameter which includes tile spacing, soil porosity, and discharge rate. Simulations are run predicting effluent concentration for various input conditions.

638

JURY, W.A.; WEEKS, L.V. Solute travel-time estimates for tile-drained fields:3. Removal of a geothermal brine spill from soil by leaching. Soil Sci. Soc. Am. J., 1978, 42:679-84.

The time required to leach a slug of saline sodic geothermal brine from the point of injection to the tile outlet of an artifically drained field is calculated. Sprinkler, complete and partial ponding leaching methods are compared as a function of drain spacing and initial location of the spill. Calculated results are presented as dimensionless parameters which scale the drainage system dimensions and the soil water transport properties. Ponded leaching required more water, but less time to leach brine out of the system for all situations except where the brine spill occurs near the mid point between tile lines. A simple calculation is proposed to estimate the leaching fluid volume required to remove excess Na from the exchange complex. Good agreement was obtained between simulated and experimental results involving a laboratory soil column with sandy loam soil. For fine-textured soils in the Imperial Valley it may require up to 30 pore volumes of leaching fluid to replace Na with Ca if saturated gypsum solution is used in reclamation. Application time per pore volume was calculated to be in excess of 1 year for all cases except ponded leaching directly over a tile line.

639

KANWAR, R.S.;JOHNSON, H.P.; SCHULT, D.; FENTON, T.E.;

HICKMAN, R.D. Drainage needs and returns in north-central Iowa. Trans.ASAE, 1983, 26:456-64.

Current status of drainage and estimated drainage needs of the soils in the Des Moines River basin as indicated from an extensive survey are presented. A high percentage of the area has the potential for receiving benefits by improving county mains and on-the-farm drainage. Analysis has been made of the economic potential for drainage improvements by comparing three drainage input levels. Benefit-cost ratios indicate drainage of very poorly drained and poorly drained soils is a good investment for corn and soybean production.

640

KAPOCSI, I. Control of the water balance of soil by mole drain deep loosening. Novenytermeles, 1980, 28:543-8.

On very heavy meadow soil tile drains were laid out at different distances apart and continuously secured by mole drains installed perpendicularly to tile drain lines at depths of 70 cm with diameters of 8 cm and distances of 65-70 cm. To maintain the favourable effects of deep tillage connected with mole-draining, this work should be repeated every third year.

641

KAPUSTINSKAITE, T.K. Biometric changes in the Pinetum ledosum under the influence of drainage. Lesovedenie, 1976, 3:20-30.

The paper presents results of biometric changes (i.e. changes in the phytomass) of Scots Pine stands on drained and undrained sites in the Pinetum ledosum in Lithuania. The un-drained sites carried 5 stands ranging in age from 28 to 120 years, and the drained sites 6 stands ranging in age from 30 to 70 years. Data are tabulated on phytomass of the trees (weights per ha of stem, branches, foliage and roots, green and oven dry), on the needle numbers per g weight and needle dimensions, and on the annual increment of the above-ground phytomass of the tree layer and subordinate layers. The results indicate that drainage caused profound changes, increasing the phytomass accumulation several-fold and bringing about an earlier culmination of current increment.

642

KAPUSTINSKAITE, T.K. Change in boggy forests as a result of drainage. Lesovedenie, 1977, 4:18-28.

A comparative study was made of forest biogeocoenoses on un-drained wetlands in Lithuania and the Kaliningrad region, and on sites drained long ago. Data are presented on the changes in the physical and chemical properties of the rooting horizon of peaty soils, and in the main components of the vegetation, as a result of drainage. The forest types compared were Pineta on oligotrophic bogs, Piceeta and Betuleta on mesotrophic sites, Betuleta and Alneta on meso-eutrophic sites, and Alneta on eutrophic sites. The results are tabulated and 5 groups of derived (anthropogenic) forest types are distinguished on raised bogs transitional bogs and fens. A scheme of desirable stand composition for drained sites is drawn up.

643

KARADI, G.; KRIZEK, R.J.; ELNAGGAR, H. Unsteady seepage flow between fully-penetrating trenches. J.Hydrol., 1968, 6:417-29.

The problem of unsteady seepage flow, including either evaporation or infiltration, in the domain between two fully-penetrating trenches is considered. The method of solution is based on matrix mathematics and has the advantage of being able to handle complicated boundary and initial conditions and to consider step by step the nonlinearity of governing differential equation. Typical results for a variety of boundary and initial conditions are presented. For a particular set, the comparison with observed data exhibited better agreement than do results calculated by linearizing the governing differential equation.

644

KARSISTO, K.; RAVELA, H. Losses of phosphorus and potassium from areas drained for forestry and topdressed at different times of the year. Suo, 1971, 22:39-46.

PK fertilizing increased the PK content of the ditch water and the K content was higher for narrow ditch spacings. Ditch-water P and K contents were not clearly influenced by fertilizer rate of placement, but application on snow resulted in higher K contents than application to bare ground. A magnitude of runoff was found at which the P and K contents of the ditch water reached a maximum value.

645

KASER, R.F. Irrigation and drainage studies of Texas Rivers J. Irrig. Drain. Div. ASCE, 1961, 87:13-36.

A new appraoch to comprehensive planning for irrigation and drainage development, involving participation and collaboration by local, state and federal levels of government, is now underway in Texas. This study will evaluate potential developments from the standpoints of water and land resources and both local and national interests. Present and potential irrigation and drainage developments are summarized for the Study Area. The possible agricultural production in the area for the next 50 years is analysed and the role of irrigation and drainage in meeting future production goals is explained. From the national standpoint there appears to be no compelling need to promote large-scale irrigation or drainage in the Study Area.

646

KASPRZAK, K.; SOJKA, J. Characteristics of the VUM plastic tube drain. Ved. Pr. Vyzk. Ustavu. Melio. Zb., 1973, 13:31-45.

The physical properties and hydraulic characteristics are given for PVC piping of the type which is formed into a tube when it is installed.

647

KAUNISTO, S. Effect of drainage intensity and fertilization on the development of pine plantations on oliogtrophic treeless sphagnum bogs. Folia Forestalia, Institutum Forestale Fenniae, 1977, 317, 31 p.

Trials were established in 1963-68 with 7 ditch spacings (5-100 m) and 3 ditch depths (30-90 cm); best growth was obtained with 5-m spacing and 90-cm depth. In one trial with comibna-tions of soil cultivation and furrowing (furrows spaced at 2-10 m), cultivation reduced early growth but furrowing (especially at 2 and 4 m) improved growth. Refertilisation with NPK in 1973 improved growth,however, refertilization with N alone reduced growth.

648

KAWASE, M. Aerenchyma formation: how plants adapt to waterlog-ging. Ohio Rep. Res. and Dev. Deptt. Horticulture Worster, 1978, 63:14-5.

In laboratory trials with 4 spp.including sunflower and bean, plants were found to adapt to flooding or waterlogged conditions by the formation of spongy aerenchyma tissue, which facilitated transport of O_2 from the leaves to the roots. The deficiency of O_2 under such conditions appeared to trigger the anaerobic formation of ethylene, causing an increase in cellulase activity which then led to aerenchyma formation.

649

KELLER, J.; ROBINSON, A.R. Laboratory research on interceptor drains. Trans. ASCE, 1961, 126:687-714.

The paper presents the results of a large scale model study concerned with the design of interceptor drains; analyses previously developed analytical relationships for determining the shape of the resultant water table drawdown curve; develops relationships for estimating the flow from drains and examples to illustrate the use of the material, have been given.

650

KELLETT, A.J. Poaching of grassland and the role of drainage. Tech. Rep. Field Drainage Experimental Unit, UK, 1978, 78/1,19p.

Poaching damage can lead to a 50 per cent reduction in yield and a 20 per cent waste of herbage when pastures are grazed in unsuitable conditions. The causes of poaching damage are outlined and the effects are discussed with reference to utilization loss, plant damage, soil damage and grass yield reduction. Methods of minimizing damage by drainage, subsoiling and grazing management are discussed. In several drainage trials on grassland in high rainfall areas mole drainage over a basic pipe system reduced damage, increased yield and extended the grazing season by 6 wk.

651

KELLETT, A.J.; DAVIES, D.B. A report on the drainage experiment at Abbots Ripton, Huntingdon, 1970-1975. Summary and discussion of 5 years' results. Tech. Bull. Ministry of Agriculture, Fisheries and Food, UK, 1977, 77-2, 11p.

Five years results from this site on Hanslope soil in a low rainfall area showed no differences in watertable control or crop yields (dwarf beans and winter wheat) between the undrained control and drains at 40 meters with moling. This is attributed to a good soil structure and the existence of subsoil channels on all plots. The results provide further evidence of the stability of mole and subsoil channels on these soils and show that mole runs longer than the standard 40 meters are practicable.

652

KELMAN, S.(CHAIRMAN) (TASK COMMITTEE ON AGRICULTURAL RUNOFF AND DRAINAGE OF THE WATER QUALITY COMMITTEE, IRRIGATION AND DRAINAGE DIVISION, ASCE), Quality aspects of agricultural runoff and drainage, J.Irrig. Drain. Div. ASCE, 1977, 103:475-95.

The ASCE Task Committee on Agricultural Runoff and Drainage has summarized the literature and research findings with respect to sources of pollutants, nature of downstream impacts, and relationships between pollutant loads and certain physical variables. Agricultural discharges from nonirrigated lands are significant contributors of nitrogen, phosphorus, and pesticides to downstream receiving waters. Management practices directed towards control of erosion and runoff, and management of fertilizer and pesticide application can reduce total pollutant load. The development of on-site discharge water quality control measures, as contemplated under PL 92- 500, must be predicted on detailed survey of a more specific nature due to the high degree of variability in the factors influencing water-borne pollutant transport.

653

KENDALL, R.G. Land Drainage. Faber and Faber Lim. London, 133 p.

The book mainly deals with drainage aspects of agricultural lands. The book containing 16 chapters also contain two appendices. In one of the appendix, some experiments on drainage have been described which help in understanding the principles. This also adds to the utility of the book for research scientists.

654

KHAN, M.Y.; KIRKHAM, D. Spacing of drainage wells in a layered aquifer. Wat. Resour. Res., 1971, 7:166-83.

A theory is presented to determine spacing of identical drainage wells that, by discharging groundwater simultaneously from a layered aquifer will lower a water table to a pre-assigned level and maintain it. The theory has been developed by solving a mathematical boundary value problem. The theory shows that the spacing depends on the thickness and hydraulic conducticity of each permeable layer tapped by the wells, the maximum allowable drawdown, the uniform vertical recharge, the radius of the wells and the well grid geometry. A table of values of a function, useful for determining spacing of wells to be tapped in an aquifer that may consists of any number of permeable layers, is given.

655

KHUDYAKOV, A.D.; KOVALENGO, N.P.; BONDARENKO, N.F. The compaction of peats and the change in their physical and mechanical properties as a result of deep drainage. Izv. Vyssh. Ucheb.Zaved, 1969, 12:5-54.

Moisture, penetration and shear resistance at 7.5, 20, 32.5 and 45 cm from the drains, and at various depths from the surface for 4 sites are tabulated.

656

KIENITZ, G. A hydrological and efficiency investigation method in connection with the establishment of the development rate of surface drainage. 5th Congr. Irrig. Drain., Tokyo, 1963, 3:15.494-15.511.

From the catchments with insignificant slopes of the Hungarian lowlands precipitation water can be conveyed into the rivers only by means of artificial conduits. A method is elaborated by which dimensions of the drainage system can be evaluated with the combined hydrological and economical efficiency. The runoff volume and inundation caused by the retardation are evaluated by channel floodwaves. An agricultural assessment work suggested will yield the net expected damages in the area and also the increased amount of returns. The comparison of increased returns at several development degrees will help the planners decide, how much drainage system of the area should be developed.

657

KIENITZ, G. The interdependence of crops and drainage. 5th Congr. Irrig. Drain. Varna, 1972, Q 27 R 7:97-104.

Effect of waterlogging duration on crop yields in Great Hungarian Plain are reviewed and the tabulated values for different crops on an average basis are made available.

658

KIM, Y.S.; CHO, M.D.; KIM, W.S. Studies on effective drainage in an apricot orchard with a heavy clay soil. J. Korean Soc. Hort. Sci. 1980, 21:152-7.

An orchard with a heavy clay soil, drained by various methods, was planted with one-year old apricots cv. Sohwa and measurements taken for the following 5 years. The systems used were 30 or 60 cm deep open drains, tile-rubble drains, tile-rubble + 30 cm open ditches, rubble only or buried pruned branches. Controls were undrained. Ground water level was reduced significantly by 60 cm open drains and tile-rubble drains with or without 30 cm open drains. The mean water table depths in these treatments being 95.4, 85.8 and 82.0 cm, respectively, compared with 64.1 cm for the undrained control. Tree growth was significantly improved in plots with tile-rubble + 30 cm open drains. Individual fruit weights and soluble solids were not affected by the treatments. In

1980, fruit yields were significantly increased with buried pruned branch drains and tile-rubble drains with or without 30 cm open ditches, these being 21.0, 19.3 and 20.1 kg/tree, respectively, compared with 9.4 kg/ha in the control. In general, however, the buried pruned branch drains were considered unsatisfactory, tile-rubble drains being far superior.

659

KINDERIS, Z. Leaching nutrients from soil by drainage water. Nauch. Tr. litov. S-kh.Akad., 1963, 10:137-46.

The salt concentration in the soil was smallest in spring and greatest in summer and autumn, but leaching was less in summer because of the small volume of runoff. Drainage water removed 196.2-238.6 kg/ha Ca annually. The NO_3 content of the drainage water increased several times towards autumn, particularly under legumes, and reached 124.7 mg/L . Drainage water removed 51.1-52.2 kg Mg, 33.2-76.0 kg Cl and 27-84.0 kg/ ha SO_4 annually.

660

KHINORI, B.Z. Manual of surface drainage engineering - volume 1. Elsevier Publishing Co., Amsterdam, London, New York, 1970, 224p.

Divides the subject matter in two parts, and an appendix. Part I (The Open Channel) - (1) Definitions and terminology (2) Hydraulics of open channels and flow regions, (3) Scour and tractive force, (4) The permissible velocity, (5) Slope of the channel walls, (6) Earth channels of composite cross section, (7) Stable earth channels, (8) the water drop. Part II (Channel Lining) - (1) Seepage from earth channels, (2) Vegetative lining of channels (grassed channels), (3) Lining channels with impervious soil, (4) Masonary lining of channels, (5) Concrete linings, (6) Lining channels with bituminous (asphaltic) material, (7) Lining with various materials, Appendix-Biological protection of water-ways and drainage channels in Mediterranean and semiarid conditions.

661

KIRKHAM, D. Physical artifices and formulas for approximating water table fall in tile drained land. Soil. Sci. Soc. Am. Proc., 1964, 5:585-90.

Two physical artifices are used to simplify the mathematics of the falling water table in tile-drained land. One artifice is to insert in the soil rigid, frictionless, infinitesimally thin, impervious membranes along the natural soil water seepage streamlines, or along approximate streamlines. The second artifice is to replace soil by gravel in the region above the level of drain tube centers. The gravel is subsequently, for final derived formulas, replaced by soil. The formulas are shown, theoretically, to be on the safe side for drainage design. The formulas agree with experimental data.

662

KIRKHAM, D. Problems and trends in drainage research, mixed boundary conditions. Soil Sci., 1972, 113:285-93.

Current problems and trends in drainage research are discussed. An analytical mixed-boundary type technique is developed which is to solve mixed-boundary type seepage problems.

663

KIRKHAM, D. Saturated conductivity as a characterizer of soil for drainage design. Proc. Drain. Efficient Crop Production Conf., Chicago, Illinois, 1965, 24-31.

Recent literature on the hyraulic conductivity k, of soil is reviewed under the following headings: Darcy's law and the conductivity k; anisotropy and k; new methods,apparatus and theory for evaluation of k; infiltration and k; calculation of k from pore-size distributions, and some recent measurements of k. Following this review, some problems in the use of the auger hole method for k are considered. The problem of the proper formula to use in reducing auger hole data receives the most attention. Ernst's simplified formula is compared in detail with the exact formula of Kirkham and Van Bavel, and it is shown that Ernst's formula (and nomographs of Van Beers based on it) are accurate enough for field work for a certain range of values of auger hole radius, depth and height of water rise given by Ernst. For accurate work, the Kirkham-Van Bavel formula should be used and detailed numerical examples illustrating its use are given. As for auger hole problems in the field, topics considered include establishment of water table level; drawdown of water table about the auger hole; capillary action; a perched water table; false artesian pressure; fluffing of soil from auger hole walls, and the problem of a tight surface layer. The paper is concluded with some suggested methods for measuring drainability.

664

KIRKHAM, D. Seepage of leachin water into drainage ditches of unequal water level heights. J. Hydrol., 1965, 3:207-24.

The theory is developed and numerical examples are worked out for the seepage of ponded water into equally spaced ditches in homogenous soil overlying a barrier. Two general cases are solved: (a) unequal depth of water in the drainage ditches and (b) equal depth. For both (a) and (b) the problem is solved for zero depth of ponded water and for finite depth of ponded water; these situations are quite different. Curves are supplied showing the surface inflow seepage velocity as it varies with distance from a ditch. Near ditches, the seepage inflow rate is very high compared with that between ditches, which helps explain why, many times, crops do better when near ditches than when moved far from them in reclaimed saline soils.

665

KIRKHAM, D. Some recent land drainage investigations at Iowa University of Science and Technology, Ames, Iowa, U.S.A. Agrokem Talajt., 1965, 14 (suppl):229-34.

A simple apparatus for measuring hydraulic conductivity for drainage, and theoretical formulae for checking field and laboratory data on falling or steady-state water tables are described. Leaching of salts by maintaining a steady ponded-water condition over saline soils is not as effective as applying a number of rinsings when the soil is underlain by an impervious layer.

666

KIRKHAM, D. Steady-state theories for drainage. J. Irrig. Drain. Div. ASCE, 1966, 92:19-39.

The paper reviews the steady-state drainage theories since 1955 for ponded; artesian, and steady rain (or equivalent excess irrigation) water, when the water seeps into drain tiles, ditches or wells, or through dams; cites reviews of the work before 1955; presents and examines the nine recently developed approximate theories for the seepage of steady rainfall into drain tile or ditches when the soil to be drained is underlain by a barrier at finite depth and brings out an appropriate theory for this problem.

667

KIRKHAM, D. An upper limit for the height of the water table in drainage formulas. Trans. 7th Congr. Soil Sci., Madison, 1960, 1:486-92.

A formula is derived for an upper limit of water table rise above a system of drains which remove rain water or excess irrigation water at a steady rate. The formula is similar to the one derived by the author earlier (1958) except that the height as earlier given must be multiplied by a factor 1/1-R/k where R is the rainfall rate and k is the soil hydraulic conductivity both in the same units. The earlier formula which is an approximation is, by use of the factor, established on a firmer basis.

568

KIRKHAM, D.; POWERS, W.L. An exact theory of seepage of steady rainfall into tile and ditch drained land of finite depth. Trans. 8th Congr. Soil Sci., 1964, 2:39-44.

Formulae are presented for the height of the water table in land, in which ditches penetrate down to an impervious layer, where rainfall maintains a steady-state water-table arch. The solution for tile drainage requires consideration of the drain radius.

569

KIRKHAM, D.; PRUNTY, L. Upslope recharge, downslope waterlogging and interceptor drains. 3rd Natn. Drain.Symp.ASAE, Michigan, 1976, 40-50.

Often-times, and in particular near river flood plains, sloped land rises from the plain, and beneath the plain and the slope there is a subsoil impervious barrier that impedes deep drainage. When sloped land receives excess rainfall in humid regions, or excess irrigation water in arid regions, the water seeps downward to the barrier and then runs more or less along the barrier to rise upward out of the soil into the plain. The seepage flow velocities for this problem have been analysed by Dupuit-Forchheimer theory and potential theory. It is found that the upward seepage velocity is greatest along a line where the sloped land intercepts the plain. The velocity of upward seepage depends on the slope. It is found, for example, by potential theory, that along the line where, the slope meets the plain, the upward seepage velocity is 7.6 times as great as at a distance 6 m on the plain away from the line. These numbers are for a slope of 10 percent, with a subsoil barrier 5 m below the plain and the sloped land extending 100 m away from the plain before there is an upslope drainage divide. For a slope of 1 per cent, the peak upward seepage velocity is one-hunderth that of a 10 percent slope. An interceptor drain should be placed at the position of maximum upward seepage velocity. The present study also shows that in arid regions where a limited amount of land on a slope is irrigated, "natural drainage" can handle excess irrigation water. The natural drainage then occurs as upward evaporation from the sides of the irrigated area, or if the barrier below the irrigated area is not completely impervious, some natural drainage may occur as leakage through the barrier. The results can apply to drainage in subhumid regions, where recharge is from rainfall and where the land might be sloped fairly steeply as in reclamation projects in strip mining. The potential theory is needed for steep strip mine slopes. Three flow nets are presented.

670

KIRKHAM, D.; SELIM, M.S. Theory of a rectangular gravel envelope in drainage design. Soil Sci.Soc. Am.Proc., 1973, 37:517-21.

A theory was developed to determine the radius of a circular drain which is equivalent to a square drain. It is shown that an equivalent circular drain tube of radius,1.172 'a' may be used to replace the square drain tube in case the impermeable barrier was at greater depths. The radius of equivalent circular drain is 1.1037 'a' when one edge of the square drain is on the impermeable barrier. 'a' in these cases is equivalent to one side of the square drain.

671

KIRKHAM. D.; TOKSOZ, S.; PLOEG, R.R. VAN DER. Steady flow to drains and wells. In Drainage for Agriculture (Agronomy No.17, Schilfgaarde, J. Van Ed). Amer. Soc. Agron. Wisconsin, 1974, 203-44.

Developments in the theory of steady-state drainage since 1965 are reviewed and tile and ditch drainage of flat land, hillside drainage, well drainage and seepage from pits and injection wells are considered.

672

KIRPICH, P.Z.; DUGAN, H.P. Province-wide irrigation and drainage planning and development in East Pakistan. 8th Congr. Irrig. Drain. Varna, 1972, Q 27 R 15:215-37.

Irrigation and drainage planning in East Pakistan is discussed. Evacuation of waters and drainage by sluicing through regulators and by pumping is recommended.

673

KISPERTH, V.; HILIGER, D. Comparison of different methods for determining the distribution of potential in drained soil. Arch. Acker-U. Pflan. Bodenk., 1971, 631-41.

Methods of solving problems of water flow towards a drain pipe above an impermeable layer in saturated, homogenous isotropic soil are compared, and electrical analogue procedures and ana-lytical solutions are discussed.

674

KJELLERUP, C.M. Experiments on drainage and improvement of soil structure of marsh soils. 1957-71. Tidsskrift Planteavl, 1974, 77:471-94.

In trials at 2 locations on marsh soils with 36-47 per cent clay, 35 per cent silt and 6-8 per cent OM, treatments imposed were various drainage depths combined with lime and gypsum application. At the 1st location (Skaerbaek) drainage depths were 80 and 115 cm. No significant differences in yield of cereal crops were found, presumably because of soil heterogeneity. At the 2nd location, (Ribe), drainage depths were 60 and 90 cm. At the 90 cm drainage depth, yields of winter wheat were significantly higher than for the 60cm depth; no significant differences in yields of barley, oats or fodder sugar beet were found. Lime and gypsum applications significantly increased yields, of all crops and were necessary on the heavy soils of the 2 trial locations if satisfactory yields were to be maintained.

675

KLAWITTERRALPHA, A.; YOUND, C. Forest drainage research in the coastal plain. J. Irrig. Drain. Div. ASCE, 1965, 91:1-7.

The paper reports the study made on the changes in the rates of growth of young slash pine trees after drainage on sandy wetlands in northwest Florida, as well as evaluation of the soil characteristics of wet pinelands in the lower plain of North Carolina which indicates (initially atleast) that an increase of approximately 6 m in site index in 50 years may be possible with adequate drainage; observe that if growth is sustained, pulpwood yields on many wetlands throughout the coastal plain can nearly be doubled by drainage.

676

KLOCK, G.O.; BOERSMA, L.; WATTS, D.G. et al. Drainage with perforated plastic tubing. Agron. J., 1969, 61:56-8.

Perforated plastic tubing placed 6.75 m apart at a depth of 50 cm provided adequate drainage in a poorly drained planosol with a mottled surface and a perched water table above the clayey sub-soil throghout the winter. Yields of lucern improved markedly and Mn levels were below toxic concentrations. When a spacing of 13.5 m was used, lucerne yields were greatly reduced and Mn concentrations were well above toxic levels.

677

KNIZE, A.A.; RUBTSOV, V.G. Evaluating the work of a forest drainage system. Lesn. Khoz., 1976, 4:39-41.

The paper presents discussions of the various aspects of the problem of evaluating the efficiency of drainage systems by the height increment of the trees. It is concluded that it is necessary to measure increment on at least 5 sample trees, and that these sample trees should be of average size at the time the drainage is carried out. The interpretation of the height measurements is discussed in terms of establishing the site class of the drained stands and the changes in site class. An example is given for a pure Scots Pine stand drained when its mean height was 5 m, which had reached a mean height of 15 m, 40 years after drainage. This example indicates that if height increment is in the range of 0.16-0.22 m/year or better, the drainage system is working normally, but if it is less than 0.16 m/year the drainage system needs to be repaired.

678

KNIZE, A.A.; RUBTSOV, V.G. New scale for determining the site class of drained stands. Lesn. Khoz., 1974, 6.47-9.

Reviews various methods of determining the site class of drained stands, and proposes a method based on current increment over the whole period since the actual drainage operation. An example is given of a treble-entry tabular scale of site classes for drained Scots Pine stands; the entry points to the scale are mean stand height at the time of drainage, number of years since drainage and mean present stand height. Field procedures and the use of the scale are described.

679

KNOPS, J.A.C. Research on envelope materials for subsurface drains. Proc. Int. Drain. Workshop, Wageningen, Netherlands (1978), 1979, 25:368-92.

Paper discusses the developments of drainage materials and their testing in the Netherlands. Two types of sand tank models commonly used are described. Studies conducted on the effect of perforations and envelope material on the entrance resistance of drain pipes have been reported. Hydraulic gradients in the immediate vicinity of the drain under various degree of drain fill and submergence have been presented. It is reported that because of insufficient data the results are inconclusive and more indepth studies are planned on the lines suggested.

680

KNOPS, J.A.C.; DIERICKX, W. Drainage materials. Proc. Int. Drain. Workshop, Wageningen, Netherlands (1978), 1979, 25:14-38.

The paper presents the overall view of the work done to date based on the 17 papers presented in the workshop. Development of envelope materials; need for envelopes; research on envelope materials; hydrological properties of drainage materials; drain design installations and maintenance; chemical and biological clogging have been discussed.

681

KOBASHI, H.; NAGAHORI, K.; OGINO, Y. The physical and mechanical characteristics of poldered paddy fields in Kojima Bay. 2. The effect of tractor operations on mole drains. Scient. Rep. Fac. Agric. Okayama Univ., 1968, 32:75-85.

A mole drain of about 9 cm diam at a depth of 35 cm was completely destroyed after a Fordson DEXTA tractor had passed three times over the area.

682

KOCHEV, K. Investigation of a front drainage in

stratified aquifers. Techniceska Missal., 1968, 5:29-40.

The paper gives analytical solutions for seepage in horizontal open trench drainages located in a four-layer confined seepage area. The seepage area consists of two alternating semi-per-meable (clayey) layers and two aquifers with different heads; test the accuracy of the analytical solutions experimentally according to the electrohydrodynamical analogy and indicates the aspects of the practical application of the solutions obtained.

683

KOCHEV, K.; YOVKOV, B. Some methods for selecting and designing of a proper drainage system in a river polder. In Polders of the World, Vol. 1. Polder Projects Land and Water Management Aspects. Int. Inst. Land Reclam. Improve., Wageningen, Netherlands, 1982, 455-63.

Research by the Institute for water problems during the last ten years into control of seepage inflow in river polders of the Danube lowlands is summarised. Aspects covered include a refined model for horizontal drainage in a semiconfined aquifer, a design model for a drainage system, and a simulation model for controlled drainage and subirrigation of the drainage system.

684

KOLENBRANDER, G.J. Nitrate content and nitrogen loss in drain water. Neth. J. Agric. Sci., 1969, 17:246-55.

The NO_3 contents and N loss in drainwater from cropped arable land and grassland were determined from data previously obtained in lysimeter and drainage trials. An average NO_3 content of 1-2mg/L was found for drainwater from grassland and an average content of 4-10 mg/L for arable land. Very little N is lost from soil with a clay content $>$ 35 per cent (particles $<$ 16 um). On sandy soils

(0-10% clay), about 60 kg/ha of N can be lost annually and about 20 per cent of the applied fertilizer N is leached out of the profile. On grassland, the leaching losses of applied N are very low during the period March-June but up to 40 per cent of the N applied in November may be lost.

685

KONING, J.C. De. Application of field drainage in Deltaic areas. ICID Bull., 1964, 13:13-20.

The drainage history and present developments in the Netherlands are discussed.

686

KONING, J.C. De.; KAMPEN, J.H. VAN. Experiences with plastic drainage in the Zuiderzee polders. 5th Congr. Irrig. Drain. Tokyo, 1963, Q 17 R 19:279-84.

The comparative analysis of cost of plastic drain and clay-tiles is made. The plastic pipe 50 mm in width and a wall thickness of 0.9 mm with 32 perforations per m (125 x 0.6 mm) without a special covering in heavy loam and with a cover of peat-litter in sand work well. Drainage with clay-tiles is cheaper than with plastic pipes. Reduction in cost of drainage with plastic pipes can be achieved by mechanization for which conditions are favourable.

687

KONSTANTINOV, V.K. Determination of moisture reserves in soil according to the depth of the groundwater. Pochvovedenie, 1966, 2:58-64.

In waterlogged, clean-felled areas of the long moss type on peaty-podzolic soils on sands and loams, the curvilinear-relation between soil moisture reserves and depth of watertable was expressed by the equation $Y = ax^2 - bx + c$, where Y = soil moisture reserves, x = depth of the water table below the soil surface and a, b and c are constants.

688

KONSTANTINOV, V.K.; POPOV, YU, A.; SKAVYSH, A.I. The intensity of draining fens in the state forest resource of Western Siberia. Lesn. Khoz., 1979, 3:20-2.

A summary account is given of Soviet research since 1974 to determine a sound basis for the designing of forest drainage systems suitable for western Siberia. Data are tabulated on the mean water-table levels on a drained fen over the period 1975-1977; part of the fen carried no stand and part carried a closed birch (Betula alba) stand. The distance between the main drainage channels was 100-760 m. The date when effective drainage occurred on the open fen, i.e. when the water-table was below 20 cm, was in mid or late May; no significant diffe-rences were caused by drainage channel spacings of 200, 300 or 400 m. The water-table during the growing season was always substantially lower (50-80 per cent lower) on the tree-covered fen than on the open fen.

689

KOSAREV, V.P. Effect of the soil water regime on the condition of drained spruce forests. Lesn. Khoz., 1981, 4:16-21.

Extensive mortality in high-yield Norway spruce forests in the Kaliningrad region (Soviet Baltic) has been observed in 1944-49 and since 1963-64, mostly on land that has been drained; the mortality appears to be caused by drought. The soil water regime and stand growth were investigated on 5 sample plots at various distances from the drainage ditches, in the Piceetum myritillosum and P. myrtilloso-oxalidosum forest types, in 45-yr-old pure stands of site class I and canopy density 0.82-1.35. The soils were mainly modertype medium podzolic gleyed clay-loams overlying a heavy gleyed boulder clay-loam. The investigations showed that mortality nad increased on all areas after the 1971 drought. The drainage network evidently had little effect on the moisture regime of the heavy soils during dry periods. Drainage is, therefore, recommended on these sites to increase yield, preferably by a dense network of shallow drains (0.7-0.8 m deep) arranged at an acute angle to the contour and not more than 50-80 m apart.

690

KOSTRZEWA, S. Influence of meteorological factors, soil permeability and drain spacing on outflow from drainage network. Unit outflow magnitude and their occurrence periods. Waid. Melior., 1968, 12:353-8.

The paper presents the results of investigations on outflow from drainage network, carried out in 1961-65. The outflow from drainage network occurred on the average in about 80-150 days of the hydrological year. Maximum unit outflows occurred as a rule in the spring-summer period as a con sequence of high precipitations.

691

KOSTRZEWA, S. Water outflow from a pipe drain system in flat, arable terrain at Czatkowice (Milicz District) and Nowawiess Ksiazeca (Kepno District). Zes. Nau. Akad. Rolncz. Wrocl. Melior., 1975, 115:203-14.

Most of the water output from the drainage system occurred in autumn and spring. Both monthly and yearly values for mean unit runoff were low (0.1 to 0.2 L/s/ha); maximum unit runoff was about 1 L/s/ha, with higher values than this occurring only a few times per year.

692

KOVDA, V.A. Origin of saline soils and their regime (Prasolov, L. I.; Ed.). Israel Program for Scientific Translation, Jerusalem (1946), 1971, Vol. I and II, 509p, 511p.

Origin of saline soils in Russia is described and brief review of measures for controlling soil salimity under irrigation discussed. Ground water control measures on the basis of entire oasis and farming units are presented.

693

KOVDA, V.A.; BERG, C.V.D.; HAGAN, R.M.(Eds.). Irrigation drainage and salinity. An international source book, Hutchinson/FAO/ UNESCO, London, 1973, 510 p.

The most comprehensive source book reviews the entire research work on the subject conducted in different countries of the world. The topics considered are: water and salt balances; soils in relation to salinity, irrigation and drainage; hydrophysics of arid and irrigated soils, chemistry of saline and alkali soils of arid zones; landscapes in relation to irrigation, drainage and salinity; quality of irrgation water; water, plant growth and crop irrigation requirements; plants in relation to waterlogging and salinity; irrigation systems and management; drainage systems and management; some effects of irrigation and drainage on soils; reclamation of saline and alkaline soils; design and operational recommendations for the establishment or improvement of irrigation and drainage projects.

694

KOVDA, V.A.; MKRTCHYAN, S.M. Control of the water-salt regime of irrigated soils by vertical drainage (as exemplified by Armenia). Pochvovadenine, 1976, 9:69-78.

A description is given of recommended practices for regulating the water-salt regime of irrigated areas by vertical drainage. An empirical formulae is proposed for controlling the water-salt regime and obtaining the desired desalinization, irrigation and salt regulating effects. Long-term field investigations showed that the recommended practices are effective.

695

KOYAMA, T.; MIYASAKA, A.; EGUCHI, K. Studies on water management in the poorly drained paddy field. VII. The effect of surface drainage on photosynthetic activity in the rice plant. Proc. Crop Sci. Soc. Japan, 1962, 30:143-5.

The yield was increased by surface drainage in the period from the end of the formation of fertile tillers to when the soil-moisture content decreased to 70 per cent. The increase was chiefly due to increases in ripe grains, ratio of ripe grains to total grains and 1,000 grain weight. On the drained plot, total leaf area and water content of plants were lower in the drainage stage and in the subsequent irrigating stage. Photosynthetic activity although depressed by drainage, increased after subsequent irrigation; this effect lasted until ripening and contributed to the yield increase.

696

KOYAMA, T.; MIYASAKA, A.; ITO, E. Studies on the water management of poorly drained paddy fields. VI. Effects of surface drainage on the tillering of rice plants. Proc. Crop Sci. Soc. Japan, 1960, 29:37-9.

Draining at the growth-revival stage after transplanting decreased the number of tillers; following development of the last fructiferous tillers, or at floral initiation, draining was more favourable for increasing tillers than was continuous submergence. Draining decreased the NH_4^+ content of the soil at every stage, decreased N absorption at early stages but promoted it at later stages. Tillering was affected more by soil temperature and N absorption than by water content and NH_4^+ content of the soil.

697

KRAFT, M.L.; MOLZ, F.J. Stochastic design procedures

for subsurface drainage. J. Agric. Engng. Res., 1972, 17:178-88.

A stochastic analysis and design procedure for subsurface drainage facilities is developed within the frame work of the well known ellipse equation. Hydraulic conductivity, rainfall rate and drainage modulus are treated as random variables. Analysis indicates that to a good approximation the probability density functions for these variables are log-normal. To apply the theory it is necessary to know or to estimate the mean and standard deviation for each random variable. By minimizing the total expected cost which consists of the system installation and maintenance costs and to risk cost, the writers outline an 'optimum' design procedure. The analysis procedure is applicable to subsurface drainage problems described by relationship other than the ellipse equation.

698

KREBS, R.D. Soil morphological characteristics and permeability- their implication in drainage design. Proc. Drain. Efficient Crop Production Conf. Chicago. Illinois, 1965, 37-9.

Soil morphological characteristics have a close relationship to soil permeability and thus can be used as a means of measuring the same. The use of such permeability values in the drainage design at Ohio (USA) have shown that the results are better predicted than with the hydraulic conductivity values obtained from auger hole. The observed data, however, differed greatly from the predicted results. It is suggested that diff-erences can be reduced by devising simple quantitative tests and their correlation with actual drain performance.

699

KRIZ, G.J. Drainage research and existing practices in USA. Harvester, 1969, 11:17-8.

Causes of appaling lack of utilization of drainage

research results obtained after spending huge sums are discussed. The possible reason of this lack of utilization are discussed. It is sugested that research on plant and its environment and unsaturated flow are essential to understand the drainage problems under field conditions. The addition of knowledge on these aspects is essential.

700

KRIZEK, R.J.; SORIANO, A.; GYUK, I. Unsteady flow to bottom drain in bounded aquifer. J. Irrig. Drain. Div. ASCE, 1973, 99:169-82.

Mathematical treatment of unsteady flow to a single drain is presented. Based on the solution and parameter study it is concluded that the drawdown caused is maximum above the drain and decreases to about one-third and one-tenth of its maximum value at horizontal distances from the drain equal to and double the depth of drainage. These ratios are almost independent of time, drain size, depth of drainage, coefficient of permeability and effective porosity. The velocity of drainage and the flow rate for a given aquifer are both controlled by the drain size in the same manner. A ten-fold increase in drain size will double the drainage flow rate. Other factors being the same, a deeper aquifer is associated with a greater flow rate and smaller drawdown above the drain.

701

KRPATA, F. Some problems of designing and execution of mole drainage. Vodni hospodarstvi, 1962, 12:102-5.

The paper draws the following conclusions as a result of research work on the subject: (1) the optimum content of water at the level of the ploughed mole drain can be characterized by the plasticity number and it should be nearer to the softness than to the fluidity limit, (2) the mole drainage should be carried out, when it is possible to assume rainfall not exceeding 25 mm for

24 hr,(3)there is no relation between the drainage depth and the underground-water level,(4)the distance of the collecting drains should be approximately 2 to 5 m.

702

KRUSTEVA, V. Effectiveness of mole-drainage of light-grey forest soils irrigated by gravity. Pochvoz. Agrokhimiya, 1980, 15:111-20.

Mole drainage to a depth of 40-45 cm on heavy, light grey forest soils, at intervals of 1.4 m, reduced surface runoff and erosion and improved soil permeability, resulting in doubled irrigation capacity and a corn yield increase of 118 per cent.

703

KULEN, M.; OZGUL, S. Measures taken for improving the salinity and drainage conditions in the lower Seyhan project. 5th Irrig. Pract. Sem. NESA, New Delhi, 254-68.

Lower Seyhan Plain with an area of 210,000 ha. consists of alluvial deposits with arid to semi-arid climate. The survey data are reported. Based on the data analysis it was observed that the most economical system for both irrigation and drainage tertiary canal construction is considered to be alternating irrigation and drainage canals each serving the land on both sides of them. One hundred hectare was taken as an average service area. The drainage canals are designed to accomodate surface flow and also to carry perennial subsurface flow. Secondary and main drains will be centrally located between secondary and main irrigation canals.

704

KULKARNI, D.G.; Problems of irrigated agriculture. Orient Longman, New Delhi, 1970, 73p.

Problems of irrigated agriculture are presented. Drainage for reclamation of waterlogged and saline soils is emphasized.

705

KUMAR, R.; MURTY, V.V.N. Investigating thickness of gravel-filter for tile drains. J. Agric. Engng., 1973, 10:1-8.

Laboratory studies were carried out to determine the optimum thickness of gravel filter below the tile drain. The gravel filter was specifically designed for the strata of the soil used. The variables measured were the sedimentation rate and the rate of water discharged into the drain tile. The studies indicate that with an impervious cover on the top one fourth of the drain and with base width of filter kept at 30 cm, the thickness of gravel filter below the drain should be around 7.5 cm. Visual inspection of soil filter interface was made and the mechanics of flow of water into the tile drain was observed.

706

KUNTZE, H. Effect of drainage on the structure of marsh soils. Z.Kult-Tech. Flurb., 1964, 5:129-34.

On bog soils unsuitable for utilization (<2 μ particles 45.2 per cent, 2-60μ particles 51.1 per cent) drains were installed in 1956 at 0.9 m depth and at 8 m intervals. In autumn 1960 the infiltration coefficient was 5.2 cm per day, 13.7 cm(265%) over the year, 28.6 cm(552%) over 2 years and in spring 1963, after a cold dry winter it was 76.6 cm per day (1476%).

707

KUNTZE, H. Iron clogging: diagnosis and therapy. Proc. Intern. Drain Workshop, Wageningen, Netherlands, (1978), 1979, 25:452-67.

Knowing the causes of chemical and biological iron

clogging and the sites where this may be expected are important for counter-measures. Both can be determined by an orohydrographic analysis of the site and field or laboratory tests on Fe^{++} content and pH in water and soil. Within a 20 year research programme it was found that Allochthonic clogging (alien origin of iron) is continuous problem. Counter-measures therefore show only short-term effects. Continuous maintenance (mechanical and chemical cleaning) remains necessary Autochthonic clogging (indigenous origin of iron)is only a temporary problem. It depends on soil ripening with time of drainage (3 to 4 years). The iron oxidation within the soil can be enhanced by subsoiling and liming. If iron clogging problems are expected, filters should not be used, except special auto-filters, and with submerged drain outlets. Plastic pipes develop adhesion forces for iron hydroxides. Drain slopes (>0.4per cent) are necessary to overcome these adhesion forces by tractive forces in partially filled drain pipes. The inlet openings should be as large as possible (>1.2 mm).

708

KUNTZE, H. Iron clogging in soils and pipes: analysis and treatment.**Bull. German Assoc. Water Res. Land Improv.**, 1982, 10,147p.

The cause of clogging in time and the local differences in iron clogging can be interpreted in terms of soil genetics and hydrology. The degree of clogging can be derived from pedological and hydrological studies of Fe^{++}content and pH. The precautions to prevent this depend on degree and nature of the clogging. For predominantly autochthonous clogging, the oxidation of iron should occur as much as possible before the gravitational water enters the drain. Useful additional drainage procedures includes aerating the soil by subsoil loosening. For predominantly allochthonous iron clogging and high iron content in the gravitational water, there are advantage in using specific materials (plastic drains),submerged outlets and special filters. The measures adopted are not completely successful, and subsequent cleaning procedures will be required. Individual drains with open collectors are recommended. Flushing shafts should be incorporated in drainage systems. High pressure flushing is a particularly important technique.

709

KUNTZE, H. Progress in drainage. ICID Bull., 1974, 23:24-6.

Drainage problems and their solutions in Federal Republic of Germany are discussed. The life of a drainage system depends upon the design and installation techniques as it is reported that whereas some systems are functioning for the last 100 years or more, some have deteriorated in 1-2 years. Secondary sedimentation and ochre clogging is presented. It is suggested that organic filter materials can delay ochre formation.

710

KUNTZE, H.; WETJEN, F.A. Subsurface drainage of a heavy alluvial soil. 4th. Natn. Drain. Symp. Proc. ASAE, 1982, Pub. 12-82:156-62.

Subsurface drainage experiment on an alluvial soil: silty loam (0-40 cm), silty clay loam (subsoil) and deeper parts silty clay. The results indicated that the closer the collector drain spacing, the longer the groundwater level remains deep enough for arable conditions. The collector drains spaced closer together have a higher rate of discharge. Both lowering of the groundwater and an increase in the rate of discharge increase grain yield. A collector drain spacing of 16 m or 32 m with mole drains spaced 2 m perpendicular to the collector drains is suitable for drainage of a heavy alluvial soil. There was no significant difference on the rate of discharge between backfill materials of dry soil, gravel and a spacial PVC-spun-yarn. Even if the mole drain disappears within a few years the loose structure around its installation keeps the permeability high and supports the drainage of the field.

711

KURYLEVA, N.I. The water and salt regime in soils of irrigated territories and its regulation. Tr. bukhar. Oblast' opyt. S-kh. Stants, 1962, 1:5-57.

The water consumption of cotton on irrigated serozems was 100 m^3/ha at the start of vegetative growth, 30-60 at time of budding and flowering and 80-100 in the period of mass flowering and fruiting. For groundwater depth of 1 m, water consumption in the field during vegetative growth averaged 10,665 m^3/ha, 48.5 per cent coming from the groundwater; the yield of cotton wool was 24-64 g per bush. For groundwater depth of 2 m the respective values were 8,804 m^3/ha, 16.1per cent and 55.6 g and for groundwater depth of 3 m, 7,800 m^3/ha, 2 per cent and 56.84 g. Cotton yields were lowest and salinization intensive with groundwater level of 1 m; over 3-4 years the soil was converted into solonchak. Yields of lucerne straw over 3 years averaged 13.4, 20.3 and 17.5 t/ha respectively, with groundwater levels of 1m, 2m and 3 m. Mineralisation of the groundwater was 5-7 g/L. The seasonal fluctuation of groundwater level on irrigated soils was 1.5-4 cm per day. The main source of salts was the surface waters. For preventing salinization, increasing drainage and lowering the drain depth to 2-2.5 m is recommended; this lowered the groundwater level from 1 m to 1.6-1.8 m with seasonal amplitude of 50-100 cm. Under these conditions cotton yields increased to 3.5-4 t/ha.

712

KUSSE, A.D. Nomograph for determining the diameter of drainage tiles. Neth. J.Agric. Sci., 1964, 12:113-22.

Based on the flow characteristics and total discharge expected from a drainage system, nomograph is presented to calculate the tile size depending upon the area to be drained.

713

KUZNETS, J.M. Drainage - irrigation systems in the Ukrainian Soviet Socialist Republic. 5th Congr. Irrig. Drain. Tokyo, 1963, Q 17 R 17:255-66.

Drainage-irrigation system also termed as double purpose reclamation system is suggested for soils of USSR. Correct observance of operational hydrometery of the system and water balance forecasting are discussed in relation to proper functioning of a large double purpose reclamation system.

714

LAGACE, R. Methods to evaluate drainable pore space. Can. Agric. Engng., 1981, 23:85-90.

After the hydraulic conductivity, the drainable porosity is the most important soil property affecting subsurface drainage under unsteady conditions. For an existing subsurface drainage system, the drainable porosity was determined using four methods viz. (1) Analysis of regression coefficients for the water table drawdown and discharge curves, (2) total water balance (out flow and soil volume drained, (3) soil moisture profile changes and (4) study of macroporosity using Vergie're soil samples. The accuracy of the first two methods is satisfactory, while that of the other two is poor. The first two are useful for experimental work but none are useful for the design of a subsurface drainage system.

715

LAGACE, R.; SKAGGS, R.W.; PARSONS, J.E. Predicting water table drawdown for two dimensional drainage. 4th Natn. Drain. Symp. Proc. ASAE, 1982, Pub. 12-82:6-15.

The product of two one-dimensional solutions were

used to approximate two-dimensional solutions for drainage problems described by the non-linear Boussinesq equation. Two cases were evaluted. The first case involved perallel drains which intersect a main (grid iron). The correlation between the product of one-dimensional solutions and the two-dimensional solution was 0.98. The maximum errors in the product solution ranged from 1.5 per cent to 4.8 per cent with an estimated error of 0.5 per cent due to the numerical procedure. The second case tested involved a square grid of drains. For this case, the maximum error ranged from 2.7 per cent to 7 per cent. In both cases, the error was within a tolerable range for engineering design. The one-dimensional solutions may be obtained from graphical plots or approximate equations. Therefore, the product solutions are much easier to compute and apply than numerical solutions to the two-dimensional equations.

716

LAGOCKI, H.F.R. Surface soil stability as defined and controlled by a drainage criterion. In Modification of Soil Structure (Emerson, W.W.; BOND, R.D.;DEXTER, A.R. Eds.), John Wiley & Sons, UK, 1978, 233-7.

The effects of sheep trampling on a pastured surface soil have been measured in relation to depth to the water table in the range 0-50 cm. Changes in the bulk density and penetration resistance of the top 0-6 cm of soil and changes in the volumetric water retention at 100 cm suction (available water holding capacity) have been determined. Provided the depth to the water table was 40 cm or more, trampling compacted the soil and increased its bearing capacity. This was accompanied by an increased water retention capacity of the compacted 0-6 cm layer. A water table level of 40 cm is recommended as an acceptable drainage criterion for this soil.

717

LARSON, C.L.; MANBECK, D.M. Tile flow and power use data for pump drainage systems. Trans. ASAE, 1962, 5:207-9.

For surface and subsurface drainage of mineral soil, it was found that a five year recurrence interval gave a drainage co-efficient of 1.9 cm/day. For organic soil with surface and subsurface drainage, the coresponding value was slightly lower. In both cases tile spacing was primarily 30 m. For organic soil and subsurface drainage, with a predominant tile spacing of 60 m, the five-year rate of flow was between 1.25 and 1.6 cm/day. Conclusions on annual power use show that it varies considerably depending upon several factors. The annual power use was about 42.5 kwh/ha for the surface and subsurface drainage systems and about 65.0 kwh/ha for those with subsurface drainage only.

718

LASSOUDIERE,A.; MARTIN, P. Drainage problems in the organic soils of banana plantations in Agoeby, Ivery Coast. Fruits, 1974, 29:255-66.

Two drainage systems were compared in peaty marsh soils. In the first, the soil was drained to a depth of 80 cm with ditches spaced at 10 m and the water level was controlled by pumping (polder system). In the second method, commonly used in the area, the drainage depth was 55 cm, ditches were spaced at 20 m and excess water was not pumped away. The growth and yield of bananas, and particularly the size and weight of the root system, were much better under the first method than the second.

719

LEE, S.Y.; CHOI D.H.; KWON, T.O.; SO, J.D.;PARK,N.P.; HAM, Y.S. The effects of mole drainage and deep ploughing with heavy fertilizer and soil amendment applications to rice on degraded fluvio-marine deposit soils. Pl. Prot. Micology, 1979, 21:39-55.

In field trials in 1972-6, rice was grown with flooding, various drainage treatments, deep ploughing, application of 10 or 20 t compost/ha or 3 t silica/ha and application of normal or high rates of fertilizer. With mole drainage root growth was increased and the root distribution in the subsoil increased. Water percolation to a depth of 6 cm increased and so the leaf chlorophyll content. With mole drainage the contents of K, Ca, Si and Mn in leaves and of N, K, Ca and Si in the leaf sheath + stem were high at heading. Mole drainage increased the ripening percentage. Application of high rates of compost and silica with mole drainage increased soil NH_4-N and OM contents and high NH_4-N content led to increases in CGR and RGR. With deep ploughing, high rates of fertilizer and application of compost, yields were low in the 1st yr but increased markedly in the 2nd yr.

720

LEEDS-HARRISON, P.B.; FRY, R.K.; CRONIN, C.J.; GREGORY, J.E. A technique for the non-destructive monitoring of subsurface drains. J.Agric. Engng. Res., 1983, 28:479-84.

A simple borescope technique for the non-destructive monitoring of mole channels and pipe drains is described. Requirements and specifications for suitable borescope and ancillary equipment are discussed as well as practical considerations for field use. The fully portable equipment is shown to be successful in both long and short term evaluation of subsurface drain performance.

721

LEES, J.C. Soil aeration response to draining intensity in basin peat. Forestry, 1972, 45:135-43.

Improvement in soil aeration was assessed following draining to a range of intensities in a raised bog consisting of lightly humified, fibrous Sphagnum/Eriophorum peat situated at Flanders Moss in central Scotland. A polarographic technique was used to measure O_2 diffusion rates in the rooting zone of lodegepole pine (Pinus contorta Dougl. var. latifolia Engelm), planted in 1965 on ploughed furrows. Soil aeration differences were related to draining intensity. Below the planting turf, O_2 diffusion rates decreased rapidly and tree rooting was restricted to the surface layers. Aeration was closely correlated with depth of water in boreholes, which provide a simple and reliable index of response to drainage. O_2 diffusion rate was a more sensitive means of assessing soil aeration than oxidation-reduction potential at the low O_2 concentrations encountered in the basin peat on the Moss.

722

LELIJ, A. VAN DER. Artificial subsoil drainage in New South Wales. In Hydrogeology of the Riverine Plain of South East Australia. Proc. Symp. Griffith, Australia, 1977 (Storrier, R.R.; Kelly, I.D. Eds.), Aust. Soc. Soil. Sci. Inc., 1978, 209-28.

The design criteria for tile drainage depends on the need of horticultural plantings, which appear to require watertables deeper than 1.2 m for average conditions. Design criteria ten-tatively determined from steady state formulas and local conditions control watertable depths to these levels, except during infrequent periods of excessive rainfall or, more frequently, excessive irrigation. This is confirmed by use of non-steady state equations to fit a model of simulated watertable levels to observed values and subsequent extrapolation to more severe conditions, as occurred in 1956. The spatial variability in hydraulic conductivity, the infiltration characteristrics and drainable porosity make the use of a safety margin in tile drainage design imperative. Measured hydraulic conductivity, which shows skew distribution, is not closely related to soil type. The proportion of total area in which drainage rates may be less than desirable may be assessed using the calculated variaance. Tubewell drainage has successfully been applied and costs of installation reduced by the introduction of PVC screens of depths of 16 m. Pumping from deeper aquifers frequently does not successfully control watertable levels because of vertical flow resistance through overburden materials.

723

LEMBKE, W.D. Observed and predicted tile flow on a lake plain soil. Trans. ASAE, 1967, 10:142-4.

Tile outflow measurements on a stratified silt loam lake plain soil are presented. It was found that actual tile outflow with a ponded layer on the soil surface was greater than that predicted by both augerholes and soil cores. Augerhole measurements, while being conservative, are recommended for measuring hydraulic conductivity in such a soil. It was also found that a tile drain embedded in a gravel filter under a ponded condition increased inflow rate during the first two years after construction. This increase was attributed to removal and re-arrangement of fine particles in and around the gravel filter.

724

LEMBKE, W.D.; DELLEUR, J.W.; MONKE, E.J. Model study of flow in steep tile drains. Trans. ASAE, 1963, 6:142-4.

Gravity flow on steep slopes was studied in 118 m model drains. The model drains were tested on slopes from one to six percent with diameters of 7.5, 10, and 12.5 cm with two degrees of roughness and with discharge rates upto pressure flow. Five different flow regimes were observed to occur at the transition from part full to full flow. Flow regimes I through IV occurred while the flow in the drain was being obstructed at the outlet. Flow regime V occurred with a free outlet. Flow regime III was found to reduce the capacity of the drain.

The range of flow rates at which this regime would occur is narrow in field drains because of high roughness. Application of flow regime V data is limited due to the difference between the model and a field drain, but high negative pressures could develop in a field tile when the flow rate becomes this large. A table of flow rates at the beginning of flow regime V is presented for common sizes and slopes of the mains.

725

LEMBKE, W.D.; DRABLOS, C.J.W.; ARNOLD, J.G.; SCARBOROUGH, J.N. A model for drainage benefits. Trans. ASAE, 1982, 25:1329-32.

A regression model for corn yield was developed which includes rainfall during two critical periods of the growing season: (a) the time during which spring tillage and planting operations occur, (b) the time during pollination. The regression model was applied to three widely different situations: (a) eight counties in Illinois that could be grouped according to soil associations with relatively similar drainage problems, (b) three University of Illinois farms (Allerton Farms) in close proximity but with different levels of drainage, (c) drainage research plots at the University of Illinois Brownstown Research Centre. Rainfall during period 1 was found to significantly affect yields on farms with inadequate drainage systems. The model has potetial to be useful in explaining yield differences between well drained and poorly drained farms.

726

LERCH, K.E. Maintenance of open drains in the Sacramento Valley. In Water Management for Irrigation and Drainage. Proc. Spec. Conf. ASCE Irrig. Drain. Div. ASCE, New York, 1977, 1:126-38.

The paper highlights many of the problems encountered in the maintenance of open drains in Sacramento Valley. Rights of way, availability of technical manpower, rapid drawdown in the drains, enlarging ditch controls, deepening ditches, drain en-croachment, improperly placed drop structures, rodents. urban area trash, siltation from sub-drains, weed control, portable pumps, pipe crossings and appretances, ditch bank stabilization are among the problems discussed in this paper.

727

LETEY, J.; BLAIR, J.W.; DEVITT, D.; LUND, L.J.; NASH, P. Nitrate-nitrogen in effluent from agricultural tile drains in California. Hilgardia, 1977, 45:289-319.

The first research phase consisted of extensive monitoring of tile drainage effluents collected in important agricultural areas in California. More intensive investigation of nitrate-N within a few soil profiles was done during the second phase. Nitrate-N concentrations were not well correlated with any measured parameter. The amount of nitrate-N contained in tile effluent was quite well correlated with total water discharge and fertilizer N application. Higher amounts of nitrate-N were removed from coarse-textured soil profiles as compared to profiles containing a layer of fine-textured material. Some alluvial soils on the west side of the San Joaquin Valley apparently contain significant native N which contributes to high nitrate-N concentrations and total mass emissions in tile effluent. It appears that the most positive approach to controll-ing nitrate-N in irrigated agriculture is through control of leaching.

728

LETEY, J. Jr.; COPPOCK, R.H. Monitoring salt levels in farmland drainage. Calif. Agric., 1977, 31:26.

In monitoring wastewater discharge for salts as a measurement of salt management, two parameters must be considered. If the primary consideration is the environmental impact of the discharge, the mass emission (total weight of pollutants) is the more appropriate parameter. If the purpose is to develop irrigation management, a measurement of salt concentration alone may be adequate.

729

LEVESQUE, M.; HAMILTON, H.A. The efficiency of a subsurface drainage system with porous filling material in a Kapuskasing clay soil. Can. J. Soil Sci., 1967, 47:55-65.

Trenches 45-cm wide were dug above existing drains and filled with either gravel or peat in an attempt to facilitate water movement towards the drains. Data on soil moisture content, water-table fluctuations and grass yield show that little advantage was gained from the trenches, except for some edge effects independent of the filling material.

730

LEVIN, M.G. Two-way regulation of the air and moisture regime of drained peat bogs. Vestn. S-kh. Nauki, Moscow, 1974, 5:80-7.

Excessively prolonged wetting of the rooting zone during sub-irrigation of peat bogs is obviated by supplying water under a head of 2.5 m along drains 4-8 m apart and by inserting in the soil vertical columns of sand spaced at intervals of 4 m and extending up to the ploughed layer. With this system, water needed to be supplied for not more than one to two days. Calculations suggest that a pressure-vacum system would decrease the total time required for the supply and discharge of water to three to four days and, in the top layer of soil, to not more than two days.

731

LEVIN, Ya, K. Investigation of flow of water into a horizontal drain when saline soils are washed. Dokl. Veses. Akad. S-kh. Nauk., 1969, 11:39-42.

In a laboratory model of a horizontal open drain mineralized groundwater entered the adjacent drain, but over most of the inter-drain area the leached salts moved with the groundwater into deeper layers and did not enter the drain.

732

LEVY, G. Effect of different drainage systems on the initial growth and regrowth of resinous species on pseudogley. In "Pseudogley and Gley". Trans. Comm. V and VI Int. Soc. Soil Sci., 1973, 763-8.

Various drainage systems (ditches at intervals of 10 m (a), 20 m (b) and 40 m (c) or ridges) were tested on (A) a pseudogley on Quaternary alluvial formations with the impervious layer at 50 cm, and (B) a pelosol-pseudogley on Keuper Marl with the impervious layer at 30 cm. On (A), methods (a) and (b) markedly lowered the water-table and, in comparison with (c), increased the height of Sitka spruce, Douglas fir and larch by 33 per cent, 48 per cent and 63 per cent, respectively, as well as the rate of regrowth of Douglas fir and Larch. Ridges were the best treatment for Chamaecyparis, but Weymouth pine showed no response to drainage. On (B), only ridges provided any noticeable drainage; they were particularly efficient for Douglas fir, decreasing its mortality from 40 per cent to 6 per cent and increasing its height by 53 per cent.

733

LEYSHON, A.J.; SHEARD, R.W. Growth and yield of barley in flooded soil; ethylene generation and Eh relationships. Can. J. Soil Sci., 1978, 58:347-55.

Barley growing on a Gleysolic soil was subjected to short-term periods of flooding (2-10 days) at four stages of development. The treatment resulted in a reduction in dry matter accumula-tion proportional to the duration of flooding. Plants 14 and 21 days at flooding were most affected immediately following inundation; at maturity, plants flooded at 28 and 35 days were the most affected in that head emergence was delayed and grain yield was reduced by as much as 55 per cent. Flooding increased the soil-produced ethylene from 0.15 to 3.78 ppm after 10 days. Ethylene production was related to a lowering of the soil Eh below 340 mv. The poor growth of barley, while in part due to oxygen deficiency, may also be a result of ethylene formation.

734

LEYTON, L.; YADAV, J.S.P. Effect of drainage on certain physical properties of a heavy clay soil. J. Soil Sci., 1960, 11:305-12.

Comparison of the physical properties of a drained and an undrained heavy clay soil revealed a higher content of water-stable aggregates, higher proportion of larger pore spaces and greater vertical and lateral hydraulic conductivity in the former. The difference generally decreased with depth; at about 38 cm there was generally little difference between the two soils. Despite possible initial differences in texture, the difference can be interpreted as being largely due to the effect of drainage on aeration.

735

LIDSTER, W.A. Investigating new equipment and materials to reduce drainage cost. Proc. Natn. Drain. Symp., Illinois, 1971, 43-6.

Drainage cost reduction through proper selection of material and equipment for installation is vividly described in this paper. Plastic tubing with automatic continuous installation machinery is suggested as the best way to reduce costs.

736

LIDSTER, W.A.; FORD, H.W. Rehabilitation of ochre-(iron) clogged agricultural drains. 11th Congr. Irrig. Drain. New Delhi, 1981, 1:451-63.

There are few satisfactory tecniques for predicting ochreous sites in advance of drainage or for removal of ochre from clogged drains. Sulfur dioxide gas is effective as a cleaning agent but it can be dangerous to both personnel and the environment. Mechanical Roto-rooters and hydraulic high-pressure jet cleaners are not entirely effective in removing more mature ochre deposits.

7

GON, J.T.; KIRKHAM, D.; JOHNSON, H.P. The falling ter table between open ditch drains. Soil Sci., 64, 97:113-8.

Equations were derived to predict the position of a falling water table between equally-spaced ditch drains partially filled with water. The derivations were based on an equation giving the shape of the water table due to steady uniformly-distributed rainfall. Comparison of the falling water table in glass-bead-glycerol drainage model with a series of water tables calculated from the theory showed that the theory was valid to within a small error, but caution should be used in applying the equation where much of the total flow into ditches is along a surface of seepage.

8

NDNER, H.; SCHAFFER, E. Effect of drain pipes with fferent parameters in a fine-grained sandy soil. ch. Acker. Pflan. Bodenk., 1980, 24:577-83.

In a fine sandy soil, comparative tests were made with: (1) corrugated PVC pipes installed without envelope material and having slots of 0.5 and 1.0 mm in length 1.0 and 2.0 mm in width: (2) polyethylene pipes ((PE pipes) covered with envelope material and with round openings of 3 mm dia. Drain pipes with slots of 0.5 mm in width showed the lowest discharge rate under equal groundwater pressure head at the drain. Slots on the crest of the drain pipes were more effective than slots in the trough. PE pipes and PVC pipe with slots of 2 mm in width showed almost equal high discharge rates under equal ground-wate pressure head at the drain in the second yea after installation. Corrugated pipes without en velope material and having slots of 1 mm in wid showed little entry resistance and high dischar rates occurred in the first year afte installation; they also showed the highest degre of siltation. Glass fibre wool (60g/m^2) used as envelope material reduced siltation of the pipe to an acceptable level.

739

LINDNER, H.; SCHAFFER, E. Prevention of drain silt tion in a medium to fine sandy soil. Arch. Acke Pflan.Bodenk., 1981, 25:727-33.

No significant difference in the degree of silt tion was observed between corrugated drain pip (63/56) covered with glass fibre fleece (60 g/m^2 in which the pipes have slits with 1.0 and 2.0 m wide openings in addition to round openings of 3 mm diam, and corrugated drain pipes without e velope material in which the pipes had slits 0.5 mm width. The latter, however, showed great increased entry resistance as the slit openin became reduced due to the presence of Fe^{+2}, H and $CaCl_2$ in the drainage water. Pipes with sli of 1.0 mm width were liable to a high degree siltation unless installed with an envelo material.

740

LINDNER, H.; SCHAFFER, E. Test results for drainpip of varying construction and for envelope materials an experimental drain box. Arch. Acker. Pfla Bodenk., 1979, 23:155-61.

Discharge rates, both as L/m and mL/cm^2 of tot entry area, were greatest for polyethylene pip having openings 3mm in diam and total entry ar of 9.3 cm^2/m, installed with envelope materia

These were followed by 63/56 corrugated PVC (hard pipes with 2 x 1 mm slots, enveloped with glass fibre fleece. Corrugated pipes with slots of 0.5 mm width on the crest gave a better discharge performance at a smaller total entry area with less siltation, compared to similar pipes with slots in the trough. Filter material was not required for slots 0.5 mm in width. Staple glass fibre was superior to glass fibre fleece for slotted pipes.

41

IST, E.J. The steady flow of precipitation to an ininite series of tile drains above an impervious ayer. J. Geophys. Res., 1964, 69:71-81.

A solution to the problem is presented. The free-surface boundary conditions are exactly satisfied, but the shape of the impervious layer is approximated. Capillary forces are neglected. The theoretical results are compared with limited field measurements, and agreement is good. The solution becomes less accurate when the drains are very close to the impervious layer. The limiting cases are shown.

42

OGAN, T.J.; SCHWAB, G.O. Nutrient and sediment haracteristics of tile effluent in Ohio. J. Soil Wat. onserv., 1976, 31:24-7.

Tile effluent from three small drainage systems (2.8 to 4.1 ha) in central Ohio are studied. Delinger and Durban farm sites were cropped to corn and received nominal amounts of chemical fertilizer. The Wedding site was in continuous alfalfa and previously received applications of dairy manure. Average nitrogen losses were 41, 21 and 0.67 kg per ha per year for the Delinger, Durban, and Wedding sites, respectively. Nitrogen concentrations tended to decrease with time during each growing season. A preplant application of 224 kg of nitrogen per ha apparently resulted in increased nitrogen concentrations on the Delinger site in 1973. Tile effluent from the alfalfa sit had consistently low nitrogen concentrations. Mos of the nitrogen lost was in the nitrate form Phosphorus losses were 0.33, 0.33 and 0.1, potas sium losses were 1.8, 3.5 and 1.2 while sedimen losses averaged 375, 52.6 and 165.8 kg per ha pe year for the three sites.

743

LONG, F.W. Drainage Design for irrigated lands. Proc Drain. Efficient Crop Production Conf, Chicago, Il linois, 1965, 71-2.

This paper briefly discusses differences in soi and climatic conditions in irrigated areas of th western states as opposed to non-irrigated areas It points out that these variations dictate dif ferences in drainage design criteria for irrigate areas.

744

LOPEZ, M. Jr. Drainage reduction program in th Wellton-Mohawk irrigation and drainage district. 10t Congr. Irrig. Drain. Athens, 1978, 4:487-510.

The U.S.A. is designing a desalting complex to im prove the quality of the drainage from the 30,35 ha Wellton Mohawk-Irrigation and Drainage Dis trict. To reduce the size of the desalting plant an extensive program of drainage reduction i being undertaken which includes retiring 4050 h with low irrigation effiency, improving irrigatic efficiency, on-farm systems improvement, and a educational program. Drainage is accomplishe through 109 drainage wells which can be in dividually controlled. Graphic scheduling, using neutron moisture probe, has greatly improved ir rigation scheduling on 7294 ha. Concrete-lined ir rigation ditches, land levelling, water contro and measurement structures, and pressure irriga tion systems have been installed on 2055 ha. mathematical model has been developed to predic future quantity and quality of the drainage fo various cropping patterns and levels of irrigation efficiency.

745

LOVAS, L. Salinity control trials by sub-soil drainage at Digod. Proc. All India Symp. Soil Salinity, Kanpur, 1971, 156-60.

Field trials were conducted at Digod, a village in right main canal of Chambal Irrigation Project, as a pilot project in 40 ha area. Soils are dark grey silty clay, light grey silty clay,yellow silty sandy clay type having pH_2 7.7-9.5 and EC_2 0.85-4.90 dS/m. The groundwater table was at 2-2.5 m in dry season and at surface during rains. EC of the groundwater was 6.65-13.11 dS/m. Treatments consisted of sub-soiling, mole drains, plastic pipe drain, A.C. pipe drains. Data collected during first three months (monsoon time) period showed that each system operated properly. EC of irrigation/rain, shallow surface drain and sub-soil drain water were 0.2-0.3, 0.3-1.2 and 3.0-9.5 dS/m respectively. Guidelines have been given for the construction and operation of different systems.

746

LOWERY, B.; KLING. G.F.; VOMOCIL. J.A. Overland flow from sloping land: Effects of perched water table and subsurface drains. Soil Sci. Soc. Am.J. 1982, 46:93-9.

Rainfall, runoff events, depth to water-table, piezometric head and matric suction data were collected from two sub watersheds in Willamette valley for 2 years. Following this in a sub-watershed of 1.4 ha subsurface drains (95 cm deep and 12.2 m intervals) were installed to determine the feasibility of this management practice for reducing runoff. Subsurface drains eliminated seepage areas and reduced the occurrence of an ephemeral base-parched water table. They also eliminated base-flow from the watershed hydrograph, overland flow continued to occur following rare storm events with high rainfall intensity or extremely long durations.

747

LUBBE, E. Situation report on land drainage in Germany. In Land Drainage (Gardiner, M.J. Ed.) A.A. Balkema, Rotterdam, Netherlands, 1982, 269-80.

Future expenditure on drainage in the German Federal Republic is likely to be limited to the maintenance and preservation of existing system rather than on major land reclamation projects.

748

LUCAS, P.E. Subsurface drainage: A comparison between field evaluation and estimated properties for fourteen soils in the midwest with restrictive layers. 4th Natn. Drain Symp., Proc. ASAE, 1982, Pub. 12-82,149-55.

A reasonable value of drain spacing generally appears to be obtainable by the Hooghoudt equation and the estimated soil permeability values as shown on the SCS soil interpretation records. If the soil interpretation records values would have been used for four of the fourteen soils, subsurface drainage design estimates would have been acceptable. Estimates of drain spacing for eight soils would have been overdesigned (spacing too narrow) and would have been underdesigned (spacing too wide) for two soils. A refined drainage performance value is obtained by field performance "drainability" studies. Only through such studies can it be determined whether the actual performance of a soil is at the low or the high end of (or above or below)values estimated by using drainage equations and estimated permeability range.

749

LUKEHURST, C.T. Land drainage and some implications for farm incomes. In Land Drainage (Gardiner, M.J.,

Ed.) A.A. Balkema, Rotterdam, Netherlands, 1982, 151-63.

Data for wheat production on the Stour Marshes in East Kent show net income per ha (1980 prices) to be $993.3 with tile drains compared with $ 197.7 without tile drains.

750

LUNDEGREN, J. Technical improvements desired in Swedish land drainage. Proc. Int. Drain. Workshop, Wageningen, Netherlands, (1978), 1979, 25: 471-6.

With an estimated 618500 ha area requiring drainage in Sweden, discussion is made of present day drainage techniques and desired improvements in these techniques.

751

LUNGU, I. Some aspects to be considered when ploughing soil containing excess moisture. Produc. Vegetala, Cer. Plante. Tecnice, 1975, 27:3-6.

In trials at 3 sites in 1958-74, breaking up the soil to a depth of 55-80 cm at intervals of 140 cm increased maize yields by 1.14-2.05 t grain/ha. Cultivation in bands 16-24 m wide formed by mouldboard ploughing increased wheat yields by 190-170 kg grain/ha. Installation of drains 0.5 m deep and 1.5 m apart increased wheat yields by 340-410 kg/ha.

752

LUPA, H. Analysis of systematic mineral soil drainage costs. Wiad. Melior., 1967, 11:247-51.

The paper examines the factors influencing the total drainage costs; draws the conclusion that the most expensive process is the manufacturing of drain pipes.

753

LUPA, H. Method of determination of economically optimal drain spacing in systematic drainage. <u>7th Congr. Irrig. Drain Mexico</u>, 1969, Q 25 R 10: 123-33.

The determination of optimum drain spacing should not only result in conditions leading to maximum crops, but also take into account operation and investment costs. The analysis carried out on investment costs and net income rate (considering exploitation costs) as the effect of drainage depending on drain spacing, permitted to estimate the time coefficient of reimbursement of investment costs. By diminishing this coefficient, the magnitude of optimal drain spacing was obtained.

754

LUPA, H. Nomogram for the hydraulic dimensioning of plastic drainage. <u>Gosp. Wod.</u>, 1969, 5:187-8.

For lack of indications of how to determine appropriate parameters, plastic has not been very widely used in drainage. The mathematical relations among the main parameters of the plastic drainage have been laid down. A specific nomogram allows to determinate the dimensions of plastic drainage upto 40 cm dia.

755

LUPA, H. Theoretical relationships between drainage system elements and execution costs. <u>Zesz. Nauk. Politech. Szczec.</u>, 1966, 69:7-16.

The paper discuses the mathematical relationships between spacing, depth, and costs of drainage at the permanent optimal mean drainage depth, carries out detailed computations for three soil types i.e. light, medium and heavy ones.

756

LUPE, I.Z.; STRIMBEI, M.; DONCA, V. Effect of drainage

of pseudogley soils on growth of wood species. Trans. 8th Intern. Congr. Soil Sci, 1964, 2:537-48.

Drainage had little effect on establishment of oaks and poplars but significantly increased their radial growth. In the undergrowth, hydrophilous species on undrained sites were replaced by mesophilous or mesohydrophilous plants in the drained area.

757

LUTHIN, J.N. Drainage analogues. In Drainage for Agriculture (Agronomy No. 17; Schilfgaarde, J.Van., Ed.). Amer. Soc. Agron. Wisconsin, 1974, 517-36.

The theory, construction and use of electric and membrane analogues, resistance and resistance-capacitance network are discussed.

758

LUTHIN, J.N. Drainage Engineering, Wiley & Sons, Inc. New York, London, Sydney, 1966, 250 p.

The book arranges the subject-matter into 14 chapters as: (1) Introduction, (2) Hydraulics and measurement of water, (3) Rainfall and runoff, (4) Soils, (5) Determining nature and extent of a subsurface drainage problem, (6) Statistics, (7) Dynamics of soil water, (8) Fundamentals of seepage analysis, (9) Measurement of soil permeability, (10) Depth and spacing of drains, (11) Subsurface drains, (12) Open ditch drains for controlling the water table, (13) Drainage wells, (14) Surface drainage tidal reclamation.

759

LUTHIN, J.N. The drainage of sloping land. Rapp. 6th Congr. CIGR, 1964, 2:169-75.

The paper presents a theoretical solution of the equation for the case of equally spaced ditch drains on a hillside, draws a graph for calculating ditch spacing for a given rate of replenishment; reports that the solution by Hele-Shaw on viscous flow model compares well with authors theoretical results.

760

LUTHIN, J.N.; FERNANDEZ, P.; MASLOV, B.; WOERNER, J.; ROBINSON, F. Displaceament front under ponded leaching. J. Irrig. Drain. Div. ASCE, 1969, 117-25.

The paper examines the leaching of tile drained soils by ponding water on the soil surface both experimentally and theoretically; presents a method of analyzing the movement of the displacement front (demarcation between saline soil water and fresh saline soil water by the added leaching water; ignore in this initial analysis the presence of precipitated salts, hydrostatic dispersion and soil-chemical reactions; shows satisfactory agreement of the theoretical results obtained by solving Kirkham's equation with the experimental values obtained on a tank; indicates that ponded leaching is inefficient in removing salts midway between the drains, intermittent leaching by applying excessive amounts of irrigation water should be more efficient in controlling the soil salinity.

761

LUTHIN, J.N.; GUITJENS, J.C. Transient solutions for drainage of sloping land. J. Irrig. Drain. Div. ASCE, 1967, 93:43-51.

It has been shown that the laws of similitude between the model (Hele-Shaw) and the prototype hold for the transient case as well as the steady state case. It presents a series of solutions for the configuration of the water table for lands with slope varying from 0 to 70 per cent and making an assumption that the drainage ditches penetrate to an impermeable layer, and ignoring capillary flow in the region above the water table.

762

LUTHIN, J.N.; HAIG, A. Some factors affecting flow into drain pipes. Hilgardia, 1972, 10:235-45.

A sand tank model experiment was conducted to find out the effect of flow into drains as affected by drain diameter, pipe segment length (no envelope), pipe segment length (with enve-lope), placement of a single line of holes on top or bottom of drains, water table height, water level in drain. The results are presented in tabular or graphical forms.

763

LUTHIN, J.N.; ROBINSON, F. Depth of drainage in irrigated areas. Trans. ASAE, 1969, 12:27-8, 31.

The field measurement on water tables and soil salinity at University of California indicated that depth of drainage, apart from capillary rise is governed by factors such as infiltration rate, soil stratification, salinity of irrigation water, cropping pattern, irrigation practices and hydraulic conductivity.

764

LUTZ, J.F. The interrelationships of soil physical conditions and drainage. N.C. Agric. Exp. Sta. Tech. Bull, 1960,139,27 p.

Bulk density, total porosity, pore-size distribution and percolation rates were determined on periodically-taken topsoil and subsoil samples from a variously managed silt loam drained with differently spaced tiles 15-30 m at 60-120 cm depth and with blasted-and V-ditches spaced 15-18 to 60-66 m. Data showed a high degree of correlation between percolation rates and percentage of large pores, and a significant inverse correlation between percolation and bulk density; these measurements serving as a good indicator of the ease of drainage. Cultivated plots had a higher percentage of large pores than had plots with no cultivation. There was evidence that tiles at 60-90 cm depths provided a more satisfactory drainage than at greater depth, especially in soils with impermeable and sticky subsoil.

765

LYON, T.; WERENFELS, L.; HOUSTON, C. Filter envelopes aid tile drainage in Sacramento-San Joaquin Delta test. Calif. Agric., 1964, 18:14-6.

The paper states that drain failure in the Delta resulted from sealing by muck soil, which reduced water movement into the tile, describes tests carried out on filter enveloping materials. It finds out from results that, (1) enveloping perforated bitumanized fibre and plastic sub-surface drain in a safflower straw sandwich gave excellent performance, and (2) 2.5 cm glass fibre mat also gave good performance, but tile drains tested without a protecting filter envelope were not successful.

766

LYU CH. T.; CHZHAN KH. CH. Saline soils of Funshu in northern Hunan and their amelioration. Acta. Pedol. Sin., 1965, 13:46-58.

There are three regions of saline soils; a region of soda saline soils; primary saline soils of the periodically inundated lowland and secondary saline soils of undulating alluvial plains. Drainage of stagnant water and washing-out of salts by means of wide shallow drains are recommended. Formation of soda saline soils is associated with swamping or closely connected with composition of the groundwater. Improvement is best effected by rice sowing and application of gypsum.

767

MACKENZIE, A.J.; VIETS, F.G. Jr.; Nutrients and other

chemicals in agricultural drainage waters. In Drainage for Agriculture (Agronomy No. 17, Schilfgaarde, J. Van. (Ed.). Am. Soc. Agron., Wisconsin, 1974, 489-508.

N,P, and trace elements and pesticides in drainage waters are discussed.

768

MACKINTOSH, E.E.; HULST, J. VAN DER. Soil drainage classes and soil water table relations in medium and coarse textured soils in southern Ontario. Can. J. Soil Sci., 1978, 58:287-301.

Water table levels were recorded for up to 5 yr on five soil catenas, each containing well, imperfectly, poorly and very poorly drained member. Water table measurements in combination with soil morphology were used to establish the zone of free water saturation and from this, periods of saturation were recorded at pre-determined depths or, alternatively, mean high and low water table levels were calculated for the periods 1 Jan. 1-31 Dec. and 1 May-31 Oct. Periods of saturation for well/imperfectly and imperfectly/very poor drained soils were singnificantly different at the 2 levels of probability, but there was no significant difference between the imperfectly/poorly or poorly/very poorly drained members. At the .05 level of probability, only two drainage classes variability raises some concern over Relyingon soil morphology to determine water table characteristics and questions the use-fulness of soil drainage classes as currently defined. The significance of the data is discussed in relation to soil mapping, tile drainage and septic tank disposal.

769

MACLEAN, N.R. Relationship between water-table height and cane yield in the Goondi mill area. Proc. Qd. Soc. SugarCane Tech. Brisbane, Queensland, 1977, 105-10.

Data are presented from 28 sites, with correlations between yield and water-table heights. Yield differences of at least 30 t/ha in adjacent areas were associated with differences in water-table heights. It was calculated that cane yields were likely to be below average if over 40 per cent of the weekly water-table measurements were within 60 cm of the soil surface.

770

MAEDA, K.; MINAMI, M. Studies on physical and chemical properties and improvement of soil productivities in heavy clay paddy fields. 3. Effect of improvement of drainage and irrigation water management on productivity of rice. Bull. Hokkaido, Prefectural Agric. Exp. Stn., 1978, 40:19-29.

On a lowland gley soil, drainage of a rice field with earthenware pipes gave yields of 4.82 t unpolished grain/ha while rice hull drainage gave 4.89 and 5.03 t depending on the method of installation. In a similar trial on upland gley soil, earthenware drains gave 5.02 t/ha while rice hulls gave 5.46 and 5.17 t/ha. In the lowland soil, yields were 6.24 t when flood water was drained off at earing, 5.7 t when it was draindcd off 7 days later and 5.36 when it was drained off 14 days later. On the upland soil, yields were 5.29 t with draining at earing and 5.02 t with draining 7 days later. On a brown lowland soil yields increased with delay in draining off flood water from 5.13 t with draining 7 days after earing to 5.62 t with draining 14 days after earing and 6.18 t with draining 14 and 21 days after earing, respectively. For both plant growth and accessibility to farm machinery a soil pF value of 2 after drainage was desirable.

771

MAGNIANT,D.; DUTIL, P. Wide variations in drainage rate during thawing in silty soils of the Perthois (Marne, France). C.R. Acad. Agric. Fr., 1970, 56:381-7.

In experimental fields on silty soil, drainage rate was 0.3-0.5 L/s/ha under usual rainfall condition, decreased to zero during prolonged frost periods and increased to a maximum of 3.8 L/s/ha within 4 hr of the soil temperature rising above freezing point.

772

MAKAROVA, V.S.; KHODAKOV, V.F. Technological improvements in the leaching of saline soils. Vest. S.kh Nauki. Moskva, 1977, 10:108-18.

In the Hungary steppe, small plots can be leached, with or without temporary drainage, by flooding with a layer of water 10-20 cm deep. Large plots can be leached by providing deep temporary drains at a depth of 1.5-2.0 m and 50-100 m apart.

773

MALAC, B.F. Research in forest fertilization at Union Camp Corporation. In Forest Fertilization: Theory and Practice, Symp. For. Fertil., Tennessee Valley Authority, Knoxville, U.S.A. (1967), 1968, 203-8.

Results of studies with 14-7-7 fertilizers applied to an old-field plantation of slash pine on Irvington loamy sand, showed that response to fertilization was strongly affected by drainage: on well-to-moderately drained soils the response was twice as great as on poorly-drained soils. Urea-formaldehyde was twice as effective as NH_4NO_3.

774

MANNERKOSKI, H. Changes in the tree cover and ground vegetation of a sedge bog after drainage. Sue, 1976, 27:97-102.

After drainage in 1950, a Scots pine stand that has been studied periodically upto summer 1976 has grown well. Changes in the composition of the ground cover appear to depend on shading and competition by the trees, and only those species with the highest moisture requirements have disappeared.

775

MANNOCCHI, F. Possible drainage systems for agricultural soils of low permeability. An.Fac. Agrari. Univ. Stu. Pergia, (1977), 1978, 32:79-100.

The traditional Italian methods of drainage by open ditches or underground pipe do not guarantee the rapid removal of surface runoff and the lowering of the water-table in impermeable soils with a high water-table. Mathematical description of drainage in such soils permits the elaboration of alternative drainage systems in which the existing system of ditches is not completely destroyed.

776

MANSELL, R.S.; SELIM, H.M.; CALVERT, D.V.; et al. Nitrogen and water distribution in fertilized sandy soil during irrigation and drainage. Soil Sci. Am. J., 1980, 44:95-102.

Spatial distribution of soil water pressure head and concentration of nitrogen in the soil solutions of a subsurface drained spodosol under citrus are measured and reported. The soil treatments included, shallow tillge and shallow liming (ST) and deep tillage and deep liming (DTL). Surface drainage and subsurface drainage (107 cm depth; 18.3 m spacing of 10 cm dia plastic tubes) were provided. Transport of water and nutrients was faster in ST compared to DTL. Water table during irrigation rose to 30 cm in DTL compared to ST in which it came up only upto 80 cm. Rates of drainage were higher for ST than DTL. Over a 2 wk period, the amount of NO_3-N in the drainage water from ST was 4 times greater than from DTL.

777

MANSELL, R.S.; WHEELER, W.B.; CALVERT, D.V. Leaching losses of two nutrients and an herbicide from two sandy soils during transient drainage. Soil Sci., 1980, 130:140-50.

Concentrations of K, NO_3-N, and Terbacil were measured in subsurface drainage from shallow-tilled (ST) and deep-tilled (DTL) plots of an acid, sandy Spodosol after 7.6 cm of irrigation. Generally, the concentrations of all these solutes were higher in ST drainage from the drains and soil solution within the soil region influenced by the drains clearly confirmed previous observations that the rates of water discharge from the ST drain were much greater than those from the DTL drain. The cumulative drainage from ST soil over the 2-week period was approximately twice that from DTL. In ST soil the water table rose to within 100 cm from the surface during the infiltration of water applied by pop-up sprinklers, but in DTL soil the water table rose to within 60 cm from the soil surface. The slower drainage, the higher soil water contents, and the higher contents of clay minerals and organic matter in the upper 60 cm of the DTL soil resulted in K, NO_3-N, and Terbacil discharges that were only 30, 37 and 14 per cent as large as those for ST soil. Thus, the deep-tilled acid, sandy soil discharged less drainage water, K, NO_3-N and Terbacil during the 2-week period than occurred from the shallow-tilled soil.

778

MANSELL, R.S.; WHEELER, W.B.; CALVERT, D.V.; STEWART, E.H. Terbacil movement in drainage waters from citrus grove in a Florida flatwood soil. Proc.Soil Crop Sci. Soc. Fl., 1978, 37:176-9.

The levels and total discharge of the herbicide, Terbacil were measured in subsurface drain water for 13 days after application at a rate of 4.5 kg/ha. Outflow water volumes were highest in surface tillage (ST) plots with deep tillage plus heavy lime plots (DTL) being lower and deep tillage plots (DT) being lowest. Terbacil concentrations in ST drains, DT drains and DTL open drain line water ranged from 75 to 100 ppb; the DTL submerged drain water values were 25-50 ppb. Maximum discharge rates ranged from 2000 mg/ha/hr in the ST plots to 250 mg/ha/hr in the DT plot. Thus, the terbacil discharge was primarily related to outflow water volume and was influenced to a lesser extent by tillage treatment. Drain treatments (open and submerged outlets), with the possible exception of the DTL plot, had no influence on Terbacil discharge. Even with the relatively high herbicide discharge of the ST plots, only 1 to 2 per cent of the applied Terbacil was detected in drainage water over the 13-day sampling period.

779

MANSON, P.W. Concrete pipe drains. Proc. Drain. Efficient Crop Production Conf. Chicago, Illinois, 1965, 49-50.

Specifications of C-412 concrete drain tile and its manufacturing procedure have been discussed. Special quality tile for acid soils and for sulphate soils are reported. It is suggested that tiles should be steam cured for high sulphate resistance.

780

MANSON, P.W. New specifications for concrete drain tile. Trans. ASAE, 1969, 12:614-5,620.

Standardisation of drainage materials is essential. Concrete drain tile standards were made in USA in as early as 1914 and now the American Society for Testing and Materials has revised and improved these standards and are designated as ASTMC 412, concrete drain tile. Special quality concrete drain tile under these standards are recommended for acid or sulphate exposure. Present improvements in physical testing of the concrete drain tile in terms of crushing strength and water adsorption limits of oven dried specimens are presented. Three edge crushing strength procedure and a 5 hr boiling absorption test are reported.

781

MANSON, P.W.; BLAISDELL, F.W. Computing energy losses at drain-tile junctions. Trans. ASAE, 1966, 9:445-8.

A laboratory study of energy losses at drain-tile junctions was conducted over a five year period. Energy loss coefficient equation is presented and design curves were drawn for different combinations. A typical example for agricultural tile drainage indicates that sum of the junction energy loss and the expansion loss for the example drain tile system where 90° junction are used is 1.00 mm in the 420 m length of the main. The loss in energy-$29x10^{-6}$m/m is so small that it can safely be neglected in the example. For agricultural drainage where the velocities are low, 90° junctions design is as satisfactory as 45° junction and that for ease of construction the lateral should join the main at most convenient angle.

782

MARCILONEK, S.; KOSTRZEWA, A.; PRZYGODZKI, M. The influence of drainage on water relations of very fine clay soils around the Goczalkowice retention reservoir. Zes. Nauk. Akad. Roln. Wrocl. Melior., 1975, 115:215-25.

A drainage system of 8 m spacing and 1 m depth was installed in 1968 in arable soils. The groundwater level varied from 20 to 160 cm from the soil surface. The mean depth during the growing season was 95-115cm. Infiltration from the reservoir did not appear to affect the water balance of the area. The total annual drainage was equal to 15 to 25 per cent of the annual precipitation, with about 70 per cent occurring in the winter six months.

783

MARCINEK, J. Effect of drainage in relation to the intensification of agriculture and forestry. Zes. Probl. Post. Nauk Roln., 1976, 177:73-157.

Changes in soil properties as a result of the lowering of the groundwater table under intensive agriculture are discussed.

784

MAREI, S.M. A Hele-Shaw model study of oscillating water tables in drained homogeneous soils. Soil Sci., 1974, 117:301-5.

A Hele-Shaw model was used to investigate water table heights in a drained soil under seasonally fluctuating irrigation intensities and the results were compared with theoretical harmonic predictions according to Childs. Mean water-table heights and their amplitudes of oscillation agreed well. For short time periods, the oscillation deviated from the simple harmonic form predicted by the theory.

785

MARTINEK, B.J. Drainage and reclamation of salt-affected soils, Bardenas area, Spain. Int. Inst. Land Reclam. Improv. Wageningen, 1978, 24,321 p.

The Bardèna areas in the Ebro basin, north-eastern Spain is a semi-arid area bounded by high mountains. The geology, geomorphology, climate, vegetation, land use, and hydrology of the area are summarized. A survey of the soils was conducted. Five main soil associations and the irrigation techniques used on the soils are detailed. Two experimental reclamation projects for saline alluvial soils were conducted. The drainage system, leaching process, and land and water management required to prevent resalinization are discussed.

786

MARUYAMA, T. The variation of soil permeability caused by underdrainage-The study of under drainage - (1). Trans. Agric. Engng. Soc. Tokyo, 1966, 16:9-13.

The paper reports the investigations and results of the variation of soil permeability, and the influences on permeab-ility of factors like the water management and planting systems and climate in non-irrigation period in cases where the underdrainage was in sandy and clayey paddy fields.

787

MARUYAMA, T. Hydraulic mechanism of underdrainage in paddy fields having underground aquifer- The study of under drainage -(2). Trans. Agric. Engng. Soc. Tokyo., 1966, 16:14-9.

It has been shown that there exists a close relationship between the ground-water table, drain flux and quantity of penetrating water and that it compares well with the calcula-ted values computed by the theory of Don Kirkham; points out that theory cannot be applied to such cases where Darcy's law is not applicable because of the large cracks in soils which remain connected with the under drains during irrigation period and a nomograph to facilitate calculations has been presented.

788

MASLOV, B.S. Employment of subsurface drainage in the USSR. 10th Congr. Irrig. Drain Athens, 1978, 4:1-15.

Subsurface drainage in the USSR mainly utilized clay pipes; PVC and other materials are used to only a limited extent. Wide use is made of filters to prevent silting. Methods for preventing ochre sedimentation in the pipes are outlined. Mole and slot drainage are of great value under certain conditions. Reclamation of heavy soils require additional measures such as land levelling, installation of absorbing wells, soil ripping, and formation of ridges and bunds in monsoon area. Use of vacuum drainage on irrigated lands has been encouraging.

789

MASLOVA, A.A. Moisture regime of derno-gleyey soil and gley-like soils drained by closed drainage in the valley of the river Yakhroma. Pochvovedenie, 1964, 9:25-32.

In these bottomland soils with the drains placed at about one m depth and 35 m apart, water entered the drains from both the plowed and the underlying horizon. Near the drains at the soil-moisture content in the A, B_1 and B_2 horizons was 60-80 per cent of the full moisture capacity (and was therefore optimal for vegetable and fodder crops) and equidistant from the drains it did not fall below 70 per cent. Groundwater was consumed during evapotranspiration.

790

MASSLAND, M. Water table fluctuations induced by irrigation. J. Irrig. Drain. Div. ASCE, 1961, 87:38-58.

The theory of intermittent recharge, as previously developed is extended and analysed. Equations are presented from which charts may be prepared for easy computation of drain spacings for irrigated agriculture. The analytical results may also be used to compute the effect of natural drainage on the height of water table as well as anticipated changes in the water table of newly irrigated areas.

791

MASSLAND, M.; DIRK, E.L.; SHERY, R.A. Falling water table between tile drains. J. Irrig. Drain. Div. ASCE, 1967, 93:9-18.

The rate of fall of a water table between drains in a Hele-Shaw model has been determined and the results with those obtained by use of the Glover equation have been compared. Considerable differences are found between the two methods. When the ratio between the drain spacing and the depth of the aquifer is decreased, the difference between the model results and the results obtained from the theory is increased.

792

MAVRIKOS, P. Drainage by plastic drains. Bull. GCID, 1968, 36-37:18-29.

The paper states that some 344,000 ha hydromorphic soils in Greece need drainage; compares plastic pipes with the tile drains for their performance and rates the pipe superior to the latter; deals with cost; maintenance and other problems regarding the pipes.

793

MAY, J.; TRAFFORD, B.D. The analysis of the hydrological data from a drainage experiment on clay land. Tech. Bull., Ministry of Agriculture, Fisheries and Food, UK, 1977, 77-1, 17 p.

The hydrologic results for 5 winters at the Drayton drainage experiment near Stratford-upon-Avon are analysed. Comparisons are made between years and between treatments in terms of overall drainage efficiency and an attempt is made to show possible changes in soil physical conditions. There are 3 different drain spacings and 3 secondary treatments, but the results suggest that the drain spacing is having a relatively small effect. Thus on this heavy clay site the treatments may be placed in the following order of merit: moling plus widely spaced pipes, subsoiling plus pipes, pipe only. Both crop yields and capital costs would also place the treatments in this order of merit.

794

MAZUR, Z. Meadow yields in relation to ground-water levels in the Wieprz-Krzna canal region. Wiadomosci Melior., 1975, 18:289-92.

On peat soils in the Wieprz-Krzna canal region of Poland, hay yields of meadows dominated by Phleum Pratense, Festuca Pratensis and cocksfoot were examined in relation to the spacing of ditches. Measurements were carried out between ditches spaced at 112-m intervals. Mean hay yields (2 cuts) for the various positions between ditches ranged from 6.64 t/ha in 1970 with low rates of fertilizer to 8.12-10.16 t/ha in other 3 years with higher rates of fertilizer. Over the 4-year period, mean hay yields were 8.44-8.45 t/ha near the ditches compared with 7.99 t mid-way between them. It was concluded that for the soil and climate conditions in the sandy area the spacing between ditches was satisfactory.

795

MCCREADY, W.; ROBSON, J.F. Development of irrigation and drainage in the Bada's Delta, Iraq. 8th Cong. Irrig. Drain. Varna, 1972, Q 27 R 1:1-16.

The Bada's delta is in the Tigris Euphrates delta and is a well established rice growing area. Designs for remodelling the irrigation system and for surface drainage are discussed. Irrigation and salinity are the major difficulties. Surface drainage through open ditches with a bed width of 1.0 m, 1 in 4 side slopes and 0.25 m minimum depth with bed graded to suit the local topography are proposed. The channels have been designed with water levels at design discharge at a minimum depth of 0.30 m below ground level. Farm management is advocated to get full benefits of the system.

796

MCWHORTE, D.B. Discussion. Drain spacing based on dynamic equilibrium. Closure. J. Irrig. Drain. Div. ASCE, 1978, 104:475-60.

The author of the original paper, published in June 1977, replies to comments concerning computations of the water-table profile following drainage.

797

MCWHORTER, D.B. Drain spacing based on dynamic equilibrium. J. Irrig. Drain. Div. ASCE, 1977, 103:259-70.

A mathematical development which is a modification of the Massland's analysis that results in a less elegent, but more practical, flexible procedure for computing drain spacing. The development uses superposition to derive a spacing method that is applicable for a periodic sequence of instantaneous recharge events of arbitrary volume and timing. The method provides a mathematically correct alternative to the pioneering procedure described by Dumm and maintains the flexibility required to handle conditions of variable recharge volumes and drainout times.

798

MEEK, B.D.; GRASS, L.B.; MACKENZIE, A.J. Chemical characterization of the gaseous and liquid environments of subsurface drain systems. Soil Sci. Soc. Am. J., 1978, 42:693-8.

Chemical composition of subsurface drainage effluents was determined in the field as a function of depth, soil texture, entry point into the drainpipe, and distance from outlet (with or without a water trap). Liquid samples were analyzed for Fe, Mn, nitrate, bicarbonate, organic carbon, EC, pH, and dissolved oxygen, and gas samples were analyzed for O_2 and CO_2. Mn ranged from 0.1 to 6.1 ppm and Fe from 0.1 to 14.2 ppm, which illustrates the wide range of oxidation-reduction conditions present. Chemical composition changed little during the season. Water entering the upper portion of the drain lines was lower in Mn, which was the result of short flow paths through soil zones with higher O_2 levels, than water moving through longer flowlines which entered the bottom of the drains. the solutions in the drain lines were undersaturated with respect to the drain line atmosphere, which indicated that O_2 flows downward through the drain trench backfill. This is an important factor in the oxidation of Fe and Mn in subsurface drain lines. However longitudinal movement of O_2 into drain lines, via the outlets, is a minor factor and does not account for the extensive precipitation of Fe or Mn within the system.

799

MEHTA, H.C.; JOSHI, R.S.; THOMSIC, M.F. Effect of pumping and surface drainage in controlling water table, waterlogging and salinity conditions on deep black cotton soil area. Proc. Symp. Waterlogging, Causes, and Measures for Prevention, Cent. Bd. Irrig. Pwr. New Delhi, 1972, 118:115-23.

Surface drains on the farm land, representing the major command area of Kakrapar Irrigation Project, were installed in 1969 and 1970 . The drainage system eliminated waterlogging in 192 ha of low lying area by removing impounded and stagnant water during monsoon and irrigation season. The tubewells, 46-49 m deep constructed at a space of 550 m, controlled water table on 40 ha of land and also supplemented canal irrigation water. The tubewells were found effective even during the years of heavy rainfall. As a result of these measures it was possible to introduce new crops such as cotton, castor, bajra, soybean, sunflower, safflower, mustard and other vegetable crops.

800

MEIN, R.G.; TURNER, A.K. A study of the drainage of irrigated sand dunes using an electric resistance analogue. J.Hydrol., 1968, 6:1-14.

An electrical resistance analogue is described, and shown to be suitable for study of sand dunes problems. It was observed that to determine drain spacing and position for a given dune cross section, water table shape be drawn according to ellipse equation for several different spacing and depths of drain to choose an appropriate combination of drainage. For dunes with a sloping impermeable layer, the theoretical solutions of Schmid and Luthin and Wooding and Chapman are good upto a slope of at least 10 per cent.

801

MELMBERG, R. Effect of subsoil loosening and mole drainage on the moisture economy of compacted soils. Sonderh. Bayer. Landw. Jb., 1967, 44:55-61.

Graphical data are given on pore distribution, moisture content and permeability of pseudogley and parabraunerde after subsoiling, and on water percolation in mole-drained pseudogley.

802

MEMMOS, N. Stratum-survey-water table investigations and soil drainage. Bull. G.C.I.D., 1983, 6:20-4.

The paper describes briefly the nature of soil drainage and points out its influence upon soil characteristics, stresses the need of stratum survey and water table investigations as fundamental steps in studying soil drainage problems, and presents an analytical study on the subject.

803

MENONEN, J.; PAIVANEN, J. Additional drainage with subsurface drains in a moled peat harvesting site. Suo, 1979, 30:17-25.

The use of subsurface drains is sometimes necessary to increase the drainage effect of open ditches on moled-peat harvesting sites. The effectiveness of different types of subsoil drains was therefore tested. These were:(1) a board-pipe drain (inner cross section 64 x 102 mm); (2) PVC pipe drains (40 and 65 mm internal diam.), either bare or covered with a filter material; (3) two types of mole drains and (4) a narrow water furrow made with a disc trencher. The board pipe drain and to less extent the PVC pipe drains had the greatest ameliorative effect on groundwater levels. Filter material did not increase the effectiveness of the pipe drains.

804

MENZ, J.A. The economic evaluation of a drainage project. J. Soil Wat. Conserv., 1964, 19:12-4.

The procedure of economic evaluation of drainage project is illustrated. Tabulated values of expected alfalfa, cotton, barley, cantaloupes and lettuce yields as influenced by water table varying from 60-90 cm to 180 cm are reported. The benefit cost ratio increases initially from 0.85 to 1.18 with increment of water table lowering from 30-90 cm but at 120 cm benefit cost ratio decreases to 1.03. It shows that water table in this project may be lowered to or beyond 90 cm but short of 120 cm.

805

MER, I.I. Method of evaluation of the capacity and trafficability of drain layers. Vestn. S-kh. Nauki., 1980, 4:69-74.

The forces and reactions acting on a machine laying tile, plastic or other drains are described theoretically. These include gravity, friction between the side walls of the hopper and the filter material, that between the french-forming tool and the soil, the force for laying the drains, preliminary backfilling, for smoothing the trench bottom, the total and resultant soil reactions etc.

806

MERGEN, Y. Operation and maintenance problems of drainage systems in Turkey. 7th Cong. Irrig. Drain. Mexico, 1969, Q 25 R 16:205-14.

About 3,289,309 ha of land in Turkey needs drainage or/and reclamation measures. In addition, approximately 2,500,000 ha of agricultural land have flooding and surface drainage problems to various extent. The author discusses the general char-acteristics of drainage problem in western and southern Turkey and describes the works carried out in Turkey. The difficulties are the lack of technical know-how on the part of contractors and inadequate availability of machinery to execute the jobs. The author then lists the maintenance and operational problems of drainage in Turkey and makes a number of useful suggestions to overcome them, including the enactment of a law to prevent damage to channels by the people and the participation of the beneficiaries in the financing of the projects.

807

MERIAUX, S. Relationship between re-moistening of the soil and drain outflow in a leached hydromorphic soil. C. R. Acad. Agric. Fr., 1977, 63:540-50

An analysis of soil water profiles showed that the depth of soil re-moistening was always greater in the region of the drainage ditch than in the soil between drains. The processes of re-moistening and water transport towards the drains were not modified by the nature of the soil profile when more rapid rates of infiltration occurred (at depth) in the drainage ditch.

808

MERLIN, S.; HERVE, J.J. Economics of drainage schemes. 10th Congr. Irrig. Drain. Athens, 1978, 5:91-107.

A analysis of the costs of drain installation using trenching and trenchless machines is presented.

809

MERVA, G.E.; SEGERLAND, L.; MURASE, H.; FAUSEY, N.R.

Finite element modelling for depth and spacing of drains in layered soils. Trans. ASAE, 1983, 26:452-6.

A finite element model was developed to solve Laplace's equation for an anisotropic, vertically nonhomogeneous soil. The model uses the soil characteristic curve for the surface layer and hydraulic conductivity of as many as three soil layers and a trench backfill to predict locations of the soil water table. The model could be used for predicting the position of the free water surface for a drainage system in Toledo silty clay soils successfully and is suggested for use in multi-layered soils.

810

METWALLY, A.I.; EL-DAMATY, A.; YANI, Y.G. Chemical changes in soils accompanying waterlogging. Beit. Trop. Landw. Veter., 1978, 16:247-54.

Under waterlogging, marked changes in the chemical properties of soils occurred up to 135 days, but only in the presence of easily reducible organic matter. They were: an increase in microbial activity of aerobes, followed by sulphate-reducers; a rapid decrease in the redox potential; an initial pH decrease, followed by a gradual increase; an increase in the formation of CO_3^{2-} and HCO_3^-, with subsequent precipitation of $CaCO_3$ and $MgCO_3$, a decrease in the content of soluble sulphate, Ca and Mg, and an increase in exchangeable Na. All these changes were promoted by higher temperatures (up to 40 degree C), but were not affected by the initial soil pH.

811

MIERS, R.H. The civil engineer and field drainage. J. Instt. Wat. Engrs., 1974, 28:21-7.

The paper presents the standards in the design of artificial drainage so as to provide optimum conditions in different types of topography for fens and marshes, upland valley and flood valleys.

812

MILIAUSKAS, V. Dynamics of the physical and agrochemical properties of drained and undrained derno-gley and podzolic-bog soils. Liet Zemdir. Moks. Tyrimo. Inst. Darb, 1964, 8:5-30.

In excessively moistened soils, drainage caused moisture deficiency in the root-inhabiting horizons during dry years. In drained derno-gley leached soils under perennial herbage, volume weight increased and porosity of the upper horizons decreased. The humus decreased in summer although in deep horizons it tended to increase; available P and NO_3 increased, but Fe content was lower than in undrained soil. To less extent, these processes occurred in drained derno-gley podzolized loamy soils. Increasing the drain intervals from 12 to 30 did not significantly affect physical or agrochemical properties.

813

MILLER, D.W.; WILLARDSON, L.S. Head loss at soil-drain envelope interfaces. J. Irrig. Drain. Engng. ASCE, 1983, 109:211-20.

Hydraulic gradients at exits from porous media are greater than those predicted by laminar flow theory. Exit gradient effects project into the soil away from the exit interface in the manner of a backwater curve. Head losses measured in the soil at an interface may be twice than those predicted. Laboratory hydraulic conductivities measured using only inflow and outflow heads may be lower than actual values.

814

MILLER, M.H. Contribution of nitrogen and phosphorous to subsurface drainage water from intensively cropped mineral and organic soils in Ontario. J. Environ. Qual., 1979, 8:42-8.

Nutrient content of drainage water was measured from six mineral soils and three organic soil sites of size ranging from 5 to 113 ha. The NO_3-N contents in mineral soil seldom exceeded 10 mg/L but when the rates of application were more than the recommended it was seldom less than 10 mg/L. Annual NO_3-N loss ranged from 4-64 kg/ha. From organic soils it ranged from 14.8-42.7 mg/L and the loss was estimated at 37-245 kg/ha. The PO_4-P content in drainage water from organic sites increased during the course of the study and averaged 18.2 mg/L on one site. It is suggested that P- adsorption capacity should be considered before organic soils areas are developed for crop production.

815

MILLIC, M.; SPASOJEVIC, M. Regulation of the water regime of a eugley soil with surface water in Donje Podrimlje. Zeml. Bilj. Ser. A, 1980, 29:1-19.

Soil layers over gravel and sand were of fine texture compact structure and low permeability. They restricted water movement resulting in considerable pressure in the water-holding layer causing a rise in water level. To reduce the pressure of the surface water, relief wells were dug at 2.6 m depth and at distances of 20 m i.e. half the distance between the drains, which were at a minimum depth of 1.1 m.

816

MILYAUSKAS, V. Effect of drainage on physical and agrochemical properties of waterlogged soils of the Lithuanian SSR. Pochvovedenie, 1963, 1:61-74.

The soil that had been drained 30-60 years ago were studied in 1949-57. Gleying was less marked in drained than in undrained soils. Total absorbed bases, base saturation and pH were lower in drained soils, and the CO_3 horizon was lower down. The content of absorbed Ca in drained soils, particularly sod-podzolic gleyey moderately podzolized soils, was lower in drained soils. Drainage either increased or did not affect the humus content of the A_1 horizon of well cultivated drained derno-gleyey and derno-podzolic-gleyey soils but decreased it in the same sub-types of slightly cultivated soils. Drainage generally increased the humus content of lower horizons. In spite of cultivation, drainage decreased the humus content of the A_1 horizon of derno-gley soils. Drainage decreased the volume weight, increased porosity, capillarly porosity and permeability to water, and improved soil aeration. Drainage of derno-gley and podzolic bog soils improved fertility more on boulder loams than on lacustrine clays and loams.

817

MINAEV, I.V. Effect of the height of the water-table on the moisture content of the ploughed layer and on yield. <u>Vestn. S-kh. Nauki.</u>, 1972, 9:81-7.

A review with 29 references. The fundamental parameters of the drainage system should be calculated from the decreases in the height of the water-table in spring and autumn, and the water balance during the vegetative period.

818

MINCHIN, F.R.; SUMMERFIELD, R.J.; EAGLESHAM, A.R.J.; STEWART, K.A. Effects of short-term waterlogging on growth and yield of cowpea (<u>Vigna unguiculata</u>). <u>J. Agric. Sci. UK</u>, 1978, 90:355-66.

When cowpea plants were grown in large pots under simulated tropical conditions, and were dependent on either symbiotic fixation or inorganic N for their total N requirements, short periods of waterlogging had adverse effects on vegetative growth and seed production. Waterlogging effects were more acute as plants experienced stressat progressively earlier stages of development. In plants depending either on biologi-cally-fixed N, waterlogging before flowering reduced vegetative growth by about 50 per cent . Although some compensation in vegetative growth occurred during the reproductive development of nodulated plants, seed yields in both cases were still 48 per cent less than in unstressed plants. Depending on the stage of plant development, treatment effects were deviated largely through changes in branch, peduncle and flower production and/ or abortion, which contributed directly to variations in the number of pods retained to maturity. Inorganic N stimulated vegetative growth but did not significantly improve seed yield compared with effectively-nodulated plants.

819

MINDERHOUD, P. A model for the design of drainage in flat agricultural lands. Agric. Wat. Mgnt., 1982. 5,95-125.

A model to determine drainage measure and costs has been elaborated for flat lands and wet tropical climates. A basic element of the model is a scheme to compute the water balance per day for a drainage parcel. Discharge, evapotranspiration, groundwater level and soil moisture storage are estimated as functions of rainfall, potential evapotranspiration, vegetation and soil characteristics and of an assumed drainage intensity. The water balance computation is performed for periods of 5-40 years of daily rainfall data, for a series of drainage intensities. The results can be subjected to a drainage criterion, from which a design drainage intensity and a corresponding drain spacing can be derived. Finally the layout of canals for a block of 4 x 1 km^2 is determined and excavation and a series of canal characteristics are computed. A summary of some applications is included.

820

MINISTRY OF AGRICULTURE FISHERIES AND FOOD. Trenchless underdrainage. Drainage Leaflets, Ministry of Agriculture, Fisheries and Food, UK, 1977, 21, 5p.

The mole drainage principles have been developed so that drain pipes can be laid without the need for soil removal. The technique is ideally suited to laying corrugated plastic pipes, but is also suitable for clayware pipes.

821

MIODUSZEWSKI, W. An attempt to estimate the effect of moisture content and frost upon the strength of tile drains of 50 mm diameter. Zesz. Nauk. SGGW, Melior. Roln., 1962, 5:53-70.

The paper attempts mathematic-statistical estimation of the moisture and frost influence upon the tile drains produced by various plants which shows no negative influence of frost or moisture content upon the tiles strength.

822

MISRA, M.K.; BEASLEY, R.P. Corrugated plastic drain tubing used as underground conduit. Trans. ASAE, 1975, 18:260-9.

In soil tank experiments no significant amount of sediment was drawn into 15 cm internal diam drain tubing even at 10% slope and a water velocity of 2.7 m/sec. In cohesive soil the amount of outflow from the pipe was low and there was no evidence of soil movement from around the test section but in less cohesive soil outflow was greater and there was considerable movement of soil away from the pipe when the test section was on a 5 or 10 per cent slope.

823

MITCHELL, J.K.; DRABLOS, C.J.W. Statistical analysis of four-inch corrugated plastic tubing deflection. Trans. ASAE, 1974, 17:685-9, 693.

Equations were developed from field data for the deflection of plastic drain tubes. The factors exerting the greatest influence were duration, moment of inertia, groove angle, and strength of tubing. Installation techniques and conditions at the time of installation were not considered.

824

MIYAMOTO, S.; WARRICK, A.W. Two-dimensional displacement into or from water-filled ditches. Soil Sci. Soc. Am. Proc., 1974, 38:723-7.

Piston type displacement of solutes for steady state saturated flow into or from water filled ditches is analyzed mathematically and by a sand tank model. The displacement front into the drains advances much faster near the drains than in the region midway between the drains. The presence of an impermeable layer accentuates this trend. The other relevant applications of theory are listed. Experimental data obtained from a sand tank model agree with the predicted fronts.

825

MIYMOTO, S.; WARRICK, A.W. Salt displacement into drain tiles under ponded leaching. Wat. Resour. Res., 1974, 10:275-8.

Steady state piston-type displacement into drain tiles is analysed mathematically when the ground surface is partially covered by an impermeable material or by ponding water away from the drains. By preventing the water from infiltrating directly over the drains more uniform leaching pattern is attained. The result is to reduce the water required for leaching to the depth of drains by as much as one half. Experimental data obtained from sand tank models support this prediction.

826

MIYAMOTO, S.; WARRICK, A.W. Two dimensional displacement into or from water-filled ditches. Soil Sci. Soc. Amer. Proc., 1974, 38:723-7.

Analytical expressions are derived describing

piston-type displacement into or from water-filled ditches for steady-state saturated flow. Parameters include the ditch size and spacing, a surface soil bank size, and the depth of an impermeable barrier. Experimental data obtained from a sand tank are compared with calculated values.

827

MIYASAKA, A. Effect of drainage on carbon-dioxide exchange and some characters related to grain yield of rice plants. Proc. Crop. Sci. Soc. Japan, 1964, 33:90-2.

Drainage tended to promote photosynthetic activity to reduce respiration and to keep leaf-blade weight high during ripening. Drainage increased grain weight and percentage of well-ripened grains.

828

MIYASAKA, A. Effects of drainage on rice growth in poorly drained paddy fields. Bull. Hokuriku Agric. Exp. Stn., 1970, 12, 80 p.

Drainage reduced the water content of leaves, increased N uptake by the roots (as a result of increased root activity), increased the photosynthetic activity during ripening and consequently increased yields. The time at which drainage took place between tillering and heading affected the distribution of carbohydrates during growth but not the final yield.

829

MOHAMED, I.S.; ATESHIAN, K.H.; WILKINS, R.A. Field irrigation and drainage techniques developed by the sugarcane industry in Guyana. 8th Congr. Irrig. Drain. Verna, 1972, Q 27 R 16:239-50.

Irrigation and drainage techniques for sugarcane cultivation in tropical climate, flat topography and the predominantly saline clay soils are discussed. The internal field layout is described in detail. The open ditch drainage system on the sugar estate is designed to remove 5.08 cm of runoff in 24 hr after saturation. The drainage system provided is hydraulically oversize and is effective not only in controlling the average ground-water table to a depth of 0.91 m below the ground but also in keeping the rootzone free of excessive accumulation of salts.

830

MOHAMMAD, F.S.; SKAGGS, R.W. Drain tube opening effects on drain inflow. J. Irrig. Drain. Engng. ASCE, 1983, 109:393-404.

Laboratory tests were conducted to determine the effects of total opening area, location of openings and gravel envelope on radial flow to 100 mm diam corrugated plastic drains. Radial flow theory was used to evaluate the effective drain tube radius, r_e and the drainage transfer coefficient for the corrugated tubes. The effective radius increased with total perforation area but the location of the drain tube perforation had little effect on r_e. The greatest increase of effective radius for an increase in perforation area occurred between areas of 38-75 cm^2/m where r_e increased from 5 mm to about 21 mm. Use of a 50 mm thick envelope resulted in an increase in the effective drain radius for a 100 mm diam pipe from 5-36 mm. Experimental results were in close agreement with predicted values for a 10 mm opening width.

831

MOHAMMAD, N.; KIRKHAM, D.; DOUGAL, M.D. Tube drainage in stratified soil above an aquifer. J. Irrig. Drain Div. ASCE, 1978, 104:209-27.

Potential flow theory is developed to give the relation of the drain spacing to the water-table height midway between equally spaced tube drains in a two-layer stratified soil overlying an artesian aquifer under pressure. The tube drains, which are located in the upper layer, receive both steady downward recharge from the soil surface and upward seepage from the aquifer. The ratio of the drain spacing for a layered soil to the spacing for a homogeneous soil is 3.8 for the following conditions:(1) the hydraulic conductivity ratio of the upper soil layer to the lower is 10; (2) the upper layer is one-fifth the combined thickness of the two; (3) the recharge is zero; (4) the ratio of the artesian head (measured with respect to the level of the drain tubes) to the desired maximum water table head above the drain centre is 2; and (5) the drain diameter is 0.02 times the combined thick ness of the two layers. If the conductivity ratio is 100, for these conditions, the drain spacing ratio changes from 3.8 to 12.7, and if 1,000, 3.8 to 41.1.

832

MOHANRAO, J.S.R. Drainage problems and design criteria for surface drains in the Nagarjunasagar project right canal ayacut area. Symp. Waterlogging Causes and Measures for its Prevention, CBIP Pub. No.118, 1972, 1:79-87.

Nagarjunasagar Project is a major irrigation project in Andhra Pradesh costing about Rs. 2,500 million to irrigate an area of about 0.831 million ha, a part of which, i.e. about 0.373 million ha is already developed and is being irrigated for the last 3 to 4 years. Salinity index maps and pre-monsoon and post-monsoon water-table prepared for each block of the ayacut to determine the degree and extent of internal surface drainage required for affording quick and effective relief to the flood affected areas of the ayacut. While there is a net-work of drains with enough fall of the country in the entire ayacut area under the right canal to effect rapid disposal of the runoff, the discharging capacities of most of the drainage courses are not sufficient to clear the floods without submersion of adjoining lands to a considerable extent, due to insufficient sections of the drains, encroachments into the drainage courses for unauthorised cultivation, formation of shoals and other obstructions in the drainage courses. While the excavation of the field drains is the job of the farmers themselves, improvments of the natural drainage in the ayacut area is to be carried out by the Irrigation Department. For this purpose, comprehensive drainage estimates are prepared for a few blocks based on detailed investigation of a few representative drains in the blocks to assess the cost. Disposal of the drainage from a major portion of ayacut lying in between branches 10 and 11 into the sea, through the adjoining Krishna Western Delta, presented another major problem, for which various alternative proposals are considered, some of which are investigated and estimated. All these studies revealed that the entire scheme of drainage for the right canal ayacut may cost as high as Rs. 600 millions in all, working out to a rate of about Rs. 1,240 per ha, a colossal figure when compared to the rate of about Rs. 2,570/ha of the project itself. The problem is how to meet this expenditure while the project estimate provides only for a nominal amount to carry out minor repairs to a few drains here and there. The possibility of any other economical and feasible alternatives to afford quick and effective drainage in the ayacut area has also to be gone into and investigated before embarking on the above scheme.

833

MOHANTY, R.B.; MISRA, P.K. Design, construction and maintenance of agricultural drainage systems. Proc. Symp. Waterlogging Causes and Measures for Prevention. CBIP Pub. No.118,1:123-37.

The paper presents general aspects of drainage based on theory. Most of the information presented is available in any text book of drainage.

834

MOLEN W.H. VAN DER. Subsidence of peat soils after

drainage. IAHS Pub., 1975, 183-6.

Considerable lowering of the soil surface occurs after drainage of peat marshes, due to subsidence. Formulae derived by Fokkens are considered to be valid except in cases where oxidation of organic matter is an important cause of subsidence. A nomographical solution is given and the effects of compaction on the water balance are briefly discussed.

835

MOLODTSOV, A.V. Salt content of soils and alteration of the groundwater level during irrigation of new land in the Hungary Steppe. Pochvovedenie, 1964, 1:12-9.

A plot under cotton was irrigated for the first time in 1962. The irrigation system was designed to minimize filtration of water away from the irrigation network. Before irrigation, the groundwater level was at a depth of 6.5-7.5 m to 10-15 m. The groundwater had a salinity of 21-30 g/L of the sulphate-chloride type and Na represented 50-70 per cent of the total cations. Irrigation raised the level of the ground water by upto 4 m during May to September but generally had little effect on its salinity. Even where the original groundwater level was at 9.4 m in spring and 4.5 m in autumn, irrigation caused considerable salinization of the top 5 cm of soil whereas before irrigation the top 50 cm was practically salt-free. Irrigation must be accompanied by drainage and washing of these solanchakous soils before they are utilized or (when the groundwater is at 10 m or more) not later than 1-2 years after utilization has begun. Washing should not be done in cold weather because of the presence of Na_2SO_4, which is only slightly soluble at low temperature.

836

MONKE, E.J.; HUGGINS, L.F.; GALLOWAY, H.M.; FOSTER, G.R. Field study of subsurface drainage on a slowly permeable soil. Trans. ASAE, 1967, 10:573-6.

The readiness of a slowly permeable soil to drain was evaluated for replicated subsurface drains placed 7.5,15 and 30 m apart. The tile drains responded within a few hr to the commencement of a dischage-producing storm and produced maximum discharge values quickly, after which long recession periods occurred. In general, the highest discharge occurred for drains which were placed farthest apart because of the likelihood of high antecedent moisture conditions. Percentage wise, the difference between peak flow rates decreased as the magnitude of flow increased, suggesting that when the soil becomes fully saturated these differences would be small. Peak rates could, therefore, be used to obtain an effective hydraulic conductivity value for the rational design of subsurface drainage systems. The response of crop production to different levels of drainage was evaluated from yield data and population stands. Corn yields were significantly different with respect to the spacing of the drains. Part of this difference was due to variation in corn population. In general, stands for the 7.5 m spacing were higher than those at either the 15 or 30 m spacing. How much of the yield differences, if any, should be attributed as a response just to drainage as opposed to the total moisture environment throughout the entire growing season is unknown. Also, before yield data can be used to justify a spacing critierion, the probability of the occurrence of yield-depressing drainage situations should be assessed.

837

MONTERIO, J.F.D.; FONSECA, V.F.D. The cultivation of the 'Machongos' (Hydromorphic soils) of the South of Mozambique. <u>8th Congr.Irrig.Drain. Varna</u>, 1972, Q 27 R 9:125-34.

Drainage and management practices to be adopted in the hydromorphic soils are discussed so as to bring them under cultivation. The secondary ditches demarcating square field area of 102 m x

102 m are provided with primary drain ditches which provide drainage to a block of 25 such field areas. Main drains with a surface width of 10 m and 1.5 m water depths are provided. Sprinkler irrigation is recommended among the other management practices in the area.

838

MONTSINGER, L.V. Influence of processing conditions and resin characteristics on polythylene corrugated tubing. <u>3rd Natn. Drain. Symp. ASAE</u>, Michigan, Illinois, 1976, 75-9.

A test was developed which evaluates HDPE corrugated tubing for brittleness properties and material distributions. Studies were made of PE resin characteristics and processing conditions. Results were substantiated with data obtained under controlled laboratory conditions. The results indicate that higher density, very broad or very narrow molecular weight distribution and high extrusion rates have a negative influence on tubing brittleness.

839

MOODY, W.T. Non-linear differential equation of drain spacing. <u>J.Irrig. Drain. Div. ASCE</u>, 1966, 92:1-9.

The paper solves numerically the non-linear partial differential equation governing the spacing of horizontal, parallel subsurface, agricultural drains, by the finite differences method; bases the differential equation on the Dupuit-Forchheimer assumptions. Since use of Hooghoudt's equivalent depth is suggested as a means of accounting for the effect of convergence, derives a simplified formula by computing that quantity. Presents solutions in the form of dimensionless curves whose abscissae are a time parameter and whose ordinates represent: (1) maximum water table height between drains, (2) rate of discharge, and (3) volume of water removed. Individual curve parameters establish the depth from drains to an impermeable barrier which can vary from zero to infinity. An example of the application, has been given.

840

MOSOLOVA, A.I.; UTKAEVA, V.F. Action time of polymer preparations in the conditioning of soils as applied to drainage-trench fill. Mosc. Univ. Soil Sci. Bull., 1977, 32:21-4.

Clay and sandy loam soddy-podzolic soils were treated in the laboratory with 11 polymers at rates of 0.05 and 0.1 per cent. Structural water-resistance was increased 2-5 times by the polymers. The effects were greater for the sandy loam soil than for the clay soil. For the clay soil, K-4, K-9, Hypan and polyacrylamide were most effective. The filtration levels of aggregates 1-3 mm in size determined in a Fadeev-Vil'yams apparatus were 6-300 times greater for the conditioned soils than for control soils, and indicated efficient drain operation under field conditions. The structural water-resistance was maintained during 50 days flooding in the laboratory. Previous tests on foam and fibreglass cloth for protection of drainpipes are discussed, and the cost of materials for these methods as well as polymer conditioners are presented.

841

MOSONYI, E.; PICHLER, J. Flood control irrigation and drainage in Hungary's Plain lands. ICID Bull., 1961, 10:50-68.

An informative account is rendered on the protection of agricultural land and the works carried out in the interests of promoting agricultural development in Hungary. Topographical and climatological conditions are described in relation to surface drainage and irrigation. A drainage canal system of 25,600 km with 240 pumping stations with a delivery of 400 m^3/sec is functioning and is described with figures. It is suggested that specific drainage capacities of the drainage basins may have to be increased to a considerable degree. Separate canal system for irrigation and drainage is recommended.

842

MOTOC, M.; TUDOR, S. The effect of tilling and drainage upon the excess moisture of homogeneous clay soils on slopes. Trans. 8th Congr. Soil Sci., 1964, 2:713-20.

In land for orchard use, breaking to a depth of 60 cm, terracing and providing each tree with stone or faggot drains are adequate measures for improving homogeneous clay soils subject to temporary moisture excess during spring.

843

MOUSTAFA, A.T.A.; IBRAHIM, M.E.; BAKHATI, H.K. Effect of depth of saline water table on cotton yield and its consumptive use. Agric. Res. Rev., 1975, 53:21-4.

In lysimeter experiments, cotton cv. Menoufy, a salt-tolerant cultivar, was sown on soil with water table depths of 40, 70, 100, 130 or 160 cm. Cotton yield was determined at harvest and daily water uptake was measured. The highest cotton yield of 220 g/plot (4 m^2) was obtained from the 160 cm water table depth treatment and the lowest (1032 g/plot) from 40-cm depth. The low yield on lysimeters with high water table was attri-buted to restriction in root growth, compared with where the water table was lower. The contribution of the saline water (10 000 ppm NaCl + $CaCl_2$) in the water to the consumptive water use of the crop ranged from 30.7 per cent where the water table was 40 cm deep to 8.8 per cent where it was 160 cm deep.

844

MUCHA, W. Problem of evolution of forest soils after reclamation. Rocz. Wyzs. Szk. Roln. Poznan., 1963,

18:91-149.

The paper discusses effect of drainage on development and course of the soil forming process in forest soils as well as on changes in the plant cover and in the organic matter; describes the physical, physico-chemical and chemical transformations connection with the above process; states that the object of the respective investigations constituted swampy soil and soil developed of swamps, usually with sandy under-ground.

845

MUIR, D.C.; BAKER, B.E. Detection of triazine herbicides and their degradation products in tile-drain water from fields under intensive corn (maize) production. J. Agric. Fd. Chem., 1976, 24:122-5.

The herbicides cyanazine, cyprazine, atrazine, and metribuzin were applied (post-emergence) to plots which were drained by separate tile drains (depth 1.2-1.6 m). Water samples were collected from the tile outlets and extracted with ethyl acetate. In the first season of monitoring, herbicides concentrations ranged from 0.30 to 1.49 kg/L for atrazine, from 0.00to 0.68 ug/L for cynazine, and from 0.00 to 0.57 ug/L for cyprazine. Similar levels of the chloro-s-triazine herbicides were found in the tile-drain water during the second year of the study. Metribuzin, which was applied during the second year, was found in the tile-drain water in concentrations ranging from 0.00 to 1.65 ug/L. The metabolites deethylated atrazine (2-chloro-4-isoprophylamino-6 amino-s-triazine) and cyanazine amide 2-chloro-4 (1-carbamoly-1-methylethylamino- ethylamino-s-triazine) and deisopropylated atrazine (2-chloro-4-amino-6-ethylamino-s-triazine) was detected at concentrations which were similar to those of the parent compounds. The discharge(g/ha) of herbicide residues in the tile-drain water was determined over a 9-month period.

846

MULQUEEN, J. Drainage of a hillside seep in Co. Cavan. Ir.J. Agric. Res., 1975, 14:337-48.

The drainage of a hillside seepage arising from a perched water table was examined and measured. A drain installed to control the seepage was not well located as an interceptor drain and only partially penetrated the upper permeable stratum. As a result, the drain only controlled the seepage over one-ninth its length. Even where the drain appeared to be well located, large amounts of water underflowed it. It is concluded that interceptor drains and ditches should penetrate into the impervious clay to intercept all downhill flow and that the drain or ditch should be located just uphill from the seepage boundary.

347

MULQUEEN, J.; KIRKHAM, D. Leaching of a surface layer of sodium chloride into tile drains in a sand-tank model. Proc. Soil Sci. Soc. Am., 1972, 36:3-9.

The effects of two recharge rates, three salinities in the salted-sand layer, and three drain spacings were evaluated. Each experiment was conducted with a steady state, arch-shaped water table maintained by recharge from battery of capillary tubes that gave a uniform infiltration over the model. All the salt was leached from the model.

348

MUSTONEN, S.E.; SEUNA, P. Influence of forest drainage on the hydrology of an cpen bog in Finland. IAHS Pub., 1975, 105:519-30.

A drainage experiment based on the control basin method was carried out in south-eastern Finland from 1936 to 1969. Two natural basins were calibrated for 22 years. Then 40 per cent of one basin was drained for forestry. The drained peatland was mostly open bog and swamp with a poor growth of pine. The soil surface settled about 20 cm during the experiment. In addition, the groundwater level dropped by about 30 cm. A drastic change in the soil moisture regime largely destroyed the wet peatland vegetation. An increase of 43 per cent in annual runoff was observed, relatively evenly distributed throughout the year. Both the spring and summer maximum runoff increased, on average by 31 and 131 per cent respectively, caused mainly by the accelerating effect of the drainage network. It is probable that annual runoff from the peatland will decrease as the new forest becomes established. Conclusions presented here do not concern peatlands drained for agriculture where cultivation changes the effect of drainage considerably.

849

MYAGI, K.A. Investigation of movement of water towards horizontal drains in slightly permeable grounds. Izv. timir. S-kh. Akad., 1962, 3:233-40.

The paper presents a mathematical treatment. When the ratio of surface runoff to subsoil runoff is ¨1, drainage will mainly accelerate removal of surface water. When the ratio is `1, drainage will mainly accelerate removal of sub-soil water. A formula is presented describing the time required to remove excess water from the arable layer.

850

MYERS, V.I.; LANGAN, L.N.; LLOYD, R.D. Interpretative soil groupings for analysis of drainage problems in western irrigated areas. Trans. ASAE, 1962, 5:103-5.

Irrigable land in western Nevada was classed in 4 groups according to degree of drainage and depth of water table. Each group was divided into 5 drainage regions depending on the nature of the subsoil, structure, filtration capacity and moisture. Further subdivision was made according to organic-matter content, porosity, presence of roots, degree of cementation and layering. Land-reclamation maps were made on the basis of the data, from which drainage recommendations could be deduced.

851

MYERS, V.I.; REKTORIK, R.J.; WOLFE, C.A. Deflection tests and trench conditions for plastic drain pipe. Trans. ASAE, 1967, 10:454-7.

Polyethylene plastic pipes were tested in the laboratory models for safety against failure. It was concluded that plastic pipes fail by deflection rather than by rupture of the wall. A 12.5 cm diam. cradle excavated in undisturbed earth, designed to support a 160° section of the pipe should support for loading upto 446.4 kg per linear m. Concave curvature of the top of the semirigid plastic pipe begins at a deflection of about 20 per cent of its diam. Almost complete collapse of 10 cm diam pipe occurs after deflection to a vertical inside diam of about 6 cm. According to these tests in which a cradle was provided, it appears that a deflection of 7.8 cm would be tolerable under field conditions similar to the laboratory. In order to minimize the load on plastic pipe it is recommended that it be placed in a narrow trench approximately 25.4 cm wide.

852

NAARDING, W.H. Developments in land drainage. Proc. Int. Drain. Workshop, , Wageningen, Netherlands, (1978) 1979, 25:512-9.

The history of drainage in relation to changing improvements in drainage pipe, cover and envelope materials and its impact on machanization of drainage is discussed. It is reported that in many cases maintenance provisions are little cared for but are essential in any successful drainage scheme.

853

NAARDING, W.H. A review of international experience with trenchless versus trenching drainage-machines. 10th Congr. Irrig. Drain. Athens, 1978, 5:39-74.

The use of trenchless drainage machines in Europe had been surveyed, and experience with trenching and trenchless machines is compared. Performance data, capacities, opera-tional costs, construction costs for drainage systems laid under Dutch conditions (clay and PVC pipes with trenching machines, PVC pipes with trenchless machines) and drainage effectiveness are compared in detail, and special problems encountered during use, for example crossing infiltration ditches and filled-up ditches, are discussed.

854

NAARDING, W.H. Trenching and trenchless drainage. Proc. Int. Drain. Workshop, Wageningen, Netherlands, (1978), 1979, 25:543-60.

General characteristics of trenchless and trenching machines are discussed. A review of the results on trenchless drainage in European countries has been made. Advantages and disadvantages of both systems of drainage are listed.

855

NAGAHAMA, K.; TEJIMA, S.; TOMITA, M. A historical consideration on the technical development of underdrainage. Trans.Agric. Engng. Soc. Japan, 1967, 21:50-4.

The paper discusses the origin of underdrainage, its technical development and how it spread in various countries including Japan. It originated in, and was popularized gradually during the Edo period (1615-1867). Describes the results of investigations on the depth and spacing of culverts and the soil drained by a culvert indicated that underdrainage was provided in planting soil in most cases.

856

NAGAHAMA, K.; TEJIMA, S.; TOMITA, M. On the under-drainage of ill-drained paddy field- substantial study on the machanism of under-drainage (2). Trans. Agric. Engng. Soc. Japan, 1968, 23:39-45.

The capacity of a pipe drainage system installed in the sub-soil of low permeability (10.6 to 10.7 cm/day) and also the ground-water table movement has been investigated and concluded that artificial means of initiating cracks in the soil be taken to ensure effective underdrainage.

857

NAGAHORI, K.; AMAYA, T.; TAKAHASHI, T. Lysimeter test for subsoil improvement of the mud soil area, Kassoka Bay polder, with particular reference to physical properties of the soil 8 years after drainage. Scientific Rep. Fac. Agric. Okayama Univ., 1980, 56:63-71.

Results of lysimeter experiments with a sea-bottom mud soil that had received surface drainage eight years previously showed that soil water relations, bearing capacity and salt concentrations had reached a satisfactory state, and plant productivity was equal to that of normal field soils.

858

NAGASAKI, A. Studies on the mole-drain under paddy fields; a simple method of under-drainage. J. Fac. Agric. Iwate Univ., 1962, 6:142-219.

Customary methods of mole drainage in Japan, the durability of mole drains, traction resistance of the mole drainer, operation of the drain and future aspects are discussed.

859

NAIK, B.; BEHERA, B.C. Adaptability of a drainage system to irrigation distribution. J. Agric. Engng.,1981, 18:74-9.

The real index of successful water management lies in designing a system of facility that would provide control of moisture regime and would suit the crop of any duty. An alternative design to elevated unlined channels for rolling terrains is explained which consists of running the channels deeper than ground level with or without separate field drains so that a network of drains is provided. The system is adopted for irrigation by locking it up with suitable hydraulic structures.

860

NAKAMURA, T. Peat of Hokkaido and overdrainage. 5th Congr. Irrig. Drain. Tokyo, 1963, 5:17.91-17.108.

In Hokkaido, there is a 200,000 ha area of peat bogs, 30 per cent of which is used for agriculture. Disadvantages of over drainage in these soils are discussed. The methods to avoid overdrainage are: dressing mineral soils to reduce permeability; use of relief wells in tile drainage systems; damming of open drainage ditches and knife drainage and rolling soil surface.

861

NAKLICKI, J.; CIESLINSKI, Z. Soil improvements as a means of supplementing drainage on chernozem soils formed from loess. Wiadom. Melior.Lakarskie, 1975, 18:255-6.

In 1974, yields of barely grown without soil drainage were 4.02 t grain/ha with subsoiling compared with 3.98 t for deep ploughing and 3.53 t for shallow ploughing. With tile drains spaced at 10, 20 or 30 m, barley yielded 4.09, 3.84 and 3.61 t/ha, respectively, in the subsoiled treatment compared with 4.04, 3.97 and 3.94 t with deep ploughing and 3.76, 3.46 and 3.44 t with shallow ploughing, respectively.

862

NARAINE, S.S. Planning and development of Irrigation and drainage projects on the coastal plains of Guyana. 7th Congr. Irrrig. Drain. Mexico, 1969, 1:C1-C18.

The historical background of coastal development in Guyana is outlined. The concept of planning for drainage and irrigation is based on the integrated plan for optimum surface water control in relation to land use in each section of the coast. This plan provides for flood control on the smaller river basins, by the storing of flood in rainy season in the reservoirs and utilising it in the dry weather for irrigation. Drainage is effected by gravity flow to the ocean as far as practicable and only if not possible pumped drainage is used. Methods of drainage, irrigation and conservancy flood relief are outlined. Basic design criteria in common use are given.

863

NATHAN, K. Depth and spacing of parallel drain tile laterals. J. Soil Water Conserv., 1961, 16:81-2.

Physical properties of Monmouth county, New Jersey were determined at eleven sites and depth and spacing of tile drains determined using seven procedures. A discussion on the comparative values has been made to illustrate the wide variations among different locations and procedures.

864

NATIONAL COMMITTEE OF ICID, Federal Republic of Germany. Use of computers in drainage systems. ICID Bull., 1973, 22:74-7.

The drainage design based on theoretical and practical drainage experience using a computer programme is presented. An example of calculations by the computer are tabulated.

865

NAUMOV, G.V. Effect of intensity of drainage on water and air regime, agrochemical properties of soil and yields of row crops. Doki. Veses. Akad. S-Kh. Nauk, 1969, 11:42-5.

In a 3 year experiment in Karelia, systematic drainage by drains 16 m apart during the vegetative period improved the water and air regime and increased the yield of Jersey kale (*Brassica oleracea* var. *acephala*) compared with drains 36 m apart or a network of open drains 40 m apart.

866

NAVEH, Z.; ELAN, S. Biological protection of waterways and drainage channels in Mediterranean and semi-arid conditions. In Manual of Surface Drainage Engineering (Kinori, B.S.). Elsevier Pub. Co. Amsterdam/London/Newyork, 1970, 180-209.

The authors discuss the problem of drainage, channels vegetation and tree and shrub planting near drainage ways in 5 heads including description of the problem.

867

NEDECO. Bhal reclamation scheme. (Govt. of the Netherlands and State of Gujarat, India). Delft Hydraulics Laboratory, Delft, The Netherlands, 1968, Vols. I-VI.

In order to suggest measures for the reclamation of the Bhal area in Gujarat, representing a former tidal flat, now abandoned by the sea, consisting of vertisols in the higher part and Halaquepts in the lower part, lying as waste land or in use as poor pasture, a pilot polder project (1500 ha) was operated by NEDECO. The results obtained in the project had been poor due to various reasons including the presence of a layer of very low permeability in the upper metre of the soil profile, swelling nature of the soil, low amount of fresh water carried by the rivers during the monsoon, irregular and short duration floods and absence of adequate sources of fresh groundwater. Discharge of water from the pilot polder could not be affected due to severe siltation of the outlet channel. Desalinization of the soil did not respond to improved drainage. The recommendations emerged from the studies were not to proceed with plans for reclaiming the entire area of saline soils in Bhal, unless after extensive research under the given circumstances, more was known about desalinisation, whether or not in combination with flooding and drainage possibilities for shortage of water (dams) for irrigation in the area west of Bhavanagar-Ahemedabad road, availability of fresh groundwater in the same area and possibilities for storage of water in the Mithapur Depression. It was further recommended to continue investigations in the existing experimental area and collect systematically hydrological and hydrographical data. Measures to check erosion, especially in the valley of the Khalkhalia and encroachment of the sea in the Sonrai creek area by low levels around the threatened arable lands, were also suggested for investigation.

868

NEGISHI, H.; TADA, A.; FURUKI, T. Studies on subsoil improvement and irrigation-drainage system for clayey paddy fields. <u>10th Congr. Irrig. Drain. Athens</u>, 1978, 4:455-68.

The physical characteristics of clayey rice fields in Japan are described. A system of pipe (PVC or clay) drains laid at a depth of 0.6-1 m and covered with filter materials, combined with mole drains at 0.35-0.45 m depth and at right angles to the pipe drains has allowed rapid drainage even in impermeable paddy soils. With proper water management, this system allows more productive use of paddy fields.

869

NELSON, R.W. Design of interceptor drains in heterogeneous soils. <u>J. Irrig. Drain. Div. ASCE</u>, 1961,

87:41-53.

The design of interceptor drainage systems is complex and often hampered by lack of methods for analyzing variable permeability. An approximate method for including variable permeability in a design procedure is presented. The analysis requires a water-table contour map for transposing the flow system in heterogeneous soil, to an equivalent system, having constant permeability throughout. In the uniform system equations derived for homogeneous soil can be applied. Reverse-transformation of the prediction from the uniform soil system to the field problem gives the expected condition after drainage facilities are installed.

870

NEMEC, J. Numerical analysis of Hooghoudt's equation for the determination of drain spacing. Ved. Pra. Vyzk. Ustavu Melior. Pr. Zb., 1978, 14:5-17.

The paper draws attention to the influence of inaccuracy in the determination of individual parameters on the calculated drain spacings.

871

NERPIN, S.V.; RYVKINA, T.F. An approximate calculation of the shape of the water-table in drained land. Vestn. S-Kh. Nauki, Mosk. 1978, 9:117-20.

Calculation of the shape of the water-table between drains in soil with a water-table of moderate height is described.

872

NESTEROVA, G.S. Effect of drainage on moisture-air regime of meadow-sod surface-gleyed soils of the southern part of the Khabarov region. Pochvovedenie, 1963, 7:84-93.

These soils are temporarily waterlogged during the

monsoon, and drainage by closed drains containing a filtering mixture of sand and gravel, or by small open hollows, improves the air-moisture regime particularly in the top 20 cm and to a less extent in the 20-60 cm layer and has no effect on the 60-100 cm layer. The period during which the air content in the plowed layer was $>$ 20 per cent was 6 times longer in 1960 and 2.6 times longer in 1961 on plots with small open hollows than on the control plots and with closed tiles longer. The period during which the air content in the plowed layer was $<$ 10 per cent was 1.7 times longer in 1960 and 4.5 times longer in 1961, on the control plots than on those with hollows. In 1960, this period was 2.8 times longer on the control plots than on those with closed drains and in 1961 the air content never fell below 10 per cent with closed drains. Open hollows combined with furrowing ridging, etc. ensure an adequate air-moisture regime.

873

NESTEROVA, G.S. Water-air regime of drained land of southern part of the Khabarovsk region. Pochvovedenie, 1966, 10:65-70.

Drainage of depressions and closed collectors in order to accelerate run-off on the soil surface and within the soil during the monsoon improved the water-air regime of the plowed, top 20 cm of soil, but had practically no effect on the 20-100 cm layer, which remained practically saturated with water. Such drainage increased yields of spring wheat and soya and green yield of maize 1.5-2.2 times.

874

NEUSTEIN, S.A.; THOMPSON, D.A. Drainage. In Rep. on Forest Res. Forestry Commission, UK, 1973, 67-8.

Experiments have shown that after ploughing there is little difference in the shrinkage of peat between planted or unplanted plots. Attempts to drain clay subsoils to improve the rooting depth of trees have been difficult. Most drying was probably caused by the tree crop and not all clay subsoils develope a structure after drying.

875

NICOLA, F.; AMER, M.H. Effect of subsurface drainage on increasing productivity of crops in the Delta area (U.A.R.). <u>ICID Bull.</u> 1969, 18:57-66.

The studies on the effect of subsurface drainage on crop yield in the Nile delta are discussed. In several studies starting since 1942, it is conclusively proved that the yields are increased. Though the percentage increase varied between studies, the report of the joint committee of Ministry of Irrigation and of Agriculture reported that percent increase due to subsurface drainage is of the order of 35, 32 and 27 percent for cotton, maize and wheat respecively.

876

NICOLA, F.; EL-MAHDY, S. Feasibility of mechanization in the execution of tile drainage in the U.A.R. <u>8th Congr. Irrig. Drain. Varna</u>, 1972, Q 28.2 R 17:23-40.

The increase in the yield of agricultural land estimated at an average of 40 per cent is found to be achieved economically through drainage. A description of the sample unit of a project is given in which the mechanization of the manufacture of tiles, the mechanized operation of trenching, pipe-laying and refilling have been accomplished. The factors influencing the choice of the type of such machines from technical, economical and mechanical aspects are given. Through mechanization cost can be reduced from $0.65/m length to $0.40/m length which is about 36 per cent less.

877

NIEUWENHUIS, G.J.A; WESSELING, J. Effect of perforation and filter material on entrance resistance and effective diameter of plastic drain pipes. Agric. Wat. Mgmt., 1979, 12: 1-9.

The flow towards drain pipes with and without filters was investigated by means of numerical solutions. Conformal mapping of the actual flow plane onto a plane with parallel flow allowed the development of a method to find the distribuition of the potential around the drain pipes. Solutions for two-dimensional flow were obtained. Although in actual cases three dimensional flow will occur, the solutions obtained may be used to find the relative influence of perforation pattern, filter thickness and filter conductivity on the effectiveness of the drain. Examples of the influence of these factors on the entrance resistance and effective diameter of some drains are given. Conditions in which the filter is partly blocked by soil material were also treated.

878

NIGAM, K.M.; PANDEY, H.N. Groundwater table and assessment of salinity hazards in major river valley project areas of Bihar. Proc. All India Symp. Soil Salinity, Kanpur, 1971, 92-8.

The quality of irrigation waters of Kosi-Sone areas from various sources like canals, rivers, borings and tubewells falls in USSL class C_1S_1 or C_2S_1. Studies on groundwater table show that there is no immediate problem of development of soil salinity in Kosi and Sone major river válley projects. Some sort of drainage system must be installed to prevent deleterious effect of rising groundwaters. Installation of tubewells at come sites shall bring proper control of soil-moisture-air balance in the soil and improve crop production.

879

NIKOLSKI, Y.N., The dependence of irrigation requirements on water-table depth in drained lands. Agric.

Wat. Mgmt., 1977, 1:191-6.

The amount of irrigation water required to maintain an optimal average water content in the root zone for maximum crop yield is dependent on the water-table depth as well as on evapotranspiration. An approximate expression is derived in terms of soil constants for the vertical soil water flux, which may be directed upwards or downwards, above the water table. This has been used in a water-balance equation in order to estimate the irrigation requirement. Satisfactory agreement is found between results calculated using this theoretical relationship and published experimental work.

880

NISKANEN, M. New techniques in the cleaning of ditches in forest drainage areas. Suo, 1977, 28:75-8.

Ditch maintenance in the forest should be primarily restricted to the main ditches, trap ditches and some of the contour ditches. Ditch cleaning units for fitting to tractors have been made more versatile by incorporating a digging stabilizer and a special scoop. A hydraulic rotary ditcher fitted to a wide-tracked swamp tractor is used for cleaning/digging in difficult terrain and ditch intersections.

881

NITU, T. Aspects of soil drainage in Spain. Circ. Agron. Moldova, 1975, 8:153-6.

the effect of drainage and reclamation of saline soils by drainage in Spain are reviewed and data are given for effects on barley, oats, wheat, Phaseolus vulgaris, sorghum, maize and cotton.

882

NIXON, M. Economic evaluation of land drainage works. In River Engineering and Water Conservation Works,

Butter Worths, 1966, 47-51.

The paper deals with the drainage objects and benefits, (direct benefits, changes of land use, intangible benefits, cost benefit ratio) and discusses economic life concept, monetary measurement of benefits, maintenance aspect, and deficiencies of economic evaluations.

883

NOJIMA, K.; TANAKA, I.; UEMURA, Y. Influence of drainage on growth of the rice plants in rice fields. I. Effect of drainage of surface water on growth, yield and nutrient uptake of rice. Proc. Crop. Sci. Japan, 1962, 30:321-4.

On drained plots, the number of tillers at tillering and booting was greater and the number of panicles at harvest fewer than on submerged plots. Uptake of N, P, K and Si and grain yield generally decreased with drainage, although nutrient uptake sometimes increased temporarily at tillering. The yield reductions with drainage were less marked with heavy fertilization and less conspicuous with granular than with powdered fertilizer.

884

NOLTE, B.H.; FAUSEY, N.R,; SKAGGS, R.W. Time available for field work on two Ohio soils. Trans. ASAE, 1983, 26:445-51.

Simulations were conducted using DRAINMOD, simulation model, to estimate the days available for field work, compute an indicator of excess water and determine a relative corn yield index. Results are based on 62 simulations for two soils, three locations, several threshold soil water contents and numerous drainage systems. Data are reported on the specific conditions simulated. The relative corn yield index includes factors for excess or deficit water and delayed planting. Simulations gave average relative yield indexes from 20 to 77 per cent. The paper deals with specific conditions and suggest that similar analysis for other poorly drained soils may result in broad generalizations to be made in this regard.

885

NORTH OF SCOTLAND COLLEGE OF AGRICULTURE UK. A systematic field approach to agricultural drainage. Bull. North of Scotland College of Agriculture, 1980, 20, 24 p.

This bulletin is intended as a guide to recognising land drainage problems and suggests remedies which are most likely to improve soil productivity at reasonable cost. Although the approach is comprehensive, the field check list at the end is provided as a quick guide for all drainage cases.

886

NOVOSAD, R.G.; IZDRIK, V.M.; MIKITCHUK, S.M. Effect of underground drains and fertilizers on yield of pea/oat mixture on low peatlands. L'viviskii sil's Kogos. Inst., 1975, 63:104-7.

In trials on low peatlands in the W. Ukraine, pea/oat mixture given 120 kg P_2O_5 + 120 kg K_2O/ha gave fresh fodder yields of 36.4 t/ha on plots with underground drains 30 m apart and 24.8 t/ha on those with ditch drains; decreasing the distance between the underground drains to 15 m adversely affected the water-air regime and decreased yields.

887

NURIJANIAN, V.N.; KHIZANTSIAN, S.M. Efficiency of chemical reclamation of soda salinized soils against the background of closed horizontal drainage. Int. Symp. New Develop. Field Salt Afffected Soils, Cairo, (1972), 1975, 499-503.

The chemical reclamation effect of soda saline soils in Ararat plain of the Armenian SSR with ferrous vitriol (in powder form) on the background of closed horizontal drainage was followed by desalinization of reclaimed soils at a depth of one m. From the estimated one m layer, the alkaline reaction has been neutrialized almost completely. In the soil exchangeable complex, active sodium has been replaced by calcium. Normally working of closed horizontal drainage ensures the outflow of leaching waters. Chemical reclamation of soda saline soils, with a extremely hindered natural outflow, must be carried out on the background of closed horizontal drainage.

888

NWA, E.U. Using the bamboo (Bambusa vulgaris) as field drainage material. Trans. ASAE, 1981, 24, 974-6.

The high cost of imported subsurface drainage materials in Nigeria, demands that efforts be made to locate relatively cheap local materials which can serve the same purpose. This study was designed to test the usefulness of the bamboo (Bambusa vulgaris) as a drainage material. It is concluded that the bamboo can be used effectively for field drainage and it is recommended that efforts be made to determine the design criteria and the drainage life of this material.

889

NWA, E.U.; TWOCOCK, J.G. Drainage design theory and practice. J. Hydrol., 1969, 9:259-76.

A review of drainage theory chosen for testing is reported. The drainage system consisted of 7.62 cm diam clay tiles with no filter at a depth of 119.94 cm on a loam soil. The spacing between drains was 40.23 m. Out of the five equations tested, all except Luthin's gave an average calculated spacing which is larger than the actual spacing. More research is recommended before one or two equations could be chosen to aid the design of drainage scheme.

890

OCHS, J.W. Controlling excess waters on agricultural land: SCS standards and Specifications. Proc. Natn. Drain. Symp. ASAE, Illinois, 1971, 50-1, 55.

Various standards of agricultural drainage used by Soil Conservation Service in USA have been discussed.

891

OCHS, J.W. Testing synthetic fabrics for use with drainage conduits. Proc. Int. Drain. Workshop, Wegeningen, Netherlands, (1978), 1979, 25:332-9.

Development of an effective synthetic material for high-speed installation of subsurface drains is essential. American Society for Testing and Materials. (ASTM) is conducting tests for evaluating these materials ASTMD-1682, ASTMD-751 and ASTMD-1682 tests are recommended for tensile strength, bursting and puncture strength and abrasion resistance respectively. A test being devised to measure the average pore size and equivalent opening size is underway and explained. Fabric filters can be used with confidence, if the silt fraction (0.05 mm to 0.002 mm) in the soil around a conduit is less than 40 per cent.

892

OGATA, H. System design for drainage management. 10th Congr. Irrig. Drain. Athens, 1978, 7:637-62.

A successful drainage system exists in the Shinkawa River district (corner of the Niigata Plain), a representative rice area. In this system, the following are computerized; the usage of information (data such as water level and rainfall), and the manipulation of drainage facilities (eg. pump and gate). Components of the system include a computer, wire-less telemeters and telecontrol devices. The system design is described in detail. The background for its need, its possibilities and limitations are explained.

893

OHLSSON, T. Redox reactions in soils; sequences of redox reactions in a waterlogged soil. Nor. Hydrol., 1970, 10:89-98.

An apparatus is described which under laboratory conditions permits sampling of water and gases from a waterlogged soil throughout the experiment period. A cultivated soil supplied with 6 per cent straw is waterlogged during 9 days and the sequence of some redox reactions is shown. After one day of waterlogging the redox potential declines and reaches negative value after six days. pH declines from 7.5 to 6.5 after 17 days and slowly increases after that to neutral again. Nitrate is the first compound to be reduced and it disappears within two days with nitrite detected during this period. After nitrate has been reduced manganese increases followed by increase of iron. Sulphur totally disappears within first three weeks, CO_2 produced during all the time of waterlogging and methane production begins only after three weeks. Nitrogen dioxide was not detected.

894

OKA, A.; NAGAISHI, Y.; IDE, K. Counter measures of drainage of paddy field for the purpose of effective usage in the rainy region of heavy marine clay. 10th Congr. Irrig. Drain. Athens, 1978, 4:469-85.

Drainage problems in the heavy alluvial marine clay of the Ariake rice area are discussed. Improved drainage is achieved by using pipe drains covered with rice chaff as filter material in combination with shallow mole drains. The importance of recent agronomic measures such as continuous production of rice and barley or wheat by direct sowing without tillage, and sowing of seed in ditches in strips is stressed. Special equipment has been developed for use with the mole to prevent clogging of the implement by straw.

Measures to prevent leakage of water from drainage channels due to drying of the rice field, and control of storage of the surface water are important aspects for future study.

895

OKAZAKI, N.A. New construction method to lay a pipe under the ground. 11th. Congr. Irrig. Drain., Grenoble, 1981, 2:789-806.

A method is described which provides for the simultaneous accomplishment of pipe-laying and refilling, so minimizing settlement problems of the ground.

896

OKOUTINE, Y.A. Drainage in land reclamation in USSR. Proc. Int. Drain. Workshop, Wageningen, Netherlands, (1978), 1979, 25:709-12.

The paper gives general information and some principal ways of development in agricultural drainage in USSR, have been discussed.

897

OKRUSZKO, H. Changes in phosphorus content of organic hydromorphic soils due to drainage. Roczn. Glebozn., 1964, 14 (suppl.):183-95.

The P content of the top layer of drained peat was three times higher than that in the lower layers; total P_2O_5 in the 0-20 cm layer was 600-1500 kg/ha, of which 25-225 kg was in mineral form and 10-50 kg in easily soluble form. About 60 per cent of the total P occurred in humic compounds, of which 1[illegible] per cent was in the hemicellulose fraction and 13 per cent in difficult decomposable lignin substances. The P:N:C ratio changed from 1:26:372 in undrained muck soil to 1:16:25 in drained peat (which approaches the 1:10:100 ratio typical humus). Results indicate that drainage of muck soils results in an increase in total and mineral P and decrease in plant-available P (due to fixation by mineral and organic compounds).

898

OLESEN, S.E. Trenchless drainage experiments in Denmark. Proc. Int. Drain. Workshop, Wageningen, Netherlands, (1978), 1979, 25: 527-35.

In four experiments trenchless drainage has been compared with trencch drainage. It is found that in each case drainage efficiency is higher in case of trenching compared to trenchless drainage. It is suggested that trenchless drainage is most suited for flat areas and sandy soils. For areas less than 6 ha, trenchless machines are not recommmended because of transportation difficulties. It should not be used on heavy, badly structured or clay soils.

899

OLSSON, T. Methods of calculating groundwater flow to drains. grundflorbattring, 1975/76, 27:53-6.

Some methods of calculating groundwater flow to a drain are presented. A relationship between groundwater drawdown and drainage yield is derived. The natural baseflow through an area is described and an equation of this baseflow is set up.

900

O'NEILL, D.G. Gravel tunnel drainage at Castle Archdale. Agric. Northern Ireland, 1977, 52:40-5.

Two experiments at Castle Archdale to compare gravel tunnel drainage with mole and pipe drainage are described. Marked increases in spring grass growth were produced by the gravel tunnel system on the difficult clay soils of this high rainfall region. Both mole and gravel systems were effective in lowering the water table; where mole and gravel drains were at the same depth, the gravel drains were rather more effective. The surface conditions of the sward were closely related to water-table level.

901

OOSTERBAAN, R.J.; WIND, G.P. Design and research. Proc. Int. Drain. Workshop, Wageningen, Netherlands, (1978), 1979, 25:3-13.

The paper is a synthesis of the 15 papers presented on the topic design and research of subsurface drainage. The paper covers mainly the work on drain spacing formulae, hydraulic conductivity, drainage criteria and drainage of slowly permeable soils.

902

OOSTERVELD, M. Disposal of saline drain water by crop irrigation. In Dryland-saline Seep Control, Proc. 11th Int. Soil Sci. Soc. Congr., Canada, 1978, 4.24-4.29.

Application of a single irrigation with saline drainwater with an EC of 9.3 dS/m did not result in a serious buildup of soil salinity. With appropriate care drainwater from many saline seeps could be disposed of by irrigating nearby crop land. The high nitrogen content often found in such drain water can be very valuable for crop production.

903

OPREA, C.V.; SANDU, G.H.; STEPANESCU, E.; VLAS, I. Reclamation of sodic soils in Rumania. Agrokem. Taljt., 1969, 18(Suppl): 321-8.

Improvement of sodic soils under meadow and pasture has been suggested through raising of Puccinellia distans, Aster tripolium, Agropyrum repens, Alopecurus pratensis, Lotus corniculatus. The most resistant crops are Agropyrum tenerum, Melilotus albas, in combination with Agropyrum cristatum. Bromus inermis is medium-salt-resistant and *Dactylis glomerata* poorly salt resistant. The deep loosening of the soil, digging of ditches spaced at 1 m and 2 m, respectively and the application of phosphogypsum and fertilizers caused significant improvement in the soil and sorghum grain yields and Sudan grass. Sodic soils with average reclamative status and satisfactory internal drainage can be reclaimed through rice growing. Combined application of hydraulic engineering, agro-technical, biological and agrochemical measures is the best approach for reclamation.

904

OSTER, J.D.; RHOADES, J.D. Calculated drainage water composition and salt Burdens resulting from irrigation with river waters in the western United States. *J. Environ. Qual.*, 1975, 4:73-9.

Drainage water compositions were calculated with a computer simulation model from irrigation water compositions, leaching fractions, aragonite and gypsum solubilities, and measured partial pressure of CO_2. The calculated compositions were compared with measured values obtained from lysimeters filled with Pachappa soil, cropped with alfalfa, and irrigated with eight syntnetic waters typical of rivers in the western U.S. Linear regression analysis of predicted vs. measured values for Na and SO_4 ion concentrations, SAR, EC and salt burden resulted in essentially one-to-one relationships. For salt-tolerant plants, leaching fractions between 0.05 to 0.1 should be safe for six of the eight waters. The predicted relative decrease in salt burden ranged from 0.32 to 0.85 for a decrease in leaching fraction from 0.33 to 0.04.

905

OSTROWSKI, J. Observations on efficacy of drainage in conditions of feeding by atmospheric waters without ground waters participation. *Zesz. Nauk. SGGW Ser. Melior.*, 1961,3:17-33.

The drained area had the characteristics of small permeability of the subsoil layers, resulting in case of intensive precipitations in overmoistening of soil. After carrying out the drainage, groundwater level was lowered by 20 cm on the average when the profile moistening nearby drain and in the centre of spacing was similar, whereas the moistening of lower soil layers was distinctly influenced by the ground water rise. It was also found that the unit outflows actually observed were considerably smaller (0.14 L/sec/ha) than the outflows accepted in the project for calibration of collectors (0.55 L/sec/ha).

906

PAFFORD, E.J.; PRICE, E.P. A disposal system for agricutural waste waters in the San Joaquin Valley of California. 7th. Congr. Irrig. Drain. Mexico, 1969, 6:26.407-26.419.

To drain the San-Joaquin Valley, 302.5 km long drain is being constructed. The drain construction follows a period of intensive study of, and public debate over the urgency for drainage disposal; the possible drainage plans; the construction, operation and maintenance costs of possible plans; the impact on economy and ecology and the possible need to treat the drain water before discharging it. Investigations on sources of nitrates in the agricultural waste waters and methods for their removal are underway in a big manner supported by Bureau of Reclamation.

907

PAHLKE, K. Investigations on structure and water content of a young sea marsh soil. Z. Kult., 1961, 2:278-96.

Tile-drain distances of 6, 9, 12 and 15 m at mean depths of 0.8 and 1.1 m in the loamy clay of a young mud deposit on the western coast of Schleswig-Holstein showed no influence on crop yields, soil structure and water content of soil.

908

PAIVANEN, J. The effect of ditch spacing and furrowing on the depth of the groundwater table and on the development of Scots Pine plantation on small-sedge bog. Silva Fennica, 1974, 8:2215-24.

Describes the results of a study of the effects of ditch spacing (10, 25 or 40 m), ditch depth (40,70 or 100 cm) and furrowing (at right angles to the ditches) in a young Pinus sylvestris plantation on drained peatland in central Finland. In general, the depth of the water table increased with narrower ditch spacing, furrowing shortened the period of highwater table, particularly where drainage was ineffective; and height growth of P. Sylvestris after five growing seasons was greater with narrow ditch spacing, though these differences were not evident in plots with furrows.

909

PAIVANEN, J.; WELLS, E.D. Guidelines for the development of peatland drainage systems for forestry in Newfoundland. Rep. Newfoundland, Forest Res. Centre, Canada, 1977, N-X-156, 44 p.

Drainage design factors include different types and forms of ditches, ditch depth, ditch spacing, drainage depth, size of main ditch,water discharge, drainage equipment, preliminary field surveys and final design of the drainage systems. The application of the drainage principles is discussed for five peatland sites in central Newfoundland.

910

PAK, K.P.; STEPANETS, I.T.; VEREDCHENKO, YU.P. Soil hydrogeological investigations in the Sarpinsk lowland in connection with rice cultivation. Pochvovedenie, 1966, 7:43-50.

When rice was grown for two years on a liman

(depression) soil whose water-table lay at a depth of 5-6.5 m in very saline, sandy-loamy layers of chocolate clay, the necessary flooding produced a perched water-table above the clay. Alkalinity due to CO_3 occurred in the perched water-table, but not in the groundwater. Flooding increased HCO_3 alkalinity in the soil. Drainage is essential for rice cultivation here.

911

PAKHUCHII, V.V. Fluctuations in the radial increment of pine stands on a drained bog. Lesn. Z., 1978, 1:15-8.

In the 1840's, a transitional bog in the Leningrad region was drained by a system of parallel ditches approx. 200 m apart. Increment cores were taken on the pines (*Pinus sylvestris*) on the drained area (125-170 yr old) and on an undrained control area (115-170 yr old), and a detailed analysis was made of fluctuations in radial increment. In the first decade after drainage, the increment on the drained plot increased by 464 per cent and the superiority in increment on the drained plot continued for 45 yr. Maximum increment was noted 15 yr after drainage (7.3 mm/yr as against only 1.2 mm/yr on the control plot). After drainage, the influence on increment of the temperature regime of the current year became of greater importance.

912

PALMER, D.B.; JOHNSON, H.P. Field evaluation of flow through blind inlets. Trans. ASAE, 1962. 5:58-61.

Blind inlets are constructed by backfilling a tile trench with various gradations of materials to increase the flow rate into the tile line and consequently reduce the time water stands in the surface depression. In a field experiment blind inlets made of soil, sand, corncobs and a combination of materials of varying sizes (graded) were tried. It was observed that statistically there were no significant differences among the four inlets. In general, however, corncobs and gradation inlets produced higher average discharge rates than soil and sand inlets.

913

PALVADI, H.K.; LAL, B. Note on susceptibility of maize to waterlogging at different growth stages. Pantnagar J.Res., 1976, 1:141-2.

In trials with maize cv. Ganga-2 given 150 kg N/ha in 3 split dressings at sowing, knee-high and tasselling stages, flooding the crop with water to a depth of 5-15 cm for 48 hr at the seedling, knee-high, tasselling or milk-ripe stage decreased grain yields by 59, 35, 63 and 31 per cent respectively, from 5.45 t/ha without flooding. The yield reductions increased with combination of flooding at 2 or more of the stages.

914

PANADIADI, A.D. Improving the quality and decreasing the cost of closed drainage systems. Vest. S-Kh. Nauki.Mosc., 1982, 2:112-8.

Technical details (degree of silting in relation to drain diameter, drain spacing, gradient of drains) regarding improved efficiency of closed drainage are examined. The need to develop new forms of drain pipes and sleeve joints is stressed.

915

PANARIN, M.N.; PANARIN, YU M. Drainage of boggy and excessively wet forest areas by aspiration vaccum drainage. Lesn. Khoz., 1980, 10:26-7.

The principles of horizontal and vertical aspiration drainage and also of vacuum vertical and vertical/horizontal drainage are explained. The scope for the application of these measures in forest drainage schemes is outlined in general terms.

916

PANDEY, S.L.; SINHA, A.K. Studies on water table position, soil properties, crop growth and yield under drained and undrained conditions. Indian J. Agron., 1971, 16:494-501.

Provision of open drains of 1 m depth and spaced at 32 m apart in waterlogged and saline area lowered the watertable sufficiently down. Rate of water table recession following heavy rainstorm was fast to avoid any stagnation of water due to presence of drains. Oxygen diffusion rates in drained soil were 10.60-37.8 x 10^{-8} g cm $^{-2}$ min^{-1} compared to in undrained soils which ranged 20-23.17 x 10^{-8} g cm^{-2} min^{-1}. Yield of sorghum and wheat during two years was continuously high in drained plots. No extra leaching water was required to remove salts as rainfall took care of the leaching requirements.

917

PANG., Z.X. Effect of tile drainage on the permeability of paddy soils. Acta. Pedol. Sin., 1980, 17:85-8.

Tile drains installed at 1.2m depth in clay and loam soils, and at 1.0 m (1.5 m for cotton) in a sandy soil with drain intervals of 10, 15 and 20 m respectively resulted in improved percentages of white roots (in relation to brown and black roots) in the total root weight observed at the tillering, elongating and booting, and emergence stage of rice.

918

PANOV, E.P.; SHISHKOV, K.N. Some problems of drainage of flooded lands. ICID Bull., 1973, 22:6-11.

Lysimeteric studies on evaluating the drainage need of crops are presented. Based on the results of the experiments aiming at creation of favourable water-air regime on humus-peaty thick soils for vegetable, industrial and grain crops, rates of drainage for under presowing tillage are 60-65 cm; under-sowing 65-75 cm, during the period from sowing to the beginning of maximum plant height 75-100 cm, maximum plant height to harvesting 100-130 cm and after harvesting 70-100 cm. For grasses the rates respectively were 30-40 cm, 40-50 cm, 50-60 cm, 60-85 cm and 70-100 cm.

919

PANOV, E.P.; SHISHKOV, K.N. Water-air regime and water supply of agricultural crops on drained humus-peat thick soils. Pochvovedenie, 1971, 10:83-93.

In the bottomland of the river Yakhron, in 1965-1970, optimal conditions of soil moisture and aeration occurred when drainage adjusted the height of the water-table to 65-130 cm for vegetable and technical crops, 60-100 cm for grain crops and 50-80 for perennial and annual herbage.

920

PANU, U.S. STAMMERS, W.N. Effect of drain boundary on subsurface drain design. 3rd Natn. Drain. Symp. ASAE, Illinois, Michigan, 1976, 22-6.

An experimental and analytical procedure is described which provides information on the effect of number, size and location of openings on drain tube inflow. For curved water table conditions that exist during drainage the results indicate that openings should be located in the lower half of the drain boundary. For 50 or more holes per meter length of drain tube the flow resistance associated with convergence toward the holes is less than that associated with 3 m of drain spacing. For a specified opening area per unit length of drain, small openings (.0016 m diam.) produced less flow resistance than large openings (.0032-0.64 m diam.).

921

PAPISOV, R.I. Combination of closed drainage with slightly raised convex ridges on heavy soils of the Kolkhida lowland. Subtrop. Kul't., 1962, 4:105-22.

On derno-meadow soils this system decreased the length of time for which standing water occurred in the rooting zone during the vegetative period from 34 per cent to 9.6 per cent. Drainage water removed 0.188 ton Ca/ha annually from the soil, 0.055 ton R_2O_3 and 0.07 ton humus.

922

PAPISOV, R.I. Effectiveness of agricultural improvement practices in tea growing in the Kolkhida lowland. Pochvovedenie, 1976, 7:84-93.

The poor soil hydrologic regime in this lowland requires improved practices. The effectiveness of profiled ridges, closed drainage, or a combination for tea growing was determined. Long-term observations showed that the highest tea-leaf-yield was obtained with a distance of 5 and 10 m between drains. Changes in soil properties are recorded.

923

PARKER, J.J.; WILLARDSON, L.S. Trench bottom shape effects on flexible pipe strength. Trans. ASAE, 1981, 24:1188-90,1198.

Flexible pipe strength is affected by the shape of the trench bottom. Laboratory tests were conducted using flat plate top loading and steel bedding forms in a universal testing machine to help quantify this effect. The steel bedding forms included a flat plate, a 90 deg V-groove, and three different diameter semicirculars. Six different pipe sizes were used.

924

PARRONDO, R.T.; GOSSELINK, J.G.; HOPKINSON, C.S. Effects of salinity and drainage on the growth of three salt marsh grasses. Bot. Gaz., 1978, 139:102-7.

Seedlings of Spartina alterniflora, Spartina cynosuroides, and Distichlis spicata were grown in a controlled environment in a nutrient solution containing different concentrations of NaCl. In a greenhouse experiment, seedlings of S.alterniflora and S.cynosuroides were grown in a 50-50 sediment-sand mixture that was constantly flooded or drained. Dry-weight accumulation by the plants indicated that D. spicata was the most tolerant of high salinity. Spartina Alterniflora and S. cynosuroides resembled each other in their response to salinity. In general, increasing the salinity of the root medium reduced shoot growth more than root growth. Roots of S. alterniflora grew better under flooded than drained conditions, while overall growth of S.cynosuroides was significantly better in drained than in flooded sediment.

925

PATERSON, E.; MITCHELL,B.D. Erosion deposits in tile-drains. Agric.Wat. Mgmt., 1978, 1:311-7.

The mechanism of silting up of tile field drains in two heavy textured surface water gley soil was examined by comparing the physical and minerological characteristics of deposits in drains with those of the clay soil directly above. The deposits consists of clay, silt and sand and are distinctly laminated. They appear to be result of internal erosion of the soil. The processes operative seems to have been: (a) inflow of soil through the drain joints; (b) sedimentation in the drain; (c) elution of very fine material. It is possible that this silting-up phenomenon is not uncommon in the tile field drains of surface water gley soils of Scotland.

926

PAUL, C.L.; VRIES, J.De. Soil trafficability in

spring. I. Simulation of the drainage process. Can. J. Soil Sci., 1983, 63: 15-26.

A mathematical-physical model developed in the Netherlands for the simulation of nonsteady flow of water in subsurface drained soils was tested by comparing its output with data collected in spring from two farmers' fields in Lower Fraser Valley of British Columbia. The ultimate objective of the tests was the prediction of soil strength and trafficability. The model was found to be useful for predicting depth to the water table,and to a lesser extent, soil water matric potential in top 15 cm in fields drained by drains at 30.5 m and 15.2 m. The two problems encountered in evaluating input data on drainage intensity and soil water flow parameters were solved by using modified Glover-Dumm transient-state drainage equation and taking heterogeneity of soil into account, respectively.

927

PAUL, C.L.; VRIES, J.DE. Trafficability in spring. II. Prediction and the effect of subsurface drainage. Can.J.Soil Sci., 1983, 63:27-35.

Effects of subsurface drainage on trafficability of farm land were assessed over a 2 yr period. When compared with undrained land, drainage by drains spaced at 30.5 m resulted in earlier trafficability by 2-4 wk in a silty clay loam and by 5 wk in a muck. Halving the spacing did not increase the benefits. Drainage intensity exerted influence on trafficability when the intensity was low, that is at large drain spacings.

928

PEDERSEN, E.F. Drainage and ground water level on marsh soil. Tidsskr. Pl. avl., 1977, 81:325-45.

Results are given of drainage trials in 1959-76 on clay soil of marine origin over sand with pH 7.6 in which drains were set (a) 80 cm deep, 24 m apart,(b)80cm deep, 18 m apart, (c) 115 cm deep, 24 m apart, (d) 115 cm deep, 18 m apart. A 4-course rotation of winter wheat, beet, barley and oats was grown upto 1966, after which 4 different rotations which included 2 yr of grass, 4 yr of clover/grass; maize and oilseed rape in addition to barley, wheat and oats were used. Data are given on soil characteristics and seasonal ground water levels. In virtually all cases, crop yields were lowest with (a) and highest for (c) and (d). Av. annual grain yields of winter wheat ranged from 5.57 t/ha for (a) to 5.78 t for (c), of barley from 5.00 t for (a) to 5.37 t for (c), of oats from 4.95 t for (a) to 5.34 t for (d) of oilseed rape from 2.04 t for (a) to 2.17 t for (c), of 4 yr clover/grass from 10.63 t DM for (a) to 10.83 DM for (d), of maize from 9.35 to for (a) to 9.94 t for (c) and of beet (roots + tops) from 17.9 t DM for (a) to 18.64 t for (d). The results indicated the importance of a good seedbed and early sowing in spring for barley and oats and that overwatering crops such as wheat and clover/grass tolerated intensive drainage best where surface water could be removed. Good drainage also increased yield dependability. In a 2nd series of trials, ground water levels were established in 40/60, 60/120 and 120/120 cm depth in summer (April-Sept) and winter, resp., and a rotation of winter wheat, root crops, barley, oats and red clover/grass established. Yields of all crops, especially barley and oats, were reduced by a ground water level of 40 cm in summer. With all crops except oats, yields with summer ground water levels of 60 and 120 cm differed little, with oats, grain yields were 10 per cent lower with the 60- than with the 120-cm level. Lack of drainage and high ground water levels in spring were particularly damaging to early sown crops.

929

PEDRERO, J.J. Cost of an agricultural piped drainage system. <u>Ing. Hidr. Mex.</u>, 1961, 15:69-85.

The paper states that in calculating costs of a

drainage system the following items should be considered: topographic chart, chart of soils and chart of subsurface water levels, soils and water's laboratory analysis, placement, course and timbering of drains, supervision, trench digging, pipes, transport costs, and distribution in the field, preparation of the gravel bed, pipe placement, covering of the same with gravel, earthfilling and covering of the ditch, supplementary works, contracts, interest and recuperation of capital.

930

PELISSIER, F. Operation and maintenance of the drainage system of the marshland of Les Baux. 7th Congr. Irrig. Drain. Mexico, 1969, Q 25, R 5:45-55.

The marshland of 'Les Baux' is a tract of lowland, at the level of the sea. A pattern consisting of double bank canals for intercepting water coming from the land outside and carrying it through the district to the outlet canal below are used for drainage. Pumping stations and subsurface-cum-seepage drains or ditches are maintained and operated by the farmers. The maintenance of drainage is done by weed cutting launches, mowing baskets and dragline excavators.

931

PELKONEN, E. Effect of damming on water table depth and ditch condition, Suo, 1960, 31:33-9.

In peatland areas, water consumption by the trees in June usually exceeded the water input as rainfall; water-tables should therefore be kept high. In late summer and autumn the water-table be deeper, owing to greater oxygen requirements of the roots and the greater probability of high-rainfall periods. In an area of 110-cm deep open ditches with 20 m spacing between ditches, the operation of dams in part of the system to maintain the water-table at a depth 15-20 cm, above that in the undammed section until the end of June did not cause over-high water-table levels during late summmer. Owing to the slight slope of the ditch network, no direct conclusions on the effect of damming on the deterioration of open ditches can be drawn. It is well known, however, that the problem of ditch deterioration is most pronounced in areas characterized by insufficient slope.

932

PELKONEN, E. The need for regulating the discharge from peat marshes drained for forestry purposes. Suo, 1976, 27:52-6.

Growth of trees in drained peaty areas is adversely affected by a high level of ground water at the end of summer and in autumn, but not in spring or early in summer. In 1972-1975, the ground water level was artificially kept near the soil surface in 2 plantations in S. Finland, in a Pine stand increase in stem girth was independent of the duration of the high level of ground water, but the longer the underflooding the greater was the increase in girth. In Spruce stands the effect of the duration of underflooding was not so marked. In 1973, there was even a decrease in girth, apparently because of an abundant yield of cones. In general, in Spruce stands also, increase in girth was directly related to the duration of high levels of ground water in spring and early in summer, but later (in July-Aug) high levels did not favour girth increment. In order to maintain ground water near the surface in spring and early summmer, it is suggested that peat dams are made in the drainage ditches, with plastic pipes for over-flow.

933

PELS, S. Disposal of saline drainage waters. In Hydrogeology of the Riverine Plain of South East Australia, (Storrier, R.S., Kelley, I.D., Eds.), Water Resources Commission, Sydney, N.S.W. Australia, 1978, 259-68.

Problems of soil salinity on the Riverine plain

are discussed in terms of inherent salt status of the soils and sediments. Regional patterns of salt distribution are shown to occur relative to geomorphological features of the alluvial environment. The repercussions of irrigation on the hydrology of the area are described and drainage measures required to overcome waterlogging and salinity problems are outlined. Difficulties in the disposal of saline drainage waters are stressed and three specific projects designed to deal with the problem are discussed.

934

PENMAN, F. Slow reclamation by tile drainage of sodic soils containing boron. 6th Congr. Irrig. Drain.,New Delhi, 1966, 2: Q 19 R 10: 113-21.

It has been found that the chloride content of the soil rapidly decreased by the construction of a tile drainage system and a 20-30 per cent deep percolation of the applied quantity of irrigation water. Boron toxicity symptoms, however, may persist for some years. Care should be taken in reusing drainage water, blended with fresh river water, on boron-sensitive crops.

935

PERRENS, S.J. Verification of a numerical model of soil water movement to a drainage ditch. In Agricultural Engineering in National Development (Choa, S.L. Bardaie, M.Z.; Saxena, N.C.; Van Vi Tran Eds.) Penerbit Universiti Pertanian Malaysia, Serdang, Selangor, Malaysia, 1981, 211-7.

A generalised numerical model for mixed saturated/unsaturated soil water systems used to analyse the non steady state drainage of water to a ditch. Results of numerical analysis are compared with laboratory experimental results for the same system.

936

PERREY, C.; VERGUET, A. An apparatus for the simultaneous measurement of rainfall and the discharge from experimental drainage networks. An. Agron., 1977, 28:671-7.

The apparatus includes one or several tanks with swinging buckets connected to a multi-channel event recorder. The mechanical and electronic parts of the apparatus are described in detail, and their performance is briefly discussed.

937

PERRIER, E.R.; MACKENZIE, A.J.; GRASS, L.B.; SHULL, H.H. Performance of a tile drainage system: An evaluation of a tile design and management. Trans. ASAE, 1972, 15:440-4.

The performance of tile-drainage system on a predominant Holtville silty clay loam highly stratified (clay loam overlying a coarse textured subsoil) had been studied, though 25 per cent of the field is of Imperial clay loam soil group. The drainage system consisted of 10.2 cm diam clay tiles 30.5 cm length with tile placed at 1.84 and 2.1 m depth and spaced at 91.4 m. Gravel envelope was used. The performance of the drainage systems as measured by the volume of outflow, showed definite dependence on the depth of the tile. Hydraulic conductivities from both systems (old and new) were similar. The piezometer method gave lower values of the hydraulic conductivity throughout the soil profile than did the tile drainage equations. Irrigation management by subsoil irrigation through these tiles can be achieved. The water level to the surface can reach in about four days and drainage in the next six days.

938

PETERS, H. Characteristics of high-speed drainage-laying machinery. 10th Congr. Irrig. Drain. Athens,

1978, 5:147-51.

The basic requirements for utilizing high capacity drainage machinery and the principal machine types are outlined.

939

PETERS H. Cost of drainage execution and economic comparison of drainage-laying machines. <u>10th Congr. Irrig. Drain</u>. Athens, 1978, 5:153-9.

The costs of drain installation using trenching and trench-less machines as experienced by the Norden Amelioration Association from 1962-75 are presented. Main cost factors are discussed and the most promising areas for economizing are indicated.

940

PETERSON, F.D. Drainage wells. <u>Proc. Drain. Efficient Crop. Production Conf</u>. Chicago, Illinois, 1965, 73-4.

Use of wells as a drainage measure is described. The drainage wells can be installed in areas with transmissivity approaching or exceeding 90 cm^2/sec. System analysis approach to the water balance of the area and ground water quality considerations in the use of drainage well water for irrigations is illustrated. The paper outlines 3 examples of drainage well systems installed in West Pakistan, Salt River Area in Arizona and Wellton Mohawk Project, Arizona highlighting the various problems of well drainage.

941

PETROV, VE G. Problems of rationally draining the Poleyes Marshes. <u>Ground Water</u>, 1980, 18:351-8.

Data for Poleyes marshes showed that optimum drainage depth was approximately equal to the height of the capillary rise above the water table. Greater or lower depths lead to lower

fields. Dokl. Vses. Ordena Lenina Akad, S-kh. Nauk. Immeni V.I. Lenina, 1981, 4:39-41.

Various types of mole drainage for flooded and non-flooded soils are examined with reference to soil salinity. Mole drainage combined with an open drainage system was highly effective in controlling the air-water, heat and salt regimes in rice soils.

953

POWERS, W.L.; KIRKHAM, D.; SNOWDEN, G. Seepage of steady rainfall through soil into ditches of unequal water level heights. Proc. Soil Sci. Soc. Am., 1967, 31:301-12.

The technique of using a finite series of orthonormal functions generated from products of trignometric and hyperbolic functions to solve a saturated water flow problem is illustrated. Several flownets showing streamlines, equipotential lines, and the water table heights for different geometries of steady rainfall seeping through soil into ditches of unequal water level heights are presented.

954

POULTION, D. Operation and groundwater hydraulics of an experimental tile drainage system at Kerang. In Hydrogeology of the Riverine Plain of South East Australia. Proc. Symp. Griffith, Australia, 1977, (Storrier, R.R.; Kelley, I.D., Eds), Australian Soc. Soil Sci. Inc., 1978, 229-46.

A description of the tile drainage installation in saline soils of the Kerang Region is given. The discharge hydrograph, water balance, 'apparent' specific yield, hydraulic head vs discharge relationship and equipotentials adjacent to the centre lateralare reported. Hydraulic conductivity in the depth range 0.5-10 m has been measured using the auger hole and piezometer methods and the values obtained are reported. Soil layers to a depth of 11 m are important in transmitting groundwater to the tile drains. A large part of the subsurface runoff is attributed to seepage of groundwater into the drained area. The amount of irrigation water applied increases, and percolation to the water table decreases, as the water demand of maize and sorghum increases during the growing season.

955

PRADHAN, S.N. Reclamation of saline inundated waste land. Proc. Symp. Sci. India's Fd. Probl., New Delhi (1967), 1971, 179-83.

Large tracts of saline inundated low lands located along the coastal belt particularly on both sides of the mouth of rivers provide adequate scope for their reclamation and bringing them under the cultivation of rice, coconut and cashewnut. The important features of such reclamation projects are (a) provision of protection embankment, (b) washing of salt concentration from top surface by flooding with fresh water received and rain water, (c) cleaning of shrubs and weeds, (d) provision of drainage, outlets and sluices and (e)layout of fields etc.

956

PREGNON, M.; REQUILLART, J.P.; FAUCHER, C. Draining by ditches and draining by underground canal systems in a small region of Central France: Limagne in Auvergne. 11th Congr. Irrig. Drain., Grenoble, 1981, 2:749-71.

Large-scale drainage in the Limagne area is discussed. Basically this consists of underground drains leading to a sink-hole situated on either side of the Allier River. Official policies with regard to techniques and financing drainage in Puy-de-Dome are outlined.

957

PRESS, H. The control of the water table in drained

soils. 5th Congr. Irrig. Drain. Tokyo., 1963, Q 17 R 20: 285-92.

Measures to prevent excessive drainage are discussed and it is recommended to use combined irrigation and drainage systems.

958

PRICE, R.K.; SLADE, J.E. A model for the design, analysis and operation of pumped drainage systems. In Polders of the World. Vol.1 Polder projects. Land and water management aspects. Int. Inst. Land Reclam. Improv., Wageningen, 1982, 464-74.

The criteria for the design of pumped systems have been greatly influenced by historic extreme events. With new modelling techniques a more refined design procedure can be used in which frequency of failure and performance are important. Two component models are described: an optimising design method to size the channels for a minimum construction cost and a simulation method to analyse unsteady flow in a pumped drainage system.

959

PROCHAL, F.; CURZYDLO, J.; NIEMIEC, J.; URBANOWICZ, A.; WITEK, T. Influence of drainage pipeline spacings on the yield of some crops. Wiado. Melior. Lakar., 1975, 18:223-8.

In trials at 2 sites in Poland in 1966-69 the relationship between the spacing of tile drains and crop yields, was examined. At Giebultow, drains were spaced at 10, 12, 13, 14 or 16 m and spring barley, winter wheat and potatoes were among the crops grown whereas at Lazany drains were spaced at 7, 9, 11 or 13 m and an oats/vetch mixture, fodder beet, spring barley and winter wheat were the crops grown. Detailed information and data on soil type and drainage characteristics are included. At Giebultow, the highest yields of barley (2.87 and 3.0 t grain/ha), wheat (2.33 t grain/ha) and potatoes (27 t tubers/ha) were obtained at spacings of 10 and 12, 12 and 16 m, respectively. These yields, however, were only 2 and 4, 1 and 3 per cent higher than those obtained at the usual 13 m spacing. At Lazany, the forage mixture and wheat both gave maxmimum yields at the 13 m spacing, whereas the yield of fodder beet roots and tops were highest (75.6 and 26.3 t/h respectively),at the 9 m spacing. These yields were 12.8 and 21.7 per cent higher than those obtained at the 13 m spacing. Barley grain yields were highest (3.3 t/ha) at the 9m spacing. This outyielded the 13 m spacing by 16.8%.

960

PROKOPOVICH, N.P. Need for better geologic input for drainage projects. <u>11th Congr. Irrig. Drain.</u>, Grenoble, 1981, 2:497-510.

Two case histories in the San Joaquin Valley, California, are used to illustrate the importance of data in the fields of geochemistry and site-specific geology. The first deals with swelling of clayey alluvium due to crystallization of mirabilite (Glaubers salt) at relatively low (usually above 0° C) winter temperature. The process resulted in severe cracking of 6.25 cm thick concrete sidelining of a major drainage canal. The second deals with numerous unexpected slope failures, squèezing, cave-ins, and costly over excavations during construction of drainage collecting facilities. The failures occurred in areas with upward gradients of shallow ground water.

961

PRUNTY, L.; KIRKHAM, D. A drainage model for a reclaimed surface mine. <u>Soil Sci. Soc. Am. J.</u>, 1979, 43:28-34.

A simplified model of the post-reclamation soil profile, employing bench-terraced geometry, at the Iowa Coal Project Demonstration Mine No. 1 is created. and used to find the steady-state water-table shape if one tube drain per terrace is installed. Dupuit-Forchheimer (D.F.) theory is used for the analysis. Water-table curves are generated for a series of alternative drain tube locations. An analysis is made to find the best position for the drain tube. For a recharge-to-soil conductivity ratio of 0.01, for terraces repeating at 45.7 m interval, the best drain-tube location is beneath the horizontal portion of the terrace 4.87 m from the toe of the backslope if the drain tubes are placed 2.44 m beneath the surface of soil underlaid by a barrier at 3.05 cm depth. The horizontal portion of the bench terraces is 38.08 m, and the vertical drop from one terrace to the next is 30.5 cm. A table of best drain locations is given for other drain depths, and recharge rates. Previous work on D.F. theory for a uniformly sloped barrier is discussed.

962

PRUS-CHACINSKI, T.M. Shrinkage of peat-lands due to drainage operations. <u>J. Inst. Wat. Engrs.</u>, 1962, 16:436-48.

Reports and amplifies a rational method devised by Ostromecki (Poland) to estimate shrinkage of peat land caused by the land drainage installations; offers worked out examples to illustrate the practical application of the method.

963

PRUS-CHACINSKI, T.M.; HARRIS, W.E. Standards for lowland drainage and flood alleviation and drainage of peat lands with special reference to the Crossens Scheme. <u>Proc. Inst. Civ. Engrs</u>, 1963, 24:177-206.

The paper compares land drainage standards in Great Britain with those applied in Northern Europe and discusses generally, the problems relating to the drainage of lowland peats; gives a brief description of the Crossens areas land drainage history and continue with the reasons for abandoning gravitational discharge, one important reason brings that no suitable excavating plant exists for working in the marginal tidal areas between dry land and full: describes the Crossens Scheme, the pumping station, the new middle channel, the enlargement of existing watercourses, the building of new bridges and the reconstruction of many existing bridges and culverts; observe that for the first time in 300 years crops can be sown with the certainty that they can be harvested, and this will lead to changes in crops, in water-table levels, in peat shrinkage and drainage standards.

964

PURVIS, E.R. Crop production on high water table soils of the New Jersey Pine barrens. N.J. Agric. Exp. Stn. Bull., 1964, 806, 15 p.

During a 8 year period on these sandy podzols mean depth to water table in the growing season in years of normal rainfall was about 40 cm at one site and 110 cm at another. Surface soils consisted of 86 per cent sand, 10.5 per cent silt, 3.5 per cent clay and retained only 12.3 per cent moisture at field capacity. Moisture content at the wilting point was 3.0% , a level not reached during the study. The narrow moisture gradient between surface soil and water table during periods of drought suggests that moisture movement in the upper profile occurs mainly as water vapour during such periods. Apparently, soils having average depths of 60-120 cm to water table in normal seasons would provide near ideal moisture conditions for crop growth during both dry and wet seasons. With occasional supplemental irrigation in dry seasons, soils of 60 cm higher elevation would probably supply sufficient moisture to crops. The soils respond readily to lime and fertilizers.

965

RAADSAMA, S.; SCHULTZE, F.E. Surface field drainage systems. In Drainage Principles and Applications. Vol.

407 Global Research on Drainage in Agriculture

IV. Design and Management of Drainage Systems. Int. Inst. Land Reclam. Improv., Wageningen, 1974, 16:67-121.

A description of the principal techniques of surface drainage and a discussion of some methods for computing design discharges.

966

RADCENKO, I.; KULHAVY, F. Soil drainage and irrigation by auto-matic control drainage system. 10th Congr. Irrig. Drain. Athens, 1978, 7:509-26.

The concept of control drainage is defined and the way in which it operates in regulating groundwater level is explained. Automatic control is the special operational benefit possible with this system. Underground retention and storage of surface water is a further important possibility. Control drainage is especially valuable in the lowland river valleys. The first automatic control drainage project was initiated in June 1976 on an area of 6.5 ha. The site condi-tions, design and principles of automatic control on this project are described.

967

RAI, A.; KHEPAR, S.D.; SINHA, B.K. Relationship between water table depth, salinity and maize yield, J. Agric. Engng., 1980, 17: 81-6.

Functional relationship between water table depth, salinity of ground water and maize yield as developed revealed that quadratic function represented the experimental data more precisely. The results showed that maize yield was more influenced by water table depth than the salinity of ground water.

68

AJ, P.A.; BAPAT, M.V. Problems of waterlogging emedied before canal inception. Proc. Symp. Waterlogging, Causes and Measures for Prev. New Delhi., Cent. Bd. Irrig. Pwr. Pub., 1972, 118,2:79-80.

The pre-irrigation soil surveys which indicate the capability of soils to various intensities of irrigation and help in predicting the problems and their remedies, have been discussed in relation to area under the Shetrunji irrigation project in Gujarat. The detailed investigations were carried out in an area of about 6000 ha near Datha village, in respect of: (1) drainage requirements, (2) boundary conditions and (3) water transmission properties of soils. For this purpose, the investigations included soil and topographic surveys, taking hydraulic conductivity values, determining outfall conditions, and the like. On the basis of the study made, the area was divided into three basins and 'open' as well as 'tile' drains were excavated. As a result of drainage, the water table receded below 240 cm and the soils could be brought under normal cultivation.

969

RAM, H.; NATH, J.; KHANNA, S.S. Effect of depth and composition of underground water and mulching on salt accumulation and yield of maize and wheat. J. Indian Soc. Soil Sci., 1981; 29:518-25.

In an experiment carried out in lysimeters, the effect of quality and depth of underground water and rice husk mulch on the growth of maize and wheat was studied. There was a gradual reduction in grain and straw yield of maize with decrease in depth of water table for non-saline under ground water without mulching and a reverse trend was observed with saline underground water. Maximum yield of wheat was obtained at 100 and 150 cm water table depths for non-saline and saline under ground waters, resp. Accumulation of salts in the surface soil layer was significantly higher for saline than for non-saline ground water treatment. An increase in the utilization of ground water with successive growth stages of these crops was observed. Mulching caused a significant decrease in ground water utilization which was significantly higher from non-saline than from saline water.

970

RAM, S.; CHAUHAN,H.S. Field studies of unsteady subsurface drainage. J. Agril. Engng. 9:1-11.

Studies of a field drainage experiment on three drain spacings are reported to compare with the performance of four chosen drainage equations. The water table profiles indicated faster and non-linear rate of drawdown for initial 12 to 15 hr and linear relationship thereafter. Schilfgaarde equation performed generally better than Glover, integrated Hooghoudt and modified Glover equations. There are indications that drain spacing of 25 m and drainage coefficients of 5 cm for surface runoff and 2 cm for tile flow for a combined system of surface and subsurface drainage may perform satisfactorily Tarai region of Uttar Pradesh.

971

RAO, B.N.; KHEPAR, S.D.; KUNDU, P.S. Investigations, reclamation of waterlogged areas, cost-benefit consideration. Proc. Symp. Waterlogging-Causes and Measures for Prevention, New Delhi. Cent. Bd. Irrig.wr.Pub. 1972, 118,2:152-9.

The paper explains various steps taken to reclaim 16 ha of waterlogged land at Ludhiana. The drainage investigations carried out under the project are topographic survey, preparation of ground water-table and isobath maps, soil chemical properties and hydraulic conductivity. Design and location of outlet, interceptor and relief open ditches for disposal of surface and part of subsurface runoff is given. Effect of drainage on watertable, crop response and economic benefit are reported.

972

RAO, K.V.G.K.; DHRUVANARAYANA, V.V. Sluice gate operation for surface drainage in the Sunderbans. J. Inst. Engrs, India (Civil Engng.), 1979, 60:162-6.

In the Sunderbans of West Bengal, the drainage of the excess rain water is being affected only by sluice gates provided in the protective bunds. There is no inland drainage network to lead the excess water to the sluice gate and the drainge disposal is mainly through overland flow. This paper presents an evaluation study on the operation of one such sluice gate for surface drainage in a typical catchment of the Sunderbans.

973

RAO, K.V.G.K.; DHRUVANARAYANA, V.V. Analysis of drainage coeffi-cients for surface drainage design in rainfed lowland rice field. Indian J. Soil Conserv., 1979., 7,2:12-20,

The design of an economic and efficient surface drainge system depends largely on the selection of a proper drainage coeffi-cient. An attempt was made to compute the drainage coeffi-cient for Sunderbans area from the rainfall data only. The rainfall data base was fifty years. The computation procedu-res involved are statistical frequency analysis of rainfall data and computation of drainage coefficient by the rainfall runoff models proposed by U.S.D.A. Soil Conservation Service, West Bengal Flood Commission, Cypress creek formula and simplified hydrological procedure. The values of drainage coefficient thus computed were compared with the experimen-tally observed values in Sunderbans. The Cypress creek formula gave relatively high values of drainage coefficient for water- sheds upto about 100 sq.km. As the watersheds in the rainfed rice cultivated areas of Sunderbans are small, the adoption of this formula results in costly and large capacity hydraulic structures for drainage. For this reason, a drainage coeffi-cient of 37.5 mm/day, computed by the simplified hydrologic procedure is seen to facilitate good drainage in 7 days from five day storm values of 5 year return period.

974

RAO, K.V.G.K.; PANDEY, R.S. Problems of subsurface drainage in coastal saline soils of Sunderbans. J. Agric. Engng., 1982, 19:85-9.

Subsurface drainage performance in coastal saline soils of Sunderbans (West Bengal India) is reported. Open drains 1.8 m deep were tried at a spacing of 15, 30 and 45 m. Profile salt was removed from surface down to 1 m for 15 m spacing and re-salinization was practically insignificant. The salt accumu-lation occurred in the other two spacings at about 1 m or shallower depth and resalinization was noticed. The rate of watertable drawdown was about two and a half times faster for the fields with a spacing of 15 m as compared to the field with a 45 m drain spacing.

975

RAO, V.R.; RAO, J.L.N. Nitrogen fixation (C_2H_2 reduction in soil samples from rhizosphere of rice grown under alternate flooded and non-flooded conditions. Pl. Soil, 1984, 81: 111-8.

Soil submergence accelerated hitrogenase activity particularly in the planted system. Microbial analysis indicated the stimulation of different groups of free-living and associative N_2-fixing micro-organisms depending on the water regime.

976

RAPP, E. Drainage requirements for agricultural crops. Agric. Forestry Bull., Univ. Alberta, 1978, 1:8-9.

Drainage requirements are determined by the effect of excess moisture on soil aeration, soil temperature, biological activity, structural stability and chemistry of the soil, and land and crop management.

977

RAPP, E.; ASHRAF, M. Deformation of corrugated plastic drains by soil surface loads. Can. Agric. Engng., 1976, 18:7-9.

Static loads were applied at the soil surface to two types of corrugated plastic drains in two soil types and at 95 and 125 cm depths. Installation was in trenches and simulated mole channels. Results were evaluated statistically by analysis of variance. Loads of 698 and 678 KN/m^2 for mole and 680 and 623 KN/m^2 for trench-installed drain tubes at the 95 cm depth in sandy loam and silt loam soils respectively were the safe soil surface loads at 30% deflection. Placing the pipes at 125 cm and simulated mole channels also allowed greater surface loads to be carried.

978

RAUT, M.S., ALI, M. Relative performance of maize germplasms to flooding and nitrogen levels. Indian J. Agron., 1976, 21: 167-8.

The average grain yields of 2 maize hybrids, 1 composite and 1 local cultivar were decreased by 16.1% due to flooding of the 9 to 10- leaf stage; flooding at the 4 to 5- leaf and 14- to 15- leaf stages produced small decreases. Composite cv. Vijay was more tolerant of flooding at the all stages than other gramplasms, but hybrid Ganga-2 gave the highest yields, both under flooded and non-flooded conditions. Increasing the N rate from 60 to 180 kg/ha increased average yields from 4.13 to 6.13 t/ha in all germplasms, except the local cultiver, which showed response to N rates upto 220 kg/ha; hybrid Ganga-2 gave the highest yields of 7.22 t/ha.

979

REEVE, R.C. Characterizing soil properties for drainage design-art or science? Proc. Drain. for Efficient Crop Production Con. Chicago, Illinois, 1965-7,21-3.

Ability to measure hydraulic conductivity and to obtain an appropriate average value to represent highly variable soil system is considered as the crux of the drainage design problem. It is suggested that tile outflow method is the best procedre to measure this though its use has been limited. Laboratory measurements are not suitable for design purposes. Extensive parameters such as conductance than the intensive quantity such as hydraulic conductivity have been suggested.

980

REEVE, R.C. Trenchless drainage. Proc. Int.Drain. Workshop, Wageningen, Netherlands, (1978), 1979, 25:571-89.

The development of drainage is discussed as well as the plough designs and performances.

981

REGAMEY, P.; JATON, J.F. Improvement of the draining capacity of a filtration system in a calcareous peat medium., C.R.Acad. Agric. Fr., 1976, 62:1271-84.

The clogging of drains is discussed in relation to transfor-mations of Fe as affected by pH and Eh. Precipitation of iron hydroxides and of humic substances is likely to occur in the case of calcareous peat soils. In laboratory studies with acid and calcareous peats the marked reduction in drainage with time in the calcareous peats was demonstrated. This was prevented by use of a filter system (woodchip), which reduced the external sedimentation around the pipes.

982

REICH, J.; HOFMANN, A.; SCHWARZ, K. The choice and construction of machinery and the technology of drainage by slotted piping and permeable envelope material. Arch. Acker. V. Pflan. Bodenk., 1977, 21:811-20.

Equipment for the installation of slotted piping and the simultaneous introduction of envelop material is described.

983

REICH, J.; UNGEWISS, R. Development of equipment for the mechanization of slot drainage and practical experience. Agrartechnik, 1979, 29:317-9.

According to the method described, some filter material, e.g. coarse sand or gravel, is fed from a hopper body of 7000 kg capacity on a wheeled chasis via a short (2.80m) longitudinal and transverse conveyer (5.51m) to a receiving hopper on the drain layer and thence into the channel formed by the machine. A layer about 10cm wide and upto 60cm deep of a coarse granular material is laid to improve drainage.

984

REICHMAN, G.A.; DOERING, E.J.; BENZ, L.C. Drainage requirement for sugarbeets grown on sandy soil. Can. Agric. Engng., 1983, 25:199-204.

Field plot and lysimeter studies were conducted to determine the optimum water table depth for sugarbeets (Beta vulgaris L.) grown in irrigated sandy soils of northern Great Plains. The water table depths (expressed as time-weighted average between 15 May and 30 Sept.) were 147,203 and 229 cm in field plots and constant water level were maintained at 46, 101, 155 and 210 cm in lysimeters. Sprinkler irrigated the field plots at levels 0.5, 1.0 and 1.5 and lysimeters at 0.3, 0.8 and 1.3 times the calculated plant water requirement. A non-irrigated (precipitation only) treatment was also included in field plots. Highest yields of fresh sugarbeet roots, root dry matter and sucrose were produced with the shallowest water table in field plots regardless of the irrigation level. In lysimeters, yields with water table 101 cm below the soil surface were greater than in lysimeters with deeper or shallower water table at 0.3 and 0.8 irrigation level. At irrigation level of 1.3, yield increased with increasing water table depth. It was concluded that water levels should be at about 10 cm to get optimum yields.

985

REKTORIK, R.J.; MYERS, V.I. Polyethylene drainage pipe instal-lation techniques and field performance. Trans. ASAE, 1967, 10:458-9, 461.

A wheel-type trencher was developed and modified for laying a continuous semirigid pipe to depths of 2.7 m in a trench only 25.4 cm wide. Polythelene plastic pipe 10 cm ID, with twenty-two rectangular slots per m was sufficiently flexible to feed to the trench bottom on a 3.6 m radius of curvature. The pipe has sufficient strength to withstand soil overburden loads at the 2.45 m depth when lateral support for the pipe was developed. A groove or cradle was formed in the undisturbed soil at the trench bottom to fit ap-proximately half the pipe periphery. The costs of installed plastic subsurface drainage systems are 10 to 15 per cent lower than other equivalent capacity systems available in the Lower Rio Grande Valley of Texas.

986

REMBEZA, L. Inflow of ground water under pressure to the drain system. Arch. Hydrotech., 1974, 21:179-99.

The Zukewski function method is used to solve the problem of drainage in soils of low permeability affected by rising groundwater using a system of parallel horizontal drains.

987

HENNER D.M.; MUELLER, C.C. Drainage system design and analysis by computer. J. Irrig. Drain. Div. ASCE,

1974, 100:255-65.

Possibility of applying techniques of computer programming to the calculations of sub-surface drains spacing and to the design layout of drainage are explored. It is concluded that computer programme can be used to find out sensitivity of system cost to drain spacing, as aid to drain line system design, forecast the extent and expected cost of future drainage system installations.

988

RESHETKINA, N.M. Operation of drainage systems in the Hunger Steppe. 7th Congr. Irrig. Drain. Mexico., 1969, 5:25.215-25.236.

The natural conditions in the Hunger Steppe are such that irrigation alone without artificial drainage inevitably results in secondary salinity of the soil. In north-eastern part most of drainage (5400 km) is open horizontal type and drain dischage amounts to 32% of the water delivery which gives negative salt balance. For reducing drainage discharge which is as high as 50% at certain locations and to make salt balance negative at others; vertical and closed horizontal drainage are provided to raise the average draining capacity to 0.23 L/sec/ha and the drain discharge to 25-30 per cent of total delivery by utilising ground water. In south Hunger Steppe closed horizontal drainage at a depth of 3-3.5 m with a length of 67 m/ha have been laid. The open collector drains 4-7 m deep are provided. In the central part where saline groundwater is 10-20 m below surface, vertical drainage wells have been sunk to maintain the natural process of gley soil formation and prevent secondary salinization. The drainage has resulted in increased cotton yields over the years.

989

RESHETNYAK, N.F. Need to lower the groundwater table

in the rice systems of the southern Ukraine. Pochovedenie, 1977, 3:98-102.

Studies over a period of 5 years showed that the reason for the low yield of rice and other crops in low areas of the Krasnoznamensk irrigation system is the presence near the surface of highly mineralized groundwaters, waterlogging, and salinization of the upper soil horizons. The groundwaters that lie at a depth of 0.5-0.8 m in these areas contains 5 to 35 g or more of salts per L. When rice is grown, the ground-waters acquire a head and discharge at the surface, damaging rice seedling and preventing the infiltration of irrigation water and the removal of salts from the upper horizons.

990

RHOADES, J.D. Drainage for salinity control. In Drainage for Agriculture (Agronomy No. 17; Schilfgaarde, J. van. editor). American Society of Agronomy, Wisconsin, 1974, 433-61.

The drainage requirement (based on leaching requirement, salt balance and capillary rise concepts) for salinity control is discussed.

991

RHOADES, J.D. Inundation tolerance of grasses in flooded area. Trans. ASAE, 1964, 7:164-6, 179.

Ability to withstand given periods of submergence for 11 grasses were tested. It was concluded that where adapted grass species such as Bermuda grass, buffalo grass, Vine-mesquite, prairie cordgrass, KSU lowland switch grass and common lowland switchgrass would be recommended when expected inundation period ranged upto 20 days; Caddo switchgrass and upland common switchgrass for 15 days; Kaw big bluestem and Woodward sand bluestem for 10 days; and eastern gamagrass, alkali sacation, El-kan bluestem, KR bluestem and weeping Lovegrass for 5 days.

RHOADES, J.D. Leaching requirement for exchangeablesodium control. Soil Sci. Soc. Am.J., 1968, 32:652-6.

A method is described for estimating the leaching fraction required to keep the sodium-adsorption-ratio (SAR) at the lower limit of the plant-rooting zone below some limiting value. The choice of the limiting SAR depends on plant tole-rance to sodium, and on soil permeability. The method is based on an equation that satisfactorily predicts the SAR values of waters draining from a calcareous soil under various leaching regimes.

993

RHOADES, J.D. Potential for using saline agricultural drainage waters for irrigation. In water Management for Irrigation and Drainage. Proc. Spec. Conf. ASCE Irrig. Drain. Div., ASCE, New York, 1977, 1:85-116.

Chemical composition of representative agricultural drainage waters in arid irrigated regions is tabulated. An evaluation of the suitability of these waters showed that many of the samples examined are suitable for irrigation of certain crops. Use of these waters for irrigation could increase food produc-tion, lessen drainage disposal requirements and improve land and water resource use efficiency.

994

RICHARDSON, J.; MAY, L.; SALTER, E.C. A fen drainage trial in central Newfoundland. Information Report, Newfoundland Forest Research Centre, Canada, 1976, N-X-139, 16p.

Describes experiments in Newfoundland to determine the effects of ditch spacing on water table levels, as part of a study to develope the optimum drainage system for the successful afforestation of bog and fen. A 6.6 ha site was drained in 1972 using a Parkgate tine plough drawn by two specially modified tractors. Measurements made in the three years after draining showed that the water table dropped considerably on the areas that had ditches spaced at 2- and 4- m intervals, compared with the unditched strip. During a wet year, however, the water table on both ditched areas was not generally below the level considered desirable for afforestation. Though this trial has shown limited success so far, the growth of the plantations established on the drained area will indicate in the long term how successful it has been in preparing the fen for afforestation.

995

RIDDER, N.A.DE., AART, R. VAN. Procedure in Drainage surveys. In drainage Principles and Applications. Vol. IV. Design and Management of Drainage Systems. Intern. Inst. Land Reclam. Improv., Wageningen, 1974, 16:383-409.

A general guide to drainage survey techniques. Procedures, data collection, processing and intention as a basis for drainage designs are briefly discussed.

996

RIDGE, R. Draining cane fields. Cane Growers' Quart. Bull., 1980, 43:55-78.

It deals with the problem of drainage of sugar cane fields under the following headings: recognizing the problem; effects on cane growth; benefits; yields; surface drainage systems (SDS); construction of SDS; maintenance of SDS; types of sub-SDS; drain types; drain design and construction; drainage in flood prone and tidal areas; and community drainage.

997

RIDGE, D.R. Spectacular cane yield response in subsurface drainage trials. Cane Growers' Q.Bull., 1978, 42:32-33.

In poorly drained or waterlogged areas cane yield losses amounted to about 0.5 t/ha for each day that the water table was closer to the surface than 0.5 m. Large increases in yields were obtained in such areas following subsurface drainage using corrugated plastic pipes placed at a depth of 1.2 m. In a trial with selection Q90, drainage increased yields from 70 to 102 t/ha.

998

RIESER, A. Standards for reclamation measures for soils water-logged due to the presence of groundwater or impeded drainage. Wasser. Boden., 1979, 31:7-10.

Investigation of the conditions according to the published standard DIN 4220 should be carried out prior to choosing reclamation techniques for waterlogged soils. Such techniques should be applied on the basis of DIN 1185 (soil water management by means of drains and subsoil management). The use of pipes, pipeless drains and subsoil improvement for soils with high water tables or impeded drainage is discussed in relation to rainfall and capital outlay.

999

RIHA, J. Prevention and causes of tile drains damage. 7th Cong. Irrig. Drain. Mexico, 1969, Q 25 R 12, 151-68.

Problems concerning the lifetime of drainage from baked tiles are outlined. The lifetime is expected at an average of 45 years from the date of beginning of operation to renew before decrease in fertility occurs. Analysis of velocity formula indicated that the formula used in the Middle European prac-tices so far, do not correspond sufficiently to the real con-ditions. To avoid silting, the minimum velocity for cross sections with diameters of 5-20 cm is assessed in the range from 0.45-0.70 m/sec. However, this method does not guarantee complete protection as zero velocity of water has to be assumed in temporarily acting drainage systems. Maintenance of tile drainage system has been discussed including classical methods of cleaning.

1000

RIMA, J. The effect of the constructional technical factors upon a drainage system. Vod. Hospad., 1966, 16:443-5.

The paper deals with the service-life of pipe drainage networks based on field research, and experience gained in other countries and enlists the deteriorating factors; hydraulic defects of pipes, sedimentation of drained soil, effect of frost, sudden strain of soil over the drainage, accidental or purposeful damaging, improper utilization of drained land, penetration of field animals in the piping, action of aggressive water, physico-chemical composition and development of drained soil, and maintenance of the network.

1001

RISSEEUW, I.I.A. Typical drainage problems of the sugarcane producing co-operatives of Peru. Bol.Tecnico, ICIA, 1977, 6:30-48.

The geomorphology, hydrology and soil properties of the sugar cane growing valleys of Peru are described in relation to the excessive groundwater levels which frequently occur, and which lead not only to poor aeration, but also to salt accumulation. Difficulties encountered in applying formulae for steady state and non-steady state flow conditions to the design of prac-tical drainage systems are discussed.

1002

RITTER, W.F.; BEER, C.E. Yield reduction by controlled flooding of corn. Trans. ASAE, 1969, 12:46-7,50.

Effect of inundation on corn yield as affected by duration of inundation and the plant stage at which flooding occurs were evaluated in a Colo flood plain soil with a permeable subsoil. The results indicated that shallow depths of flooding after silking stage will not cause any noticeable amount of damage. 4-5 days of inundation at initial growth stages may completely kill the plants. High nitrogen rates could to some extent alleviate the damage caused by flooding at early growth stages.

1003

ROBINSON, M.; BEVAN, K.J. The effect of mole drainage on the hydrological response of a swelling clay soil. J. Hydrol., 1983, 64:205-23.

The results of a plot experiment to determine the effects of a mole drainage on the hydrological response of a swelling clay soil under pasture are described. It is shown that the response is dependent on antecendent moisture condition. Recession discharges and total water yields are almost always higher from the drained plots. Mole drainage may be expected to reduce the magnitude of winter flood peaks, but may increase the magnitude of summer and autumn flood peaks.

1004

ROBINSON, M.; BLYTH, K. The effect of forestry drainage opera-tions on upland sediment yields: a case study. Earth Surface Processes and Landforms, 1982, 7:85-90.

Forestry is generally considered to be a land use giving low sediment yields. However, in Britain most forestry is on land that requires extensive drainage, and it was found that the drainage of a small upland catchment resulted in sediment yields over the following five years equivalent to nearly half a century's load at pre-drainage rates. Subsequently sediment yields did not decline to pre-drainage levels, but remained about four times higher, as a result of erosion of the drains.

1005

ROCHESTER, E.W.; KRIZ, G.J. Ditch drainage of anisotropic porous media: a model study. Trns. ASAE, 1970, 13:626-8.

The critical depth of an impermeable layer below equally spaced parallel drains was determined for a homogenous and a two layered isotropic soil. The results were then mapped mathematically to the anisotropic case. The effect of the impermeable layer was determined by observing the rate of fall of the fluid free surface at the mid point of Hele-Shaw model. The results for open ditch indicated that (1) for homogeneous, isotropic soils the critical depth to impermeable layer is one ninth the spacing. Same is true for two layer isotropic system when ditches do not penetrate the second layer of lower hydraulic conductivity K ($0.5 \leq K_2/K_1 \leq 1$). Formulae are presented for homogeneous anisotropic and two layered anisotropic soils.

1006

ROCHESTER, E.W.; KRIZ, G.J. Model study of boundary effects on ditch drainage. J. Irrig. Drain. Div. ASCE, 1968, 94:493-504.

The studies on a Hele-Shaw model to evaluate the influence of the depth to an inmpermeable layer on transient subsurface drainage into equally spaced open drains, indicate that a horizontal impermeable layer affects drainage significantly only if the layer is closer to the ditch bottm than one-twelfth the ditch spacing, and the initial depth of water above the ditch bottom does not affect this criteria; add that midpoint recession of the free surface is exponential if the flow is not affected by the impermeable layer and find that some dimensionless groupings are valid only if the ditch bottom extends to the impermeable layer.

1007

ROGERS, J.S.; HARRISON, D.S. Crop response to drainage in the Flatwoods. Proc. Florida State Hort Soc., 1975, 87:193-5.

Data are presented on the yields of several crops, including Phaseolus beans, capsicums and oranges, grown on Flatwoods soils with controlled water table at 20 or 60 cm. The highest yields and returns were obtained with the later.

1008

ROGERS, J.S.;HENSEL, D.R.; CAMPBELL, K.L. Subsurface drainage and irrigation for potatoes. Proc. ~~Soil Crop~~ Sci. Soc. Florida, 1975, 34:-7.

In a field plot experiment with potatoes, water furrows used for irrigation and drainage were replaced by sub-surface drains spaced 18.3 m apart. The drains discharged through flexible pipes that were raised or lowered to control the water table which was normally held about 20 cm below the furrows between rows. Average tuber yield was increased from 15.5 to 24.2 t/ha, partly due to the extra rows planted in the spaces previously occupied by irrigation furrows.

1009

ROGERS, J.S.; STEWART, E.H. Hydraulic gradients near drains in modified soil profiles. Soil Crop Sci Soc Florida, 1974, 33:148-50.

Large hydraulic gradients were found near the interface between a gravel envelop and a soil. An analysis of the physical system for flow indicated that the hydraulic conductivity of a soil may be as much as 30 times that of a thin restricting layer.

1010

ROGERS, J.S.; STEWART, E.H. Watertable elevation-drain outflow relationships in a mixed and unmixed Spodosol. Proc. Soil Crop Sci. Soc. Florida, 1977, 36:94-6.

Computer simulations of flow to drains for a Spodosol after surface tillage, or after deep tillage involving mixing of the soil indicated that flow was restricted near the drains in the mixed profile. The relationship between watertable elevation and drain outflow rate for both soil profiles was constant for 5 years. This indicated that the hydraulic characteristics of neither profile have changed significantly. Thus, the restriction to flow near the drains developed very soon after the drains were installed.

1011

ROGERS, J.S.; STEWART, E.H.; CALVERT, D.V.; MANSELL, R.S. Water quality from a subsurface drained citrus grove. 3rd Natn. Drain Symp. ASAE, Michigan, 1976, 99-103.

A 9 ha citrus grove on flatwoods soils was drained with subsurface drains at a 18.3 m spacing and a 105-cm depth. Three initial tillage treatments [surface tillage (ST), deep mixing to a 107-cm depth (DT), and deep mixing to a 107-cm depth with 54 mt/ha of line (DTL)] were used in the study. Drain outflow was monitored for its NO_3-N and PO_4-P content. Volume of drain outflow was in the order ST $<$ DT = DTL. NO_3-P concentrations were in the order ST$<$DTL$<$DT. When heavy rainfall followed fertilizer applications, the NO_3-N concentration in drain outflow from ST exceeded 10 ppm. PO_4-P concentrations were in the order ST$<$DT=DTL. The PO_4-N concentration for ST at times exceeded 1 mg/1. Drainage water from the grove alongwith surface runoff was collected in a ditch and pumped into the drainage district canal. Water samples at the pump discharge never exceeded 5 mg/1 NO_3-N and 0.5 mg/1 PO 4-P.

1012

ROGOWSKI, A.S.; SHRADER, W.D.; JOHNSON, H.P. et al. Tile drainage experimentation on Webster soils: results of years 1954-1963. Iowa St. J. Sci., 1966,

41:25-40.

In experiments involving continuous maize, continuous lucerne or rotation of maize-maize-oats-ley on silty clay loams, average crop yields showed no difference with tile spacings of 30 and 60 m or with the cropping systems. Differences in the permeability of substrata rather than with differences in treatment, were observed.

1013

ROKITA, A. Contribution to investigations on draining of cultivated soils on the Chelm Experimental Farm, near Krakow. Zes. Nauk. Wyz. Szk. Roln. Krak. Melior. 1972, 68:127-49.

20 plots totalling 1000 m^2 and supporting a wheat/oats.,/potato/clover rotation were drained by laying 2 rows of tile drains (5 cm in diam) at a depth of 1 m, 10 m apart through the middle of the plots. Yield data are given for 3 years before drainage and 2 years after. Yields declined in the year following drainage as a result of damage caused during drain laying, but in the subsequent year yields increased markedly and in some cases even exceeded those obtained before drainage. Drainage enabled work to begin earlier in spring and allowed potatoes to be grown even in very wet years without tubers rotting.

1014

ROOK, B.; SIPKES, K.H. Drainage of the Beemster Polder. 8th Cong. Irrig. Drain., Varna, 1972, Q 27 R 20:297-314.

The paper discusses the deficiencies of drainage of 7200 ha polder in the north-western part of the Netherlands and suggest improvement measures. The drainage should provide minimum depths of 0.90 m for arable lands orchards and vegetable farming and 0.65 m for pastures. The maximum drainage depths may not exceed 1.50 m and 1.15 m for the two cases, respectively. The water balance analysis indicated that a discharge capacity of 30 mm in 3 days and 54 mm in 5 days equivalent to a design discharge of 0.14 m^3/sec per sq km for the main drainage system may be sufficient. Pumping stations and necessary structures are recommended for achieving good drainage.

1015

RUDD, A.V.; CHARDON, C.W. The effects of drainage on cane yields as measured by water-table heights in the Macknade mill area. Proc. Qd. Soc. Sugar Cane Tech., Brisbane, Queensland, 1977:111-7.

Data are presented on the relationship between yield and mean water table level, and between yield and number of days when the water table was $<$0.5 m from the surface, assessed from studies on 10 sites. It was calculated that a reduction in yield of the order 0.46 t/ha could be expected for every day the water table was $<$0.5 m from the surface.

1016

RUEDISILI, L.C.; LOGAN, S.M. Locating drainage systems by magnetometer surveys. J. Irrig. Drain Div., ASCE, 1978, 104:261-73.

The feasibility of using a portable, proton precession, ground magenetometer in the location of subsurface drainage systems is analyzed. Autumn, winter and spring surveys of a field drained by clay tile and plastic (PVC) pipe and of a specially constructed test plot on agricultural land are presented. Magnetic anomalies over the drainage lines resulting from the contrast between the drain material, the voids in the drain and the surrounding soil are identified. This magnetic contrast is not affected by the different seasons or the moisture content of the soils. The technique is inexpensive after the initial machine purchase, it is relatively quick, it can be used with or withour crop cover, it can be used in any season other than winter, it requires only minimum manpower and limited technical skills by the magnetometer operator and data interpreter.

1017

RURAL RESEARCH IN CSIRO. Plastic agricultural drains. CSIRO Rur. Res., 1964, 47, 22 p.

Plastic pipes, mainly from PVC, in agricultural drains, are light, easy to handle, cheap to transport and install, and efficient. Savings in work varied from 20 to 60% according to soil types. Slot size is critical. They should be 2-2.5 cm long and 0.6 mm wide.

1018

RYABUKHA, A.S. Effectiveness of draining pine stand in the Ukrainian Poles'e. Lesn. Khoz., 1977, 7:28-31.

Investigations are described on the increment response to drainage of pine (**Pinus sylvestris**) stands 43-110 years old, on the basis of data from 7 plots near drains made in 1960-61. The results indicate that in the 10 years since drainage, the stands produced as much wood as had been produced during their entire previous life before drainage (33-110 years). The pattern of movements of the water-table in the period after drainage is discussed, in relation to the increment response, and some general conclusions on drainage policy are drawn.

1019

RYABUKHA, A.S. The groundwater regime in Scots pine stands of the Ukrainian Polesye drained by a sparse and dense draiange network. Lesoved., 1980, 1:64-70.

An account is given of studies on the ground water level on several areas of raised bogs and transitional bogs drained in the early 1960s by network of ditches. Data are tabulated on the seasonal and annual fluctuations in the water table for 9 yr following drainage, at various distances from the ditches. 1.6-1.8 m deep achieved rapid lowering of the water-table on transitional bogs, but not on typical raised bogs. It is recommended that for forest drainage purposes the ditch depth be limited to 1.2 m, and the spacing between the ditches be adjusted according to local conditions.

1020

RYKBOST, K.A.; ZWERMAN, P.J.; LUCEY, R.F. Surface drainage research: I. Bedding as evaluated by corn response under two fertility levels on a very fine textured Aeric orchraqualf. Agron.J., 1967, 59:324-7.

Beds were constructed 24.4 m wide, crowned to 50 cm, and orientated at right angles to an existing system of plowed beds 9 m wide, with 15- to 20- cm elevation. The latter had no outlets, with the result that standing water occurred on the field as late as mid-June. Soils were 40% silt loam (aeric orchraqualf), 50% clay loam (mollic haplaquept) and 10% fine sandy loam (aeric haplaquept). During a normally wet autumn, surface drainage by bedding gave greater assurance of harvesting the silage and permitted earlier planting in the spring. Non-bedded and bedded treatments showed similar yields. There were marked silage, grain and maturity responses to fertilising on varying position on the bed.

1021

SABO, E.D. Distinguishing zones where forest drainage is highly effective. Lesn. Khoz., 1979, 3:17-20.

The factors affecting the effectiveness of forest drainage in the USSR are discussed. The bioclimatic potential (BCP), which characterizes the biological productivity of the climate, differs by regions and has values from below 0.5 to 2.5. Mean forest increment in m^3/ha has been found to depend on the % bogginess of the forest resource, and expressions have been derived to estimate increment from % bogginess at different values of BCP. An expression has also been derived for determining the coefficient of productivity according to the type of forest growth conditions (based on moisture and nutrient status). The influence of the area of treeless bogs in the territory, and the influence of the species composition off the stands can also be calculated. On the basis of these studies it is possible to define the zones where forest drainage will be most advantageous, and figures are presented illustrating this for the non-chrnozem region of the RSFSR.

1022

SABO, E.D. Drainge of forests and the conservation of natural resources. Lesn.Khoz., 1974, 9:43-6.

Discusses in general terms the enormous scope for drainage in the forest lands of the USSR (possibly 200-250 m ha), and the effects of large-scale drainage operations on forest increment in adjacent dry areas, on inland fisheries, and on game populations. Research and observations to water indicate that drainage will have only an insignificant adverse effect on the increment of forest in adjacent dry areas. Fish production in rivers and lakes decreases with increasing forest drainage operations in the catchments, unless special measures are taken to counteract this. Drainage adversely affects the populations of waterflow but increases the populations of game such as hazel-hens and mountain hares.

1023

SABO, E.D. Economic effectiveness of drainge in forestry. Lesn. Khoz., 1977, 4:15-9.

In the context of the Soviet socio-economic system, various approaches are discussed for calculating the economics of drainage operations in forestry. Formulae are derived for determining the pay-back period, the absolute economic effectiveness, and the coefficient of economic effectiveness of capital investments in drainage.

1024

SABO, E.D. Expediency of forest drainage on potentially fertile soils. In Improvements in Forest Growth on Peat-Bog Soils of the USSR Forest Zone and Tundra, Jerusalem, Israel. IPST, 1965, 72-85.

In the Moscow region, in a 60 year pine-spruce-birch plantation with individual 130-year old pines on peaty-gleys, the effect of drainage since 1890 suggests that the younger a stand the more economic is drainage. The calculation of the economic effectiveness of drainage is discussed.

1025

SADLER, L.D.M.; TAYLOR, S.A.; WILLARDSON, L.S.; KELLER, J. Misci-ble displacement of soluble salts in reclaiming a salted soil. Soil Sci., 1965, 100:348-55.

Fresh water was applied to the surface of a saline soil, and the change in concentration of ionic species was measured in the effluent from the tile drain. Both hydrodynamic dispersion and ionic diffusion actively contributed salt to the effluent; salt immediately over the line of the drain was removed largely by velocity flow and dispersion. Diffusion was of increasing importance in removing salt from soil more distant from the drain.

1026

SAHOO, N.; SINGH, J.; PANDE, H.K. Planning an integrated drain-age and irrigation system in humid rice farming-A case study. J. Agric. Engng, 1982, 19:37-47.

Twenty seven years (1951-77) of weekly rainfall and evaporatioon data were analysed to determine the irrigation and drainage requirements for the Central Rice Research Institute Farm in Cuttack, Orissa. Such requirements were determined at 70% and 60% chances of success, by analysing the data probabilistically. Subsequently, for an integrated irrigation and drainage plan, the cost of the system and the expected benefits were estimated. The calculated benefit: cost ratio of 2.09 indicated that such a plan would be feasible.

1027

SAHU, N.N. Studies on the design criteria for surface drainage in Hirakud Command Area. J. Agric. Engng., 1982, 19: 49-51.

Trials on surface drainage were conducted at the Regional Research Station, Chiplima, Orissa to determine suitable longitudinal slopes of the field for increasing irigation efficiency. Under high rainfall and high water table situations, the maximum yield was obtained at 0.3% slope (1244 kg/ha) and minimum at 0% slope (1103 kg/ha). Higher slopes give rise to scouring and silting problems. The study indicated that a 0.3 per cent slope was optimum for surface drainage.

1028

SAKKAS, J.G. Generalized nomographic solution of Hooghoudt equa-tions. J. Irrig. Drain. Div.ASCE, 1975, 101:21-39.

Hooghoudt's equations for the equivalent depth and drain spacing are put into dimensionless form. A generalized nomogram for the direct solution of the drain-spacing formula was also prepared.

1029

SAMANI, Z.A.; WILLARDSON, L.S. Soil hydraulic stability in a subsurface drainage system. Trans. ASAE, 1981, 24:66-9.

Clogging of drains with sediment is an important problem in the design and maintenance of sub-surface drainage systems. Soil stability and density, represented by plasticity index and hydraulic conductivity can be used to estimate the internal water flow gradient at which hydraulic failure of the soil structure occurs.

1030

SAMI SELIM, M.; KIRKHAM, D. Screen theory for wells and soil drainpipes. Water Resourc. Res., 1974, 10:1019-30.

A solution is given for steady flow in a confined aquifer to a well tube having equally distributed screens. The solution may also be applied to seepage flow to the joints in soil drains. It is shown, quantitatively, by use of Darcy's law and Laplace's equation, that more flow results if a given total length of screen is utilized in a number of sections over the full length of the well tube than if the given length of well screen is all used at the bottom of the tube.

1031

SANDHU, H.K.; KAHLON, A.S.; KHEPAR, S.D. Economic aspects of reclamation of water-loggged areas in the Punjab: A case study J. Res. PAU, 1969, 6:1001-9.

A techno-economic evaluation of the drainage and reclamation project in village Habowal Ludhiana showed a decrease in ground water level by 45 to 75 cm, and reduced salt content. The annual field reclamation cost/ha was estimated at Rs.28.10 and the ave-rage net return/ha was Rs.65.65 in 1966-67. The cost benefit ratio more than unity in the first year justifies the establishment of the project and its take-up in other water-logged areas.

1032

SANDHU, J.S.; CHAUHAN, N.S.; BEHAL, S.P.; PANSHI, B.S. Cropping patterns for waterlogged areas. Proc. Symp. Waterlogging Causes and Measures for Prevention, 1972, CBIP New Delhi, 118(1):161-75.

Comparative production capacity of different kharif and rabi crops under varying degrees of waterlogged conditions has been studied and a suitable cropping pattern for such areas for maximum production is suggested. Investigations have proved that the total cropped area, total foodgrain area and total foodgrain output are higher in high water-table areas. Rice and wheat are not affected adversely in areas having water-table depths upto 0 and 0.6 m. The performance of maize, bajra, cotton are normal beyond 1.12 m depth of water-table. Sugarcane withstands water-logged conditions beyond 0.6 m depth. Gram is sensitive up to 0.9m depth. Barley cannot withstand stagnation and is suitable beyond 0.9 m of depth of water-table. Efficient development of these waterlogged areas with improved agronomic practices by using the land according to its capacity have proved a boon in bringing green revolution in the country.

1033

SANDU, G.; CORNEL, R. Pedological problems of drainage and management of the Calmatzui in view of the amelioration of the saline and alkali soils. Int. Symp. New Develop. Field Salt Affected Soils, Cairo, (1972), 1975, 585-94.

The saline alkali soils in Calmatui valley, are characterized by shallow and highly mineralised groundwaters, high EC and ESP. In the area of soluble salt transition-migration and accumulation of soda (Lipia-Albesti), the drainage should provide water reserves necessary for leaching. In the area of reduced soluble salt migration and accumulation (Albesti-Ciresu) the drainage by pumping well had no influence on mineralization and chemical composition of water. On the contrary, tile drainage, completed by other reclamation works, contributed to a high demineralisation of ground water; at the end of soil reclamation works their waters may be used in irrigation after a calcium pretreatment. In the area of high accumulations in soluble salts (Ciresu-Lisco-teanca) ground waters are highly mineralised and consequently demineralization by drainage does not assure the qualities necessary for waters to be used in leaching. In lightly drained and salt migration area (Liscotenca-spiru-Haret), under the influence of drained Danube waters the degree of ground waters mineralization is smaller. Drainage works conducted in the four areas are presented.

1034

SANDU, G.; ISPAS, E.; YAKOB., E. Reclamation of mole-drained saline and solonetzic soils of the Rumanian plain. Pochovedenie, 1978, 59-66.

Field experiments on solonchak-solonetzes and crustal solonetzes are described. Leaching irrigation over mole drainage reduced the content of soluble salts in the physiological root zone. Cultivation of the soils using special reclamation practices promoted the continuation of desalinization and desolonetzization. High crop yields were obtained.

1035

SAMDU, G.; RAUTA, C. Reclamation of the saline-alkali soils of the Calmatzui valley. Irrig. Drain. Paper,FAO, 1972, 16:17-30.

Reclamation of the Calmatzui valley could be achieved through several stages. First of all river training works, should be executed. During this period, the ground water regime and the soil salinity should be observed. The next step should include drainage works with a view to improve the drainage conditions over the entire area; these works are in correlation with the hydro-technical and land reclamation works carried out in the Romanian Plain. When both the groundwater regime and the salt levels in the soil have been stabilised, the actual reclamation works may start; these shold be carried out differently according to the prevailing soil and water conditions.

1036

SARKAR, T.K. Drainage system design of Research farm at Indian Agricultural Research Institute New Delhi. Design Manual IARI, New Delhi (India), 193p.

The manual is a compilation of the design work carried out to alleviate waterlogging problems of the Indian Agricultural Research Institute farm at New Delhi(India). Manual is divided into eight chapters dealing with general aspects of the area, project formulation, hydrologic design of farm drainage system, hydraulic and structural design, economic analysis of the plan. The glossary of terminology used has been given in a separate chapter.

1037

SARKAR, T.K.; BHATTACHARYA, A.K. Proposed viscous model for the design of drainage system of Madanpur, Khadar, Asola, Ali etc. Waterlogging Areas. Symp. Waterlogging Causes and Measures for Prevention, CBIP New Delhi. 1972,118, 1:143-8.

The general description of the area and the problem are given in the paper "proposed combination drainage system for Madanpur, Asola and Ali area". An attempt has been made to simulate a viscous (Hele-Shaw) model for the above mentioned problem. The importance of the model becomes more eminent as it gives the visual solutions of the complex problem for which analytical solutions do not exist. The survey of the area is in progress. From the preliminary survey it is being proposed to have a combination of tile drain and open ditches. This model will help in designing the tile spacing and diameter for steady state equilibrium condition with a steady rate of replenishment. In this paper, the complete description of the model along with the scales are given. As soon as the field data from the survey report will be available, then by plugging in these data the model can be constructed. Also a stepwise procedure is given here for how to proceed further for the experiment to obtain the required drain spacing and its diameter.

1038

SARKAR, T.K.; MAJI, C.C. Integration of drainage decision with irrigation management and crop planning -- an application of concave programming. J. Agric. Engng, 1978, 15(2): 60-8.

The study attempts to highlight the interdependence among drainage decision, crop planning and reservoir management and suggests that drainage decision should be made in a wider context of overall resources management. The conventional procedure of basing decisions solely on the available fund may be inconsistent with the optimal use of scarce inputs and hence investment in drainage development may not generate commmensurate benefit to the society. A concave programming model which seeks to maxomimize the return from the command area by ensuring optimal use of available resources was developed. The nonlinear model was then transformed into a linear programming model by the method of Partitioning of Variables. The applicability of the model was demonstrated under four hypothetical resource situations.

1039

SASANABE, S. A study on drainage in apple orchards. Bull. Fac. Agric. Hiosaki Univ., 1974, 23:83-192.

An orchard drained by plastic pipes was compared with an undrained orchard. The drained orchard had a soil temperature in the top 40-60 cm which was 0.8° C higher on average than that of the undrained orchard. Fruits of cv. Ralls Janet from the drained orchard had 1-2% more sugar, were 5% larger and were of better quality than those from the undrained orchard.

1040

SAULMON, R.W. Need for envelope material above drain tile. J. Irrig. Drain. Div. ASCE., 1971, 97:661-3.

Results of a laboratory study on the need of envelope above the tile indicated that elimination of this portion of the envelope material can cause a decrease in flow rates to the tile.

1041

SAVESON, I.L. Surface drainage of flatlands. Misc. Publs. U.S. Dep. Agric., Agric. Res. Serv. 1967, 1062, 10p.

Two main systems of field drains and land forming are described: the cut crowing system, used for the production of sugar cane, vegetables and pasture, and the precision land forming system, used in the production of all crops except rice and pasture.

1042

SAVILL, P.S. Drainage/ploughing experiments on deep oligotrophic peat. Forester, Belfast, UK, 1976, 15:46-8.

All combinations of two drain depths (0.6 and 1.2 m) and three drain spacings (7.5, 15 and 30 m) were superimposed on the basic ploughing treatment of alternate deep single and shallower double mouldboard ploughing when Sitka Spruce (Piecea sitchensis) was to be planted on oligotrophic peat on two sites in Northern. Ireland in 1965 and 1967. In 1975, best growth was found in the plots with the deepest drain at the closest spacing. Two types of control plot had been provided, (a) with closely spaced deep single mouldboard ploughing alone, and (b) with shallow double mould-board ploughing alone, growth was as good on (a) as an the most intensively drained plots, whereas on (b) it was similar to the least drained plots. These results appear to be related to the quantitity of peat thrown up in the 'ribbons' for each plant to exploit as a source of N.

1043

SAVILL, P.S. The effect of drainage and ploughing of surface water gleys on rooting and windthrow of Sitka Spruce in Northern Ireland. Forestry, 1976, 49:133-41

Fielf studies and surveys were made in established Piecea sitchensis plantations on both peaty and mineral soils in several poorly drained areas. Water levels were periodically recorded in bore-holes 0-16 m distant from 1-m deep drainage ditches which were dug at 30-m intervals on experimental plots in stands 2-5 m high. Drainage and soil type had little influence on water-table levels and therefore on rooting depth. Root distribution was measured (in sample trenches) in plantations 3-12 m high on (a) (strip) ploughed sites and (b) turf-planted sites. Root development was markedly restricted by the furrows made by ploughing or turfing, the furrows being at intervals of 1-2 rows in (a) and 4 rows in (b). This is related to the much higher incidence of windthrow on ploughed sites (23-32%) than on turf-planted sites (0.4-1.1.%) in a comparative trial. The use of double-mouldboard ploughs or other measures, to reduce the number of furrows per units area, is recommended to allow wider root-plate development and reduce the risk of windthrow.

1044

SCHAFFER, E. et al. Use of plastic materials for drainage and groundwater control by subsurface pipe systems. Fortsch. Landw. Nahrungs, 1976, 14, 54p.

The design production and use of pipes, fittings and filters for drainage and groundwater control are reviewed. Some emphasis is given to future trends in research, development and testing. The performance of plastic pipes with regard to silting up, deformation, impact resistance, hydraulic efficiency etc., as well as their increasing use in land improvement, are reviewed at length.

1045

SCHAIK, J.C. VAN.; MILNE, R.A. Reclamation of a saline-sodic soil with shallow tile drainage. Can. J. Soil. Sci., 1962, 42:43-8.

Leaching studies showed that shallow soils that have become saline-sodic can be reclaimed with tile drains installed at a depth of 75 cm with a 90 m spacing. At 30-60 cm below the surface occurred a slowly permeable glacial tile layer. Only after the application of 180 cm of water, the saline-sodic soil was sufficiently reclaimed to allow normal plant growth. Some decrease in salt concentration was found below the tile drains. There was no difference in salt movement with respect of proximity to the tile lines. The development of non-saline sodic soils was not evident during the trial. Most of the gypsum applied during the trial remained in the surface 15 cm after the final water application.

1046

SCHEFFER, B. Drain flushing in field tests in Bokern. Z. Kult. Flurb., 1982, 23:110-14.

Drain sludges can be eliminated by flushing, especially high-pressure flushing. About 66 per cent of the drain sludges were eliminated after one high-pressure flushing. Repeated flushing during the discharge period gave still further improvement. After flushing, about 17% of the blocked openings were re-opened. Flushing showed little effectiveness, however, against incrusted and old iron cloggings, which change from ferrihydrite to goethite. Drain pipes subject to iron-clogging should be flushed after a short period of use, and before the content of drain sludge becomes too high.

1047

SCHEFFER, B.; KUNTZE, H. Organic drain filters to prevent clog-ging with ochre. 3. Drain sludge and filter quality. Z. Kult. Flurb., 1979, 20:146-51.

In an experiment with two drain filters rich in tannin, the average discharge in 1973-1976 from drains laid in 1973 in a sandy Gley-Podzol was the same with Antoc and Duo filters, al-though the Fe content of the filter and clogging of the drain openings were greater with the Duo filter. In autumn 1976, the percentage contents of HC-insoluble material and wash in the sludge in the drains were higher, and the percentage content of Fe in the sludge was lower with Antoc than with Duo. Application of pig slurry to the soil increased the amount of sludge in the drain and the Fe content of the sludge. The slurry promoted decomposition of Antoc.

048

CHILFGAARDE, J.V. Design of tile drainage for falling ater tables. J.Irrig. Drain. Div. ASCE, 1963, 89:1-2.

The paper derives an equation for the spacing of the drains in terms of the rate of water-table drop, the depth of an impervious layer below the drains, the hydraulic conductivity of the soil, and an assumed constant drainable pore space. It treats the depth of the water-bearing stratum as a variable, so that it is not restricted to cases in which the water-table fluctuation is a small percentage of the aquifer thickness, proposes the use of a graphical correction, adopted from the tabulation of "equivalent depths" by Hooghout, to account for the convergence of flow towards the drain when drain spacing is not relative to aquifer depth; shows that the solution is in terms of an incomplete beta function which can either be evaluated numerically from a simple graph or determined within 3% from an approximate equation, makes comparison with an earlier, widely usd, solution by Glover to illustrate the significance of the improvements.

049

SCHILFGAARDE, J.V. Design for drainage systems an rainfall distribution. Proc. Drain for Efficient Cro Production Conf. Chicago, Illinois, 1965, 66-8.

A constant water table or an assigned rate of fal of water table is inadequate in principlethough a the present state of technology, it appears as th only practical criterion available for drainag design. A new drainage criterion based on prob ability distribution of water table height i described which is based on predicting water tabl height is from long term precipitation records a influenced by drainage-system parameters. Limita tion of the theory can be overcome and it is sug gested that this methodlogy can be refined to mak it practically useful.

1050

SCHILFGAARDE, J.V. Drainage yesterday, today and tom morow. Proc. Natn. Drain. Symp. Chicago, Illinois 1971, 2-4,7.

Drainage history which dates back from second cen tury before Christ has been presented. Th development of tile drainage system on agricul tural lands, drainage material and machinery an its future needs have been discussed.

1051

SCHILFGAARDE, J.V. (Ed). Drainage for Agriculture. Agronomy Monograph, No.17, Ameri. Soc. Agron, Wiscon sin, 1974, 700 p.

The previous monograph in this series devoted t drainage and entitled: Drainage of Agricultura lands, was published in 1957. This new monograp includes developments since 1957 as well as cover ing subjects that were not dealt with in detail i the earlier book. It is in 11 sections: drainag and crop production, current drainage practice (in the USA and in flat areas of humid regions o Europe), materials and methods, saturated flo theory and its applications, unsaturatd flo theory and its application, salts and water movement, quality of drainage waters, models and analogues for the study of groundwater flow, determining soil properties, and water management systems.

052

CHILFGAARDE, J.V. Limitation of Dupuit-Forchheimer heory in drainage. Trans. ASAE, 1965, 8:515-6, 519.

Dupuit-Forchheimer (D-F) theory has been commonly and success-fully used not withstanding repeated attacks on its validity. The paper intended to bring out that the limitations of the theory are kept in mind, the D-F theory can provide answers that are more than adequate in terms of accuracy, especially for engineering design when estimation of parameters such as porosity and hydraulic conductivity results in errors far greater than those associated with the theoretical limitations.

053

SCHILFGAARDE, J.V. Progress and problems in drainage design. Proc. Int. Drain. Workshop. Wageningen, Netherlands, (1978), 1979, 25:633-44.

Drainage design problems and progress made in this field, are presented. The drainage evaluation guidelines for different cases are reported. It is suggested that in the rainfed areas, the effectiveness of drainage probably should be evaluated in terms of the soil water content of the rootzone and/or the surface soil, rather than the height of the water table. For irrigated areas, the primary concern is removal of excessive salts. In the latter case, the interaction between irrigation management (and cropping pattern) and drainage requirements must be recognized. Furthermore, in both cases, the drainage facilities must be designed with due regard for the natural drainage rate. Finally, the impact of drainage is not restricted to the farm; in some cases, the off-site effects are importannt and may even dominate.

1054

SCHILFGAARDE, J.V. Theory of flow to drains. <u>Adv. Hydro-Sci</u>., 1970, 6:43-106.

The review paper in general discusses the equations of motion, steady and nonsteady saturated flow towards drains, partially saturated flow and discusses the measurements of soil properties (Hydraulic conductivity) which is used in drainage design. Application of flow theory to drainage design is discussed. The paper contains 82 references.

1055

SCHILFGAARDE, J.V. Non steady flow to drains. <u>In</u> Drainage for Agriculture (Agronomy No.17, Schilfgaarde, J. V. Ed). Amer. Soc. Agron, Wisconsin, 1974, 245-70.

Equations describing non-steady flow to drains and their applications in the field are discussed.

1056

SCHILFGAARDE, J.V.; WILLIAMSON, R.E. Studies of crop response to drainage. I. Growth chambers. <u>Trans.ASAE</u>, 1965, 8:94-7.

Studies on drainage requirements of crops in growth chambers indicated that response of plants to water table height is dependent upon day time temperatures and thus this factor must be carefully considered. To establish a useful criteria of drainage for crops aeration status of the root-zone must be specified than alone the water table as plants do not respond to water table <u>per se</u>.

1057

SCHMID, P.; LUTHIN, J. Drainage of sloping lands. <u>J.Geophys. Res.</u>, 1964, 69:1525-29.

The paper analyses problem of water table in equilibrium with rainfall on sloping land by solving Boussinesq's equation; assumes that hill-side is drained by parallel ditches which are found in impermeable layer; presents a nomogram for calculating ditch spacing as function of rate of replenishment, soil hydraulic conductivity and maximum height of water table above impermeable layer.

1058

SCHNEIDER, C.B.H. Drainage of peat grassland. Bedrifsontwik-keling, 1973, 4:441-7.

In this review it is pointed out that the traditional high water levels in the drainage ditches do not ensure an improved water supply to grassland in summer but do result in 'invisible water-logging' in winter. Water levels 50-60 cm below field level were usually required in order to combat treading damage.

1059

SCHOIZ, A.; HILLIGER, D. Filter-less perforations in the base of drains for groundwater regulation in soils with a sandy layer at drain depth. Arch.Acker-U.Pfl. Bodenk., 1974, 18:177-87.

Modifications to the arrangement and shape of the perforations in plastic drains are described that prevent silting up in the absence of filters.

1060

SCHOLZ, A.; WERTZ, G. Investigation of groundwater regulation in peat soils in the GDR. IAHS Publ., 1975, 105:47-54.

Systems of two-way groundwater regulation are required to ensure high and stable crop yields on sand, sand-underlined turf and thick peat soils. The transition to industrialized methods in agriculture is highly conducive to the replacement of open drainage with closed drainage. To this end, novel plastic drain pipes of large diameter are being used in GDR in both sand and peat soils, in the latter case in combination with mole drains. Depending on how extensively the grasslands are utilized, ground-water regulation is carried out with due regard to water saving, utilisation of the soil waterholding capacity and differential water provision.

1061

SCHOTHORST, C.J. Effects of different drain depths in peat soils in the Netherlands. Meded. Inst. Cult.Waterhuishouding, 1974, 149, 16p.

The eutrophic peat soils used as pastures in the western part of the Netherland are drained by ditches customarily situated at 0.40 m depth. With the intensification of dairy farming the bearing capacity of the soil is insufficient and, in wet period, grass is lost by trampling and sod layer damage. Bearing capacities of the sod layer are sufficient if the penetration resistance is 6kg/cm^2 or higher. To meet these requirements the groundwater level in wet periods should not rise higher than 0.30 m below the surface. The investigations showed that at a mean distance between the drains of 35 m, and with good soil permeability, a lowering of the open water level from 0.40 to 1.0 m below the surface always met the drainage and bearing capacity requirement.

1062

SCHOTHORST, C.J. Subsidence of low moor peat soils in the Western Netherlands. Geoderma, 1977, 17:265-91.

Three experimental fields were laid out to study the effects of deeper drainage on the process of subsidence. It was found that 6 years after ditch-water levels had been lowered, surfaces had subsided 6-10 cm. Because the deep peat is affected only very slightly by deeper drainage, it may be assumed that compression of it caused little or no subsidence over the past 1000 years in areas with high ground water levels. The bulk density comparisons, N-supply, rate of loss of organic matter for shallow and deeper drainage are presented to have a plausible explanation for the surface subsidence of the low moor peat soils in the Western Netherlands.

1063

SCHUCH, M. Regulation of water regime of heavy soils by drainage, subsoiling and lining and water movement in this soil. Proc. Int. Drain. Workshop, Wageningen, Netherlands, (1978), 1979, 25:253-67.

The amont of runoff from two drain fields on pseudogleys of loess and clay with increasing drainage distances combined with vertical soil drainage and loosening of the subsoil (0.95 m depth, 0.75 m vertical distance) were compared with the amount of runoff from a conventional drainage without sub-soiling. The highest yield was obtained at a drainage distance of 80 m. This mode of drainage is not only good for farming but also for environment. By using different materials for filling the drain ditch, it was observed that not gravel but excavated topsoil gives the greatest runoff. The water movemennt in soil to a drain was observed by measuring the natural electric streaming potential.

1064

SCHULIZ, E. A model to determine optimal sizes for the drainage system in a polder. In Polders of the World. Vol.1. Polder pro-jects. Land and Water Management Aspects. Int. Inst. Land Reclam. Improv, Wageningen, Netherlands, 1982, 664-74.

A method has been developed to determine optimal values for the main components (sluice or pumping station, canals, main ditches, ditches, trenches or subsurface drainage) in the drainage system of a new polder, with reference to conditions in the Netherlands. The method combines hydrologic and economic computations, followed by an optimalization routine.

1065

SCHUSCHKE, F.; BISCHOFF, H.M. Studies on the influence of floods on grasses. 1st communication; the influence of the duration of flooding. Arch. Acker-U. Pflan. Bodenk., 1980, 24:521-9.

Pure stands of meadow fescue, timothy, tall fescue and reed canarygrass and mixtures including these spp., Alpecurus pratensis, Poa pratensis and perennial ryegrass were subjected to artificial flooding for up to 30 days and flooding tolerance, yields and changes in chemical composition were studied. Plant injury increased with flooding duration. Young stands were more flood-tolerant than older ones. A.pratensis, reed canarygrass and timothy were suitable for sites flooded more frequently and over a longer period of time; perennial ryegrass, meadow fescue and tall fescue were only suitable for less frequently flooded areas. Mixtures behaved in accordance with the flooding tolerance of the individual components. Flooding caused large increases in ash content and reduced CP and crude fat content. CP content decreased with duration of flooding in the autumn of the establishment yr and increased with duration of flooding in the spring of the 1st utilization yr.

1066

SCHUSCHKE, F.; BISCHOFF, H.M. Studies on the influence of floods on grasses. 2nd communication: the influence of the level of flood water. Arch. Acker. Pflan Bodenk., 1980, 24:531-7.

In pot experiments, Italian ryegrass, perennial ryegrass, timothy, meadow fescue, tall fescue, reed canarygrass, Alopecurus pratensis and Poa pratensis were subjected to flooding to depths of 0, 20, 40 or 80 cm for 28 days at various stages of development. Tillering capacity plant ht., DM yield, regrowth capacity and soluble carbohydrate content decreased with increasing water depth. The response to differences in water depth was greatest when flooding began at the start of emergence and the plants were tall enough not to be completely submerged by flooding to 20 cm depth. All depths of flooding at earlier stages significantly reduced plant ht. and DM yield.

1067

SCHUSCHKE, F.; BISCHOFF, H.M. Studies on the influence of floods on grasses, 3rd communication: the effect of special cultivation measures. Arch. Acker-U. Pflan Bodenk., 1980, 24:539-44.

Plots of parennial ryegrass, Italian ryegrass and mixtures including Alopecurus pratensis, Poa Pratensis, meadow fescue, timothy and reed canarygrass were uncut or cut to a ht. of 8 cm in autumn 1977 and were flooded to a depth of 30 cm from 15 March to 13 April 1978, with or without subsequent loosening of the sediment deposited. Both treatments improved plant density and yield, especially in spp. with rapid early development. Cutting in autumn had more effect than loosening of the sediments, but the combined treatment gave the best results.

1068

SCHUURMAN, J.J. Root and crop growth of oats as affected by the length of periods of highwater-table. Neth. J. Agric. Sci., 1980, 28:20-8.

The influence of the duration of a high water-table on root and shoot growth of oats in an early growth stage was studied. Root wt., depth of rooting and number of model roots with a length of 10 cm reacted favourably on a low water-table. In this experiment shoot growth, however, hardly reacted to better root growth. This meant that the plants with restricted root growth could absorb water and nutrients as well as those with larger root systems. The fertilization of the top 0-25 cm of soil could have been an important factor in this report.

1069

SCHWAB, G.O. Age effects on surface and subsurface drain flow. 3rd Natn. Drain. Symp. ASAE, Michigan, 1976, 10-3.

Results from field drainage experiment, installed in northern Ohio in 1957, were analysed to determine the effectiveness of drainage systems with time. Peak tile flow and tile flow volume did not change significantly over a 15 yr period (1960-1975) with comparable crops and water applications. A slight, although not statistically significant, drainage improvement was measured for the combination of surface and subsurface drains.

1070

SCCHAB, G.O. Procedure for determining hydraulic capacity of tile drains in humid areas. Trans. ASAE, 1969, 12:766-8.

Knolewdge on drainage requirement of crops in terms of volume of water to be removed in dropping the water table in the desired time is essential to evaluate the hydraulic capacity of tile drains. An equation is developed to find out the average drainage rate. The proposed procedure of evaluation is illustrated by two examples on Toledo silty clay soil and Tedrow fine sand. It is suggested that because of limited data available on crop drainage requirement in terms of duration of the water-table level and the initial and the final height, more research on these aspects may be conducted.

1071

SCHWAB, G.O. Tile or surface drainage for Ohio's heavy soils? Ohio Rep. Res. Develop, 1976, 61:19-21.

In comparisons of surface drainage, tile drainage

and a combination of both types on poorly-drained lake bed soil with a heavy impermeable subsoil, the yields of maize, soyabeans and oats were measured. For maize, in undrained surface, tile and combination plots, with conventional tillage, grain yields were 50, 79, 100 and 106 bu/ac, respectively and with no tillage were 35, 50, 63 and 83 bu/ac. Oats, over 2 years, gave 0, 24, 53 and 62 bu/ac and soybeans gave, 4, 42, 50 and 50 bu/ac, respectively. The yearly variations in yield were decreased with the combined drainage system. Surface drainage was shortlived reaching nil during the subsequent 3 years. Nutrient losses/ha with conventional tillage over 5 years were 4, 24 and 20 lb N, 2.4, 1.2 and 1.9 lb P and 42.8, 30.5 and 40.5 kg K for surface, tile and combination systems respectively.
(1 bu/ac = 50.44 kg/ha).

1072

SCHWAB, G.O.; DEBOER, D.W.; JOHNSON, H.P. Effect of openings on design of subsurface drains. J. Irrig. Drain. Div. ASCE, 1969, 95:199-209.

The paper presents theoretical effects of joint spacing on flow into drain tile and effects of perforations in a form useful for the design engineer and also the influence of these openings on the rate of water table drawdown between parallel drains and observes experimental field data being supported by the theoretical results.

1073

SCHWAB, G.O., DESMOND, E.D.; OCHS, W.J. Ultraviolet standard for corrugated plastic tubing. 4th Natn.Drain. Symp. Proc. ASAE, 1982, Pub. 12-82:87-92.

A parallel plate impact test was developed to determine brittle-ness of 102mm dia corrugated plastic tubing when exposed to ultraviolet (UV) radiation. Artificial exposure was obtained with an Atlas Weather-Ometer Ci 35 using a xenon-arc lamp. Proposed exposure time was 200 hr which was selected to simulate two years or more of outdoor exposure. Samples from ten manufacturers were tested to evaluate proposed procedures. All samples passed the artificial UV exposure test, but three failed the parallel plate impact that due to a manufacturing problem at the mold seam and not due to embrittlement.

1074

SCHWAB, G.O.; FAUSEY, N.R.; WEAVER, C.R. Tile and surface drainage of clay soils. II. Hydrologic performance with field crops (1962-72). II. Corn, Oats, and soyabeans yields (1962-72). Res. Bull. Ohio Agric. Res. Develop. Centre, 1975, 1081.

For all drainage systems, the flow was nearly the same for conventional and no tillage. For the combination tile and surface drainage system, seasonal tile flow was more than double the surface runoff for conventional tillage, but nearly the same for no tillage. For all drainage systems, the stands and yields were significantly higher in the conventional than in the no tillage system. The stands and yields increased in the order of no drainage, surface tile and the combination tile and surface. The conventional and no tillage management treatments showed similar patterns of yield and stand for the four drainage treatments. The adjusted corn yields for conventional tillage were 50.4, 78.5, 100.1 and 105.9 bu/acre, respectively, for the four systems. The average oat yields for 1966 and 1972 were 0, 24.0, 53.3 and 62.1 bu/acre, respectively, for the four systems. The annual coefficients of variation of corn yields were 47.8% for no drainage, 50.1% for surface, 20.7% for tile, and 21.3% for the combination tile and surface drainage system. By pairing treatments, the variance of any two systems was not significantly different for either yield or stand.

1075

SCHWAB, G.O., FOUSS, J.L. Tile flow and surface runoff from drainage systems with corn and grass cover. Trans. ASAE, 1967, 10:492-3, 496.

Flow from surface drainage, tile drainage (level with no surface runoff), and a combination of tile and surface drainage systems was measured over a 5-year period (1960-64) from a replicated field experiment. For the 7-month growing seasons the total outflow from the tile-drined (C) plots and the comibination-drained (D) plots was considerably higher than from the plots with only surface drainage (B). The combination tile and surface drainage system (D) gave a higher total outflow than the plots with only tile drainage (C). The surface runoff volume and peats were higher (also percentage of water removed) with fescue than with corn. Conversely, the tile flow was greater from corn than from fescue. For the conditions studied, surface drainage and tile drainage systems complemented each other. When the meadow crop was sprinkled with 9.9 cm (flow corrected to a 7.6 cm application) at 0.50 cm per hr, surface drainage reduced tile flow by 53 percent. For similar soil moisture conditions the following year, with 7.6 cm of water sprinkled on corn, the percentage flows were nearly reversed, that is, surface drainage reduced tile flow 31 percent and tile drainage reduced surface runoff 51 percent.

1076

SCHWAB, G.O.; MANSON, P.W.; LUTHIN, J.N.; REEVE, R.D.; EDMINISTER, T.W. Engineering aspects of land drainage. In Drainage of Agricultural Lands (Luthin, J.N. Ed). Wisconsin, 1957, 287-394.

Drainage in arid and humid areas are presented. Design and installation of open drains for large areas and tile drainage systems have been discussed. Problems of drainage of irrigated lands are reported.

1077

SCHWAB, G.O.; MCLEAN, E.O.; WALDRON, A.C.; WHITE, R.K.; MICHENER, D.M. Quality of drainage water from a heavy-textured soil. Trans.ASAE, 1973, 16:104-7.

Sediment, dissolved solids, nine chemical elements or ions, five pesticides, EC, pH, and EOD in tile drainage effluent and in surface runoff were measured on Toedo siltyclay soil near Sandusky, Ohio, for a 3-year period, 1969-1971. Measurements were taken from both conventional tilled and no tilled plots in continuous corn. Irrigation water was applied each year to simulate heavy rainfall. Sediment losses were about the same in the tile effluent and in the surface runoff. Nitrate nitrogen losses were generally low (about 14.6 to 21.3 kg/ha per year). These loses were about the same from the tile as from surface runoff, but slightly higher for conventional tillage than for no tillge. Except for the storms in June 1970, phosporous losses were less than 1.7kg/ha per year, but they were greater in the surface runoff than in the tile effluent. Potassium losses in the surface runoff were 11.2-14.6 compared to 4.5 kg/ha per year in the tile effluent. HCO_3, Cl, and Na losses were higher in the tile effluent than in the surface runoff. Average strazine losses from rainfall were low (2.8 to 6.9 per cent of amount applied), and dicamba (2.0 to 7.3 percent). Aldrin and heptachlor losses were insignificant being much less than 0.01 per cent of the amount applied. The average EC of the tile effluent and the surface runoff from rainfall were about the same 729 to 1062 ms/m. The BOD of the drainage water was insignificant, being less than 3.4 kg/ha.

1078

SCHWAB, G.O., MICHENER, D.W.; ROPAK, D.E. Spacing for subsurface drains in heavy soils. Res. Bull. Ohio Agric. Res. Develop. Centre, 1982, 1141, 20p.

Hydraulic conductivity (K) and drainable porosity (f) were evaluated for eight slowly permeable soil types by saturating the soil to the surface at each field site by sprinkler irrigation. Water table measurements were taken at three distances from parallel drains and corrections were made for evaporation to compute drainable porosity and K/f. For the existing procedure developed by Skaggs, from these K/f values, drain spacings were computed for drawdown rates from the soil surface of 10, 20 and 30 cm/day. Drain spacings ranged from 7 to 18 m for a drain depth of 0.8 to 1 m.

1079

SCHWAB, G.O.; TAYLOR, G.S.; FOUSS, J.L.; STIBBE, E.;Crop response from tile and surface drainage. Soil Sci. Soc. Am.Proc., 1966, 30:634-7.

The effects of various drainage systems and different levels of N treatments on corn yields were evaluated for a fine textured lakebed soil during a period of 3 years. The drainage treatments consisted of undrained plots, surface drained, tile drained (90 cm depth, 12 m spcing) and a surface and tile drained combination. The three drained treatments gave significantly higher yields than the undrained treatment, and N increased corn yields significantly at all drainage levels. An econnomic-evaluation of the drainge system showed that the benefit/cost ratio of surface drains was 10:1, tile drains 6:1 and the combination drains 4:1 when compared to undrained. Surface drainage gave the greatest benefit per dollar invested on drainage.

1080

SCHWIETERMAN, C. Causes of premature failure and proper installation of large (25 cm to 75 cm subsurface drainage mains. 4th Natn.Drain Symp. Proc.ASAE, 1982, Pub. 12-82:93-104.

A contractor's view on failure of subsurface drainage systems is presented. Because of expected increase in agricultural machinery 25 years hence, wider trenches, better bedding, longer length of tile & tile with greter crushing strength are recommended. Continuous on site inspection and annual maintenance enforced by rules are essential to increase the life of drainage systems. For plastic drains installations care should be taken to ensure correct contour of the bottom of the trench, eliminate strectcing as much as possible, use of friable material bedding and avoidance of large machinery before natural setting.

1081

SCOTT, R.S. The effect of mole drainage and winter pugging on grassland production. <u>N.Z. Grassl. Ass. Proc.</u>, 1963, 25:119-27.

Trials in foothill country on a silt loam overlying clay suggested that, where winter grazing was not practised, drainage is of doubtful value for increasing pasture production. Pasture production may suffer considerably under winter grazing and drainage can alleviate this damage to some extent.

1082

SCOTTISH AGICULTURAL COLLEGES. Drainage of soils of low permea- bility. East of Scotland College of Agriculture. Edinburg. Scottish Agric. Colleges, U.K. 1983, 8:(Revised), 19 p.

A discussion of drainage requirements of Scottish soils from which it is concluded that to maintain the necessary condition for the rapid removal of excess water from the less permeable soils more effective methods of drainage are required and that some form of subsoil modifications, such as moling or subsoiling, will have to be accepted as part of the routine cultivations of the farm.

1083

SEGEREN, W.A.; SMITS, H. Drainage of newly reclaimed marine clayey sediments, peat soils, and acid-sulphate soils. <u>In</u> Drainage Principles and Applications. Vol. IV. Design and Management of Drainage Systems. Int. Inst. Land Reclam. Improv., Wageningen, 1974, 16:261-95.

A discussion of the specific problems encountered in draining three types of soils: recently emerged marine clayey sediments, peat soils and acid-sulphate soils.

1084

SEGINER, I. A model for surface drainage of cultivated fields. J. Hydrol., 1971, 13:139-52.

A computer model for the simulation of the drainage pattern and surface storage capacity of cultivated fields is presented in which furrows and random roughness are superimposed over the general topography of the field. Results show that storage capacity increased with increased depth of contour furrows but that the depth of furrows become insignificant as the furrows deviate more and more from the contour. Storage capacity of furrows running obliquely to the contour seems to be a function of the slope along the furrow only and independent of the general slope of the field.

1085

SEKHON, G.S.; GILL, S.S. Effect of drainage on desalinization in a Ludhiana silt loam soil. Proc. All India Symp. Soil Salinity, Kanpur, 1971, 178-81.

Drainage with natural precipitation alone considerably reduced the pH and ESP. Exchangeable Ca increased generally during winter. ESP decreased with proximity to the relief drain. Effectiveness of the drain was only when located upto 67m distance. The intercepting drain on the eastern side considerably helped the adjoining fields. Fluctuating water table closer than 60-90 cm was associated with high pH and ESP.

1086

SEL'CHENOK, V.P. Automation in the control of the groundwater level in drainage-irrigation systems.

Dokl. Vses. Ord. Len. Akad. Sel. Nauk. Im., Lenina, 1980, 12:35-7.

Automation in controlling the groundwater level in drainage-irrigation systems is analyzed and described by a series of mathematical equations.

1087

SEPPALA, K. Effect of drain spacing on the volume and growth of peatland-forest stands. Suo, 1969, 20:55-60.

Stand volume and annual volume increment were decreased by > 20 percent by increasing drain spacing from 20 to 60 cm. The decreases were greatest on less fertile sites.

1088

SEPPALA, K. Relationship between the stem diameter of trees at time of draining and their post-drainage increment. Suo, 1976, 27:55-60.

The post-drainage radial increment of Norway spruce and Scots pine trees of similar age was, over a period of about 50 years, more or less independent of tree diameter at the time of draining. However, the post-drainage basal area increment of similar aged trees was related to the size of trees at draining.

1089

SEWEL, J.I. Land grading for improved surface drainage. Trans. ASAE, 1970, 13:817-9.

Land improvements including a drainage canal, drainage ditch, levee, land grading and drainage pump are discussed for West Tennesse Experimental Station so as to result in improved surface drainage. Cost of land grading is $320/ha whereas total cost comes to $557.5/ha. In the East Tennesse, grading forms the major part of land improvement for surface drainage and the cost is around $325/ha.

1090

SHALHEVET, J.; ENOCH, H.; DASBERG, S. Response of sugar-beet to soil drainage and aeration. I. J. Agric. Res., 1969, 19:161-70.

The experiment, conducted with fine-textured soils (vertisols) in large containers under polyethylene cover, showed that crop yields increased with watertable depth, most of the increase occurring between depth of 20 and 60 cm. Compared with conditions without a watertable, a high watertable resulted in a great decrease (60%) in the Fe content and an increase (2-fold) in the NO_3 content of leaves, as well as a 35 percent decrease in the Na and K contents of roots. Rates of O_2 diffusion and redox potential increased with distance from the watertable, and provided indices that correlated well with yields, whereas composition of soil air was a less sensitive index of soil aeration.

1091

SHARMA, A.K.; FEHRENBACHER, J.B.; JONES, B.A, Jr. Effects of gypsum, soil disturbance and tile spacing on the amelioration of Huey silt loam, a natric soil in Illinois. Proc. Soil. Sci. Soc.Am., 1974, 38:628-32.

In a study of the effect of various rates of gypsum, various soil disturbances and tile spacings on a natric soil in Illinois, the most effective treatment combination in removing sodium and increasing corn yields was 62 t/ha of gypsum thoroughly mixed with soil to the 90 cm depth and tile spacing of 10 m or less.

1092

SHARMA, D.R.; PRIHAR, S.S. Effect of depth and salinity of ground-water on evaporation and soil salinisation. Indian. J. Agric. Sci., 1973, 43:582-6.

In a controlled study with soil columns, the

evaporation from a sandy-loam soil with watertable at 50 cm or more was not influenced by an evaporation of 8.8-16 mm a day from a U.S. Weather Bureau class A open-pan evaporimeter. The maximal evaporation from soil, Y (mm) was related with the depth of the watertable, X (cm), by the equation of X = 149.2-150 Y. The effect of groundwater salinities of EC 8 and 16 dS/m and groundwater depths of 50, 75 and 100 cm on soil salinisation during a 13 week rainless summer period was also investigated. The effect of groundwater salinity on the salinisation of the top 40-cm soil layer was more pronounced with shallower watertable than with deeper watertable.

1093

SHARMA, H.D.; CHAWLA, A.S. Use of interceptor drain as an anti-waterlogging measure. Proc. Symp. Waterlogging Causes and Measures for Prevention, Cent. Bd. Irrig. Pwr. New Delhi, Pub 118:1972, 1:101-10.

An analytical solution has been developed for the problem of seepage flow from a canal to the interceptor drain and to the main drainage, location in homogeneous and isotropic soil formation of depth extending to infinity. Equations have been obtained for the seepage discharge from the canal into the interceptor drain and main drainage, and for determining the phreatic surface under steady condition of flow. The solutions of these equations have been plotted in the form or dimensionless curves for facility of design. The computed results have been found to be in close conformity with those observed in the field in case of Ganga Canal.

1094

SHARMA, R.P.; SEHGAL, S.R. Design criteria for drainage projects in the Punjab. Proc. Symp. Waterlogging Causes and Measures for Prevention. Cent. Bd. Irrig. Pwr. New Delhi, Pub. 118, 1972, 1:111-22.

The applicability of use of Boston Society Formula to condition in Punjab (India) has been examined. The value of constant C in the Boston Society formula $Q = C.A^{1/2}$ as influenced by rainfall and area have been reported. The observed data show that usually recommended values of C can be drastically reduced for different rainfall conditions. The value of C for rainfall less than 50.80 cm can be reduced to half from its present value of 200 whereas between 50.80 cm and 76.20 cm it can be reduced from 480 to 200. Similarly the value for rainfall between 76.20 and 101.60 cm can be taken as 400 instead of 2000. For rainfall exceeding 101.60 cm the value can be taken as 2500 instead of each case to be considered on actual study recommended earlier. The effect on C values because of change in area has also been discussed.

1095

SHARMA, T.C. Response functions applied to a drainage system. J.Hydrol., 1980, 45:279-87.

Mathematical expressions have been developed for the response functions based on the discrete dynamic representation of drainage system. The response functions, viz. impulse, step and frequency functions could serve as performance parameters of the drainage system.

1096

SHARMA, T.C.; IRWIN, R.W. Drain discharge from a level agricultural watershed. J. Irrig. Drain. Div.ASCE, 1977, 103:429-43.

Data for the Merlin watershed on the north shore of Lake Erie were analysed. A hypothesis was developed that watershed discharge from rainfall, for nonfreezing periods from a tile drained level watershed, was mainly tile flow. Daily discharges at the watershed outlet were due to depletion of two parallel linear reservoirs. A fast reacting reservoir, represented by the physical shallow ploughed layer, constituted nearly 40% of the total runoff volume. The slow reacting reservoir, represented by the subsoil zone, contributed nearly 60% of the total runoff volume.

1097

SHARMA, T.C.; IRWIN, R.W. A model to determine a subsurface drainage coefficient for flat land soils., Can Agric. Engng., 1976, 18:46-8.

A model was developed to predict drainage rates from a flat tile drained basin for non-freezing periods, using a probability analysis of drainage rates for the 11 year period. Probability analysis is a sound way of choosing a drainage coefficient for designing and evaluating tile drainage systems. With satisfactory return period selected, based on submergence tolerance of crops, probability analysis will provide a sound basis for the selection of an appropriate drainage coefficient.

1098

SHEIKH, G.S.; PERVEZ, K. Modelling of farm irrigation and drainage structures. Agric. Mechanization Asia, 1980, 11:51-5.

Prediction equations were developed using the principles of similitude for the performance of pumping machines, open channels, porous media, underground drains, pipes and sewers. It was found that viscous forces for pumping machines, gravitational forces for open channels and pressure head with gravity for underground drains are the most important parameters.

1099

SHIH, S.F.; CASCHO, G.J. Soil temperature related to water table depth. Proc Soil Crop Sci. Soc. Fl., 1980, 39:98-102.

Tenperature was measured in organic soil where shallow (SWTD), 30 cm, and deep water table depths (DWTD),90 cm, were maintained in a lysimeter system. Tenperatures were measured at depths of 2, 15, 30, 60 and 90 cm below the soil surface. The data were analysed based on three different time periods, i.e. annual, initial growth (March, April and May), and 'grand growth' (April through October) for sugarcane. Soil temperatures in DWTD were about 1.5°C higher than in SWTD for the annual period, 1.8° C for the initial growth period, and 1.6°C for the grand growth period. The lower temperatures in SWTD can reduce the annual subsidence rate by about 10 percent based on computations from three subsidence models. The high water tables in the initial and grand growth periods significantly reduced sugarcane production. Thus, a high water table in organic soil reduces subsidence, but some sugarcane yield reductions are expected. However, a high water table should be maintained during periods when cane growth or sugar accumulation is not adversely affected.

1100

SHILESKY, D.M.; WALSH, J.J. Applicable experience land disposal sites: drainage-leaching-monitoring. In Risk and decision analysis for hazardous waste disposal. Hazardous Materials Control Research Institute, Maryland, USA, 1981, 106-16.

Drainage and leaching are discussed in relation to current US monitoring requirements.

1101

SHKINKIS, C.N. The effect of drainage on water regime of coarse textural loamy-sandy soddy-podzolic soils in the Latvian SSR. Trans. 10th Int. Cong. Soil Sci, Moscow, 1974, 10:260-6.

The most effective reclamation and the best economic results have been achieved by application of deep systematic drainage in combination with mole drainage or deep lossening of soil. The average amount of moisture lost on artificial drainage in light loamy and sandy loam soils of

Latvian SSR ranges from 150-200 mm, in excessively wet year from 400-500 or more. If drainage is 1.3-1.5 m deep, it is recommended to decrease the normative spacing between drains by 2-4 m and to increase depth by 0.1-0.2 m.

1102

SHKINKIS, C.N. Subsurface drinage results for overwetted soils of the Latvian SSR. Agric. Wat.Mgmt., 1979, 2, 55-65.

In the Latvian SSR, experiments on hydrological drainage action have been carried out for some 16-17 years on 200 drainage fields. It is found that the average annual removal of excess water is 150-250 mm and, in particularly moist years, 400-500 mm and more. Drainage results in a considerable reduction of the duration of overwetting and over flooding of the active soil layer. The importance of drainage in the control of the soil water regime is the greatest in the winter and spring periods. The most effective draining and the best economic indices have been achieved by applying deep systematic drainage. For sandy and loamy soils, a 1.3-1.5 m deep drainage installation and use of larger diameter drain pipes are recommended. It is expedient to determine the drain spacing by combining the hydromechanical and the empirical methods, making the most of the data of many years of observations of the action of drainage systems.

1103

SHKINKIS, T.N. Effect of soil conditions on the hydrological action of drainage. Pochvovedenie, 1961, 11:98-103.

With drains at a given depth and a given distance apart, the height of the water table decreased more slowly in derno-podzolic-gleyey soils than in dernopodzolic soils. The maximum modulus of drainage discharge sometimes exceeded 4L/sec/ha from dernopodzolic-gleyey soils and 10L/sec/ha from dernopodzolic soils. The maximum hydraulic pressure of the groundwater at which drainage began was 10-30 cm for dernopodzolic soils and 50-70 cm for dernopodzolic-gleyey soils.

1104

SHKINKIS, T.N.; DAUGAVIETE, M.V. Covered drainage for forest land. Vestn. S-Kh. Nauki. Mosk., 1977, 7:72-83.

The lowering of the groundwater table in Latvian forests was greater and more uniform by covered tile and plastic drains than by open drains. Variations in drainage discharge during the course of a year were less variable from forest than from agricultural land although total annual drainage from forest and agricultural land was similar. Preliminary data indicated that total evaporation from the two types of land was also similar. Three years after drains had been laid drainage improved the growth of 40-50 year old trees.

1105

SHOPSKI, N. Duration of the effectiveness and technology in the construction of mole drains. Pochvoz. Agrokhim., 1973, 8:9-16.

Data for a system of unlined mole drains on the flood plain of the Danube showed that almost all the channels were sound after three years but that only 1/3 of the channels were actually filled by irrigation water when the system was used for subirrigation.

1106

SHOPSKI, N.; ENEVA, S. The salinization of soils and groundwater in the Stara Zagora irrigation system. Pochvoz. Agrokhim., 1975, 10:50-61.

Salinization data are presented for the period 1968-1973 and it is suggested that the groundwater level should be controlled at a depth of greater than 130-150 cm in the summer.

1107

SINE, L. The effect of land amelioration on soil productivity. Ann. Gembl., 1963, 69:642-59.

Data are reported on the seasonal movement of groundwater and on the botanical composition and yields of grassland on some drained and undrained soils in Belgium.

1108

SINE, L.; GASPAR, S. Evaluation of aeration porosity in the calculation of the descent of watertable in drainage. Pedol., 1965, 15:177-95.

The "equivalent drainable porosity" characterizes the air content of soil above the watertable. Curves of watertable descent in drained grassland soils show a constant air content in loamy and clayey soils and no effect of capillary rise on the trend of watertable descent.

1109

SINGH, B.; O'CALLAGHAN, J.R. Effect of interflow on soil drainage. J.Agric. Engng. Res., 1978, 23:397-415.

A sand tank study was carried out to compare the performance of diferent drainage configurations on the rate of removal of water in the steady state and on the rate of drawdown of the water table in the non-steady state. A shallow layer of high permeability overlying a deep layer of lower permeability was used to encourage interflow. It is observed that when the ratio exceeds 5, most of the water moved horizontally through the upper more permeable layer. A finite element solution was used to examine the effects of different drain spacings and ratios of soil permeabilities on drain performance.

1110

SINGH, B.; O'CALLAGHAN, J.R. Non-steady drainage in a layered soil. J.Agric. Engng. Res., 1978, 23:417-427.

A sand tank study was carried out to compare the performance of different drainage configurations both on rate of removal of water and on the rate of drawdown of the water table in a layered soil where a permeable layer was overlying a less permeable sub-soil. Narrowly spaced shallow drains in combination with a widely spaced deep drain was the most effective combination. The initial rate of discharge of shallow drains either alone or in combination with a deep drain was twice that of deep drain alone.

1111

SINGH, B.; MICHAEL, A.M. Predicting discharge from tile drains. J.Agric. Engng., 11:22-6.

Applicability of theoretical drainage equations in predicting tile discharge has been verified by conducting field tests. Prediction of tile discharge by United States Bureau of Reclamation equation compares well with observed discharge.

1112

SINGH, J. Ground water studies in rising water table area. J. Res. HAU, 1982,12:667-74.

Ground water fluctuations and quality trend over an area of approximately 1, 46, 875 ha in rising water table area have been analysed from June 1977 to June 1980 which includes a drought year. There was rising trend of water table and an average annual rise worked out to be 28.9 cm. The subsurface inflow and outflow were almost in equilibrium with each other and the rise was mainly due to the losses from irrigation application and conveyance system. Nearly 70 percent of the under-ground water had EC exceeding 3 dS/m. There is need for increased drainage through tubewell and its subsequent utilization in conjunction with surface water to arrest the rate of rise of water table.

1113

SINGH, J.H.; WILKINS, R.A. Implementing techniques for efficient drainage on Ogle estate. 10th Congr. Irrig. Drain. Athens, 1978, 6:149-66.

The decline in sugar cane yields in Guyana prompted attempts, late in 1974, to improve the drainage systems on estates, and by the end of 1975 substantial improvements resulted. Perhaps the best and most clearly defined drainage improvements resulted at Ogle estate, and some of the methods and techniques applied for effecting these improvements are described.

1114

SINGH, K.K.; MISHRA, A.P.; OJHA, T.P. Ponded leaching on sloping land with tile drains. Harvester, 1976, 17:29-31.

The paper deals with leaching performance of saline soils with different slopes of soils and arrangements of tiles in a sand box model under ponded conditions. The tile spacing in the prototype are 10, 15 and 30 m. The studies indicated that leaching efficiency and salt distribution patterns are very little affected with the spacing of tiles and bed slopes.

1115

SINGH, O.P. Salt and water balance for evaluating effectiveness of drainage in saline soils. Curr. Agric., 1982, 6:9-16.

Using the salt and water balance technique for evaluating the effectiveness of drainage practice for the reclamation of saline soils, it is concluded that (1) the depth of ground water outflow is always higher than from the undrained area for different seasons, (2) the provision of drainage improves the distribution of salts in the soil profile creating favourable environment for crop production, (3) the increase in the salts within the soil profile due to use of saline irrigation water in rabi season is reduced through leaching due to rainwater during monsoon season, (4) the accumulation of salts as a result of soil water evaporation is higher in case of undrained area as compared to drained area. It is further concluded that due to provision of subsurface drainage, the barren saline soils can produce good yields of wheat, barley, mustard and bajra, if managed properly.

1116

SINGH, R.N.; RAI, S.N. On subsurface drainage of transient recharge. J. Hydrol., 1980, 48:303-11.

An approximate closed form solution of the Boussinesq non-linear parabolic equation for a time varying recharge has been obtained. It is shown that for unit value of a non-dimensional recharge parameter the water table profile is a function of time whereas in the uniform recharge case it was time independent.

1117

SISSON, D.R. Envelope materials, their use in agricultural drainage. Proc. Drain for Efficient Crop Production Conf. Chicago, Illionois, 1965, 51-54.

Use of envelope materials in agricultural drainage is important as it serves wide-ranging functions in the well functioning of a tile drainage system. The review of recent research on this aspect shows that ratio of 50 percent of the grain size of filter materials to the 50 percent grain size of the base soil material should be between 12 to 58. Same ratios for 15-percent grain size should vary from 12 to 40. Filter stability ratio defined as ratio of 15-percent fine size of filter to 85-percent fine size of filter should not be less than 5. In addition, the gradation curves of the base soil and the filter material should be parallel and the maximum size of the filter should not exceed 38 mm and not more than 5 percent of the filter material should pass a No.200 sieve. Present use of filter and likely future development in this field have been discussed.

1118

SISSON, D.R.; JONES, B.A. Filter materials for tile drains in a medium sand - A laboratory comparison. <u>Trans. ASAE</u>, 1962, 5:54-8.

A uniform medium sand known to cause silting of drain tile was selected as a base and seven filter or blinding materials (corn cobs, sawdust, straw, topsoil, gravel fiber glass mat over the top three-fourth of the drain tube and fibreglass over the top three-fourth and vinyl film under the bottom one-fourth of the drain) were compared in a laboratory drain-tank models. 5 cm lucite tubing was used to simulate drain tile. The results indicated that with no filter material, sand filled the draintile quickly. Corncob provided only a small protection whereas fibregrass with plastic,straw and sawdust in that order provided the best protection. Gravel, fibreglass and topsoils followed in that order. Top soil though sixth provided a much higher protection than no filter. Negative pressure within filter materials during ponded water flow is considered to result in low rate of soil movement.

1119

SKAGGS, R.W. Effect of drain tube openings on water-table drawdown. <u>J. Irrig. Drain. Div. ASCE</u>, 1978, 104:13-21.

Conventional methods assume that the drain tube is completely open and offer no resistance to the entry of water. An approximate method of determining the effect of drain tube openings on water table drawdown is presented, based on the use of Hooghoudt's equivalent depth to account for convergence near the drain during water-table drawdown. Data from the literature are used to define an effective drain tube radius, r_e, for drains with a finite number of openings. The r_e value is then used to define the Hooghoudt equivalent depth which is in turn used in solutions to the Boussinesq equation for water-table drawdown. The use of an envelope wrap material will permit a somewhat wider drain spacing for all cases with a larger increase in spacing for the deeper profile. The allowable increase in spacing for both cases was relatively small and the total system cost with the envelope wrap was about 6% higher than without the wrap for the shallow profile and about 7% lower for the deep profile.

1120

SKAGGS, R.W. Evaluation of drainage-water table control systems using a water management model. <u>3rd Natn. Drain. Symp. ASAE</u>, Michigan, 1976, 61-8.

A water management model was used to continuously simulate the soil water regime over a 20 yr period for alternative designs of drainage and water table control systems for two soils. Water management systems consisting of surface and subsurface drainage, controlled drainage and subirrigation were considered in the analysis. The systems were evaluated in terms of their effect on trafficable conditions for seedbed preparation in the spring, adequate drainage for crop growth during the growing season. The results are presented on the basis of a five-year recurrence interval. It was found that a 43 m drain spacing will satisfy conventional drainage requirements for the Wagram soil. Closer spacings are not desirable as they will increase soil water deficiencies during dry years unless water table control practices are used. Drain spacings required for the Bladen soil varied from 12 to 20 depending on the quality of surface drainage. Subirrigation can be used to good advantage on the Wagram soil with a 30 m drain spacing and good surface drainage but was found to be infeasible for the tight Bladen soil.

1121

SKAGGS, R.W. Field evaluation of a water management simulation model. Trans. ASAE, 1982, 25:666-74.

The water management simulation model DRAINMOD was tested using field data for over 5 years of record from three locations. Each site had field scale drainage systems with provisions for subirrigation and controlled drainage. Comparison of predicted and measured water table elevations were in excellent agreement with the daily water table depths having standard error of estimate ranging from 7.5 to 19.6 cm. The average absolute deviations between predicted and observed water table depths for 21 site years of data was only 3.1 cm. It is concluded that DRAINMOD can be used for predicting the effect of drainage system on design of water table elevations.

1122

SKAGGS, R.W. A water management model for shallow water table soils. Report, Water Resources Research Institute of the University of North Carolina, 1978, 134, 178 p.

A study was conducted to develop and test a water management model DRAINMOD, for shallow water table soils that require artificial drainage either surface or subsurface, for efficient crop production. The model has the capability of simulating on a day-to-day, hour-by-hour basis the water table position, soil water content, drainage, ET and surface runoff in terms of climatological data, soil properties, crop parameters, and the water management system design. The model is given in full in an appendix to the report. Tests of the validity of DRAINMOD were conducted on three field sites with a total of five water management treatments over a five year period of record. DRAINMOD can also be used to determine hydraulic loading capacities for systems for land application of waste water, an example was given to demonstrate this use of the model. Finally, one example was given to show how DRAINMOD can be used to determine the effect of rooting depth limitations on the number of days and the frequency that a crop suffers from drought stress.

1123

SKAGGS, R.W.; GILLIAM, J.W. Efect of drainage system design and operation on nitrate transport. Trans. ASAE, 1981, 24:929-34.

The computer simulation model, DRAINMOD, was modified to predict nitrate movement from artificially drained soils with high water-tables. Nitrate cocncentrations in surface runoff, subsurface drainage and seepage waters were assumed to be constant and independent of the drainage system design for a corn-soybean rotation on these high water table soils. Total annual nitrate outflow was determined for alternative drainage system designs and operational procedures. The results showed that trafficability and crop protection requirements can be satisfied by several different drainage system designs. For the poorly drained soil considered in this study, there was a threefold difference in NO_3-N outflow among systems that satisfied drainage objectives. The amount of nitrate that leaves the field through drainage waters can be reduced by using controlled drainage during the winter months and during the growing season. However, if the controlled drainage systems are not used properly an overall increase in nitrate outflow may result.

1124

SKAGGS, R.W.; HARDJOAMIDJOJO, S.; WISER, E.H.; HILER, E.A. Simulation of crop response to surface and subsurface drainage systems. Trans. ASAE, 1982, 1673-78.

The water management simulation model, DRAINMOD, was modified to quantify stresses on corn caused by both excessive and deficient soil water conditions and to weigh those stresses according to the stage of the growing season and crop susceptibility at the time they occur. An approximate crop response model was used to predict annual crop yields in terms of stress day indexes for deficient and excessive soil water conditions and the delay in planting date. The methods are demonstrated in the paper by simulating the performance of several subsurface drainage designs for a Tomothy sandy loam soil near Wilimington, NC. The results showed that, in the absence of irrigation, the long-term average yield will be maximized by a subsurface drain spacing of 20 m or less for good surface drainage. For poor surface drainage a subsurface drain spacing of 12 m would be required to obtain the same yield. The effect of surface drainage on yields is particularly important when subsurface drainage is poor. For example, good surface drainage, on a field with a 90 m subsurface drain spacing, will more than double the long-term average corn yield over that obtained for poor surface drainage on the same field. Potential yield increases due to irrigation will not be realized if adequate drainage is not provided.

1125

SKAGGS, R.W.; KRIZ, G.J.; HERNAL, R. Field evaluation of transient drain spacing equations. Trans. ASAE, 1973, 16:590-5.

Field experiments were conducted on a subirrigation-drainage system to determine the applicability of four drain spacing equations for a shallow sandy loam soil. The water table drawdown following subirrigation and rainfall events were measured for drain spacings of 7.5, 15 and 30 m. The total rise and response time of the water table will depend on its initial depth, the rate and amount of rainfall, and the soil hydraulic properties. When the effective hydraulic conductivities were determined from water table drawdown measurements for the 15 m spacing, the order of preference of the four equations tested was Hammad, van Schilfgaarde Bouwer and van Schilfgaarde, and Glover. However, it was concluded that if the conductivity is determined from soil cores or other independent measurements, the Hammad equation will not give a reliable prediction of the drain spacing for shallow soils. In either case, the reliability of the other three equations could probably be improved for shallow soil if a better procedure for compensating for convergence near the drain tube were available.

1126

SKAGGS, R.W.; TABRIZI, A.N. Drainage system for land treatment of waste water. J. Irrig. Drain. Div.ASCE, 1982, 108: 196-211.

Simulation methods for design and evaluation of drainage systems for waste water land treatment are presented which are based on water management model, DRAINMOD. Several drainage design and application strategies are evaluated to determine their effects on the amount of waste water that can be treated and how that amount changes from year to year with weather conditions. Effects on water table depth, out flow volumes and required wastewater storage are also evaluated. When wastewater application is limited by drainage conditions of the site, more water can be treated by applying small amounts frequently than with larger amounts on a less frequent basis. Results showed that DRAINMOD can be used to optimize the drainage system design for a given land application site.

1127

SKAGGS, R.W.; TABRIZI, A.N.; FOSTER, G.R. Subsurface drainage effects on erosion. J. Soil Wat. Conserv., 1982, 37:167-72.

A mean of evaulating the effects of combination subsurface/ surface drainage systems on erosion was developed by coupling the drainage simulation model, DRAINMOD, with the CREAMS model for simulating erosion. Use of the coupled model was demonstrated in an analysis of the effects of artificial drainage on erosion for Goldsboro sandy loam soil in eastern North Carolina. The simulation showed that erosion on Goldsboro soil with an average slope of 2 percent could be reduced appreciably with subsurface drainage. For example, predicted average annual rates of erosion declined by a factor of 10 (9 to 0.9 metric tons per hectare) by changing the drainage system from one with good surface drainage and poor subsurface drainage to one with poor surface drainage and good subsurface drainage. Subsurface drainage, therefore, should be considered as a possible best management practice for reducing erosion on relatively flat land.

1128

SKAGGS, R.W.; TABRIZI, A.N. Optimizing drainage system design for corn.
4th Natn. Drain. Symp. Proc.ASAE, 1982, Pub. 12-82:50-61.

The simulation model DRAINMOD as modified to estimate the effects of soil water stresses on corn yields was used to determine the effect of surface and subsurface drainage on corn production for a Rins sandy loam in eastern North Carolina. The simulation and analyses indicated that (a) surface and subsurface drainage have a good long term effect on corn yields and an increase to a extent of 63% is possible by providing better subsurface drainage compared to conventional drainage ditches at 100 m spacing; (b) Drains (subsurface) should be spaced 20 m apart for good surface drainage and at 17 m for poor surface drainage. Maximum profits occur for spacings of 24-27 m depending upon surface drainage intensity; (c) The results are not good and depending upon the hydraulic conductivity of soils within the series, drain spacing may vary to a factor of as high as 2.5.

1129

SKAGGS, R.W.; TANG, Y.K. Effect of drain diameter,

openings and envelopes on water table drawdown. Trans. ASAE, 1979, 22:326-33.

Numerical solutions to the Richards equation for saturated and unsaturated flow to parallel drains were used to determine the effects of drain tube dia openings and envelopes on water table drawdown. Conventional drain tubes with finite number of openings in the walls were represented as completely open tubes with an effective radius. The effect of drain tube size and the use of envelopes on water table drawdown was determined for different soils, drain spacings and profile depths. It was concluded that envelopes will allow an increase in drain spacing without reducing the design drawdown rate. However, the allowable spacing increase is smaller than previously reported in the literature. The midpoint drawdown was found to be relatively insensitive to change of one or two sizes in the drain tube diam. It was shown that the effect of drain tube diam. on drainage rates is dependent on drain spacing and profile depth as well as hydraulic properties of the soil.

1130

SKOGERBOE, G.V.; WALKER, W.R.; AYARS, J.E. Drainage for reducing salt pickup from grand valley. 3rd Natn. Drain. Symp. ASAE, Michigan, 1976, 87-90.

Irrigation return flows from the Grand Valley in western Colorado contributes to the serious salinity problems in the Colorado River Basin by carrying large salt loads which result from the contact of subsurface return flows (seepage and deep percolation)with local saline soils and the underlying Mancos Shale formation. This paper discusses the role of drainage in reducing the salt load in the Colorado River. The results indicate that while field drainage is effective in skimming water that is at least 30 percent less saline than underlying ground water flows, the high costs emphasize the need to reduce seepage losses by lining canals and minimizing deep percolation losses through improved on-farm water managaement practices.

1131

SECHTA, Y. Groundwater flow for systematic drainage on slopes. <u>Sb. UVTIZ-Melior.</u>, 1978, 14:115-24.

If a drainage network is located on an impermeable sloping substratum, in a homogeneous and isotropic uniformly saturated soil medium, asymmetric flow may occur. The solutions of dif-ferent equations of flow to drains on a slope under the given conditions provided equations which describe the pattern of phreatic surface between twc drainage networks on a slope. The application of the equations to three examples is presented.

1132

SLUIS, P.M. VAN DER. Lay out of drainage systems in irrigated areas. <u>Landb. Tijdschr.</u>, 1964, 76:888-99.

The paper reviews the requirements of suitable drainage systems for irrigated soils in arid or semi-arid climates; states that the drainage systems should permit the drainage of excess rain water, deep percolation water, waste water and tailwater; further states that the macro-topography, the slope of the land, the hydrological configuration of the subsoil, the prevailing percolation pattern, the results of drainability surveys and the depth of the laterals influence the selection of the lay-out; discusses the advantages and drawbacks of open drains versus closed systems; observes that in Egypt the use of 5 cm drains instead of 10 cm pipes would mean a tremendous saving of investment.

1133

SMART, P. Mole drainage of difficult land. <u>Proc. 9th Conf. Int. Soil. Tillage, Res. Org. Osijek, Yugoslavia</u>, 1982, 644-9.

A theory for the design of mole drainage schemes, when a few small collapse-induced features are present, is presented.

1134

SMEDEMA, L.K. Drainage criteria for soil workability. Neth. J. Agric. Sci., 1979, 27:27-35.

Drainage requirements for workability are formulated on the basis of the soil moisture content at the lower plastic limit, and the relationship of the latter value to field capacity is investigated. Current drainage design criteria are analysed with respect to soil workability requirements; it is concluded that there is little technical and even less economic scope for improving soil workability by adopting stricter criteria, leading to narrower drain spacings. However, a case can be made for increasing the drain depth in medium and heavy textured soils beyond the commonly applied depth of about 100 cm, in the Netherlands. The effects of soil management on soil workability are briefly discussed.

1135

SMEDEMA, L.K. Evaluation of field drainage discharge from heavy land in the tropics. J. Hydrol., 1984, 67:325-37.

Drainage discharges were measured in a sugarcane field in Tanzania. Field drainage consisted of an in-field system of shallow ditches intercepting drainage flow in the cane furrow (row drainage). Observed hydrographs are compared with discharge prediction of three relevant rainfall discharge models, i.e. linear storage model, a travelling time model, (rational formula) and curve number synthetic unit hydrograph model of US Soil Conservation Service. Very good agreement is obtained with the linear storage model although, in this case, the linearity between discharge rate and dynamic storage appeared to be somewhat incidental being promoted by the particular row drainage layout of the fields.

1136

SMEDEMA, L.K.; MEER, K. VAN DER. The role of adapted farming in the solution of drainage problems of basin clay soils. Agric. Wat. Mgmt., 1979, 2, 257-69.

Adequate control of excess water in fine textured, low permeable soils often requires an uneconomical, highly intensive drainage system. Highly intensive drainage systems (spacing ` 15-20 m) may help control waterlogging but cost involved is enormous. Solutions to this problem can be found by using a less intensive drainage system that removes only part of the excess water, complemented by adaptations in farming practices. Two cases are presented to illustrate this principle. The usefulness of soil moisture regime studies in the planning of such solutions is also demonstrated.

1137

SMEDEMA, L.K.; RYCROFT, D.W. Land drainage: planning and design of agricultural drainage systems. Batsford Academic, London, 1983, 376 p.

A modern comprehensive text on land drainage which discusses the diagnosis and solution of agricultural drainage problems, based on the understanding of the physical principles involved. All major problems on drainage such as groudwater drainage and water table control, surface drainage of sloping and flat lands, shallow drainage of heavy land, drainage for salinity control in irrigated land; drainage and reclamation of polders; drainage and seepage control; and main drainage: design discharges, canal design and outlets. Theory is backed up by practical application. A handy book for professionals and a useful text at university level.

1138

SMITH, L.P.; TRAFFORD, B.D. Climate and drainage. Tech. Bull., Ministry of Agriculture, Fisheries and Food, U.K. 1976, 34, 199 p.

The use of climatic data for drainage planning and for detailed design is discussed with particular reference to definition of the date of return to field capacity, the overall drainage need, duration of winter season, likel range of variation and the mole drainage/subsoiling. The methods used in climatic analysis are described and the use of data is demonstrated by a number of examples. Data on area rainfall, transpiration, soil moisture deficits, field capacity and related characters, together with notes on the type of farming practised, are presented for the 52 agro-climatic areas in England and Wales.

1139

SNEYD, A.D. Effect of semi-circular arches at drain gaps on drain conductance. J. Hydrol., 1976, 30:293-4.

A modification of the formula for drain conductance is given for the case where erosion leads to the formation of stable semi-circular arches at the drain gaps.

1140

SNIADOWSKI,A. Silting of drain pipelines and its prevention ways. Wiad. Melior., 1970, 7/8:22-25.

The paper presents a detailed analysis of causes of silting drain pipelines and the remedial measures applied for its prevention.

1141

SOJKA, J. An analysis of the load on installed drain pipes. Ved. Pr. Vyzk. Ust. Melior. Zb., 1973, 13:21-30.

A theoretical analysis of the pressures on drainage pipes exerted by the soil or backfill.

1142

SOLIMAN, N.F.; ANTER, I.M.; SOLIMAN, M.F. Effect of depth of water table on growth and yield of tomato. Agric. Res. Rev., 1978, 56:17-21.

In lysimeter trials with the tomato cv. Ace, the water table was maintained at depths of 40, 70, 100, 130 or 160 cm. A significant yield increase occurred when the table was lowered from 40 to 70-100 cm. Further lowering of the water table gave only slight yield increase over the 70-100 cm depth.

1143.

SOLIMAN, M.M.; EL-DAYEM, M.S.A.; AMER, M.H. Recent technique for tile drainage design in the Nile delta. 10th Congr. Irrig. Drain. Athens, 1978, 4:543-8.

A new water-table shape equation has been developed under Nile Delta conditions. Based on this a numerical analysis was set up to find the relations between the different tile drainage parameters and hence the spacing between tile drains.

1144

SOLOV'EV, V.V.; DRUZHININ, N.A.; DANILOV, M.A. Growth and structure of young pine stands on drained peatlands of the Central Urals. Lesn. Zl.,1978, 6:29-33.

Investigation was made in 1973-76 in the Perm and Sverdlovsk regions to determine the changes in the mensurational indices of Pinus sylvestris stands developing as a result of natural regeneration on areas drained in 1930-40 for peat harvesting. Data were analysed from 235 sample trees from 21 plots established on different type of peatland. Data are tabulated on the tree dimensions and stand structure on drained and undrained areas of transitional bog and of fen. The resutls indicate that stand formation and development are better on drained than on undrained areas. The best situation for the formation of high-yielding stands is where the water table is lowered gradually, in accordance with the growth and development of the trees; in selecting trees for thinning, the increased growth vigour of subordinate trees should be remembered. In winter, the drainage channels can be used as skidding trials for the extraction of thinnings.

1145

SOMMERFELDT, T.G. Outflow from various subsurface drainage materials. Trans. ASAE, 1975, 18:85-8.

In lysimeters, the discharge from 55, 65 and 105 mm internal diameter perforated plastic drains and 105 mm clay tile drains with and without gravel or fibreglass envelopes was determined in clay loam and fine sandy loam. In the field, discharge from fibreglass envelopes, in a coarse textured soil was determined. In the field, discharge from the plastic drain was greater during the first 12 months but during the second year there was little difference between the plastic and clay tile drains. Relative discharge from drains in the lysimeter experiments also changed with time.

1146

SOMMERFELDT, T.G. Soil moisture and texture effects on mole drain stability and on force requirements for installation - A laboratory study. Can. Agric. Engng., 1983, 25:1-4.

The effect of soil moisture and texture on force required to install mole drains and the stability of drains after installation were studied in the laboratory, and the results were used to explain the differences in mole drain performance, at two field sites in southern Alberta. For effective durable drains, the soil should contain sufficient clay to be cohesive and plastic (texture: clay loam or finer) and the moisture content should be at or above that of the plastic limit (50% of field capacity) during installation. Below the plastic limit, less force was required in the loam than in the finer-textured soils but reverse was true when moisture levels were above the plastic limit. Failure of the moles to function as drains in a clay loam at one field location and in a clay loam at another, indicate the limitations in such soils.

1147

SOMMERFELDT, T.G.; HENDRY, M.J.; CHANG, C. Soil properties impeding subsurface drainage of a soil near Vauxhall Alberta: A case study. Can. Agric. Engng., 1983, 25:205-8.

Chemical and physical properties of a fluvial soil near Vauxhall, Alberta, were studied to determine the causes of poor subsurface drainage. The results indicated that salinity and alkalinity should not restrict the hydraulic conductivity of the soil. Poor performance of subsurface drainage was attributed to a high percentage (63.8-73.8) of very fine sand (<0.125 mm) plus silt which restricted water flow through the soil and plugged the drain filter. The reclamation measures suggested include construction of new drains with gravel or styrofoam bead cover extending to the soil surface. The new drains have improved the drainage of the land.

1148

SOMMERFELDT, T.G.; PLUYM, H.V.; CHRISTIE, H. Drainage of dryland saline seeps in Alberta. In Dryland-saline Seep Control. Proc. 11th Int. Soil Sci. Congr. Canada, 1978, 4.15-4.23.

Drainage at three locations is discussed and pointed out that it is location specific. It appears that minimal drainage, to remove temporary excesses of water, can be used to supple-ment natural conditions in maintaining the water table at safe depth.

1149

SOOVIK, E. Determination of drainage discharge from the degree of gleying of the soil. Pochvovedenie, 1967, 9:100-7.

In south-west Estonia, the rate of discharge of drainage water from the tile drains in particular plots of dernopodzolic soils could be deduced from a subjective estimate of the degree of gleying in the plots.

1150

SOOVIK, E.A. The soil-fertility method for determining the spacing between drains. Dok. Vses. Ord. Len. Akad. S-Kh. Nauk. Ime. V.I., Lenina, 1979, 1:44-6.

Tabulated data describe the effects of halving the distance between tile drains on boggy soils on yields of field crops in 1969-1974. A nomogram is presented for calculating the spacing between tile drains in sandy-loamy and loamy soils from the filtration coefficient of the soil, the degree of gleying of the soil, the depth at which the drains are laid and the degree of drainage required.

1151

SORIBE, F.I.; FOUSS, J.L.; SCHWAB, G.O. Comparison of strength test methods for corrugated plastic drainage tubing. Trans. ASAE, 1972, 15:445-7.

A comparative study of the load deflection test methods given by U.S. Soil Conservation Services (Sand bed method) and by the British Ministry of Agriculture (Parallel-plate test) had been conducted on corrugated plastic drainage tubing. Either of the two methods can be used for testing the strength of corrugated plastic drainage tubing and the study shows that there is theoretical justification for the same, at least upto 10-cm diameter tubing.

1152

SOWA, E. Analysis and simulation of ground-water dynamics in drain systems by using physically substantiated block models. Arch. Acker. Pflan. Bodenk., 1980, 24:561-9.

A block model is introduced for transforming the variable quantities: precipitation or evaporation and water level in the drain into the output: groundwater level at the observation pipe'. The approximation level of the model is tested. The model may be used to work out criteria for the determination of drain spacing when field trials for drainage and groundwater regulation are evaluated.

1153

SPENCER, W.F.; PATRICK, R.; FORD. H.W. The occurrence and cause of iron oxide deposits in tile drains. Soil Sci. Soc. Am. Proc., 1963, 27:134-7.

Red sludge-like deposits in tile drains were due chiefly to iron bacteria which oxidize and precipitate reduced iron in the drainage water. Much of the material was organic probably bacterial cells and waste products. The inorganic portion was chiefly hydrated Fe and Al oxides. Gallionella, Leptrothrix, Thiothrix, Thiobacillus, Crenothrix and Cladothrix were mainly responsible. The deposits occur in clay tile and bituminous pipe whether installed in sawdust filter or wrapped with fibre-glass. The amount of deposits was related to the amounts of reduced Fe extracted from the soils with NH_4Cl solution.

1154

SPOOR, G. Soil disturbance with deep working tined implements in field drainage situations. Proc. Int. Drain. Workshop, Wageningen, Netherlands, (1978), 1979, 25:561-70.

Two types of soil disturbances may result at depth when using trenchless drainage ploughs, subsoilers and mole ploughs. The first, common at shallow depths, a loosening, fissuring disturbance ideal for the first two operations and secondly at great depths, a compressive, compacting disturbance suitable only for the latter operation. The working depth at which the transition between the two types of disturbance occurs, known as the critical depth, is dependent upon implement geometry and soil conditions. Increase in tine width and inclining the tine forwards increases the critical depth. Loosening the surface soil layers prior to deep tining increases the chances of loosening at depth. The use of shallow leading tines immediately ahead of the deep tine increases the critical depth without increasing the draught.

1155

SPOOR, G.; FRY, R.K. Field performance of trenchless drainage tines and implications for drainage system efficiency. J. Agric. Engng. Res., 1983, 28:319-35.

The soil disturbance produced by a range of trenchless drainage tines, working under different soil conditions, is described and the effect of different types of disturbance on subsequent drainage performance considered. Few problems are likely to arise with trenchless tines working above critical depth. Below critical depth impeding layers may develop alongside the pipe reducing drainage efficiency. Soil conditionns most susceptible to damage are highlighted and methods suggested for avoiding; alleviating and minimizing potential problems.

1156

SPOOR, G.; FRY, R.K. Trenchless drainage pipe installation and its implications for subsequent drain performance. 4th Natn. Drain. Symp. Proc. ASAE, 1982, pub. 12-82:105-14.

Problems of trenchless drainage are enumerated and improvements to improve drainage efficiency suggested.

1157

SPOOR, G.; LEEDS-HARRISON, P.B.; GODWIN, R.J. Potential role of soil density and clay minerology in assessing the suitability of soils for mole drainage. J. Soil Sci., 1982, 33:427-41.

The failure mechanisms causing mole channel deterioration or collapse, which are controlled mainly by the shear, swell/ shrink and apparent viscosity properties of soils, are shown to be sensitive to soil density and clay minerology. These two properties have, therefore, a potential role in helping to assess the suitability of soils for mole drainage and to define the particular failure mechanism likely to be active in given situations.

1158

SPOOR, G.; LEEDS-HARRISON, P.B.; GODWIN, R.J. Some fundamental aspects of the formation, stability and failure of mole drainage channels. J. Soil Sci., 1982, 33:411-25.

The various stresses which can arise in the vicinity of a mole channel from its formation to ultimate collapse are examined. Possible failure mechanisms contributing to channel collapse or deterioration are proposed, together with soil properties associataed with them. Evidence for the proposals, obtained from laboratory and field experiments, is presented.

1159

SREEDHARAN, C.; VAMADEVAN, V.K. Drainage requirements of rice. Oryza, 1974, 11:119-25.

The various aspects of drainage of ill-drained rice fields are discussed. The productivity of ill-drained soils can be improved with suitable drainage methods. Adequate rate of percolation should vary in relation to the amount of toxic substances and nutrient-holding capacity of soils. In swampy rice fields abundant in toxic substances, a percolation rate of 3 cm/day is required and surface drainage should be practised if hard pan or rock is present and percolation rate is $<$2.5 mm/day. The late tillering stage is the best time for drainage of such soils. Percolation is beneficial to soils with abundant production of toxic substances and also to fields given large amounts of green manure or FYM.

1160

STAMM, G.G.; GRAY, H.R. Current drainage construction of the Columbia basin project is based on past experience. 7th Congr. Irrig. Drain. Mexico, 1969, 6:233-246.

The Columbia basin project in northwestern United State being developed to serve 400,000 ha of agricultural land is based on re-examination of policy because previous experience showed early drainage problems. Early land classification did not identify subsurface conditions at depths greater than 1.5 m. Current policy requires intensive and integrated land classification, drainage and economic studies. Largely cohesionless soils have led to use of shield-type excavation equipment in the construction of 800 km buried pipe drains at 3 m depth to control watertable at 1.2 m. A comprehensive forest of drainage requirement and the cost is essential prior to large scale irrigation development and present development works are on these lines.

1161

STAPEL, Z. Backfill effect on drain pipeline efficiency. Wiad. Melior., 1970, 10:291-3

The paper discusses the relationship between water flow-off intensity from the soil and its reception by the drain; argues that water reception ability of a drain pipeline depends on permeability of soil around the drain and total area of slots per length unit; finds slot area of 9-10 cm^2 ensuring an adequate water reception ability by the drain.

1162

STAPEL, Z.; COLASZOWSKI, M. Increase of effectiveness of the drainage network by producing an underpressure in it. 10th Congr. Irrig. Drain. Athens., 1978, 4:147-1.

A device consisting of a well and self-sucking syphon is described which produces an underpressure in drains. In an experiment on heavy soil, at an underpressure pFh_2 = 0.30 m of the water column, an increase in water outflow from drains of 26 percent has been obtained. This method could be applied to small areas, upto 2 ha in size.

1163

STARIKOV, KH. N. Drainage regime for low-moor bogs in various zones of the Ukraine. Vestn.S-Kh. Nauki., 1972, 10:107-13.

Permissible variations in the height of the water-table for various crops are tabulated for the left bank of the wooded steppe as compared with Polesia and the right bank of the wooded steppe.

1164

STAVEREN, J.M. VAN. Surveys and their sequence. In Drainage Principles and Applications. Vol.III. Surveys and investigations. Intern. Inst. Land Reclam. Improv., Wageningen, 1974, 16: 1-12.

Three phases of study, at the reconnaissance, semi-detailed and detailed levels, are recommended, and factors to be considered during the project survey are discussed.

1165

STEINHARDT, R.; BENYAMINI, Y.; GANANI, A.; DASBERG, S.; MALIACH, J.; NADLER, A.; ROSENZWEIG, D.; ROSNER, M. The effects of subsurface drainage, compaction and irrigation on the soil-water air relationships of a Grumusol in a cotton field. Spe. Pub. Div. Scientific Pub. Bet Dagan, 1980, 176, 50 p.

The effects of subsurface drainage on physical soil conditions were monitored in a cotton field, irrigated by sprinklers and subjected to machine compaction. Subsurface drainage reduced the settling of the tillage layer during the winter, increased the rate of soil drying in spring (thereby increasing the air content and reducing the compactability of the tillage layer) and reduced the effects of a water surplus during irrigation. Additional effects of subsurface drainage were: accumulation of salts in the vicinity of drain; compaction of the tillage layer in the absence of traffic throughout the growing period; accumulation of subsurface water entering the field from adjacent higher areas in the vicinity of the drains; and small changes in clay percentage, exchangeable sodium and water-holding capacity. Although traffic-induced compaction resulted in appreciable increase in volumetric water content and soil water storage, these did not cause any change in soil water depletion, but only a reduction in air content. The rate of water application increased by 10 percent to 15 percent at 6-8 m from the sprinkler line compared to 2-4 m. This resulted in the swelling of the tillage layer without compaction and its shrinkage under heavy compaction.

1166

STEINHARDT, R.; TRAFFORD, B.D. Some effects of subsurface drainage and ploughing on the structure and compactability of a clay soil. J. Soil Sci., 1974, 25:137-52.

In a field experiment to determine the direct and

indirect effects on soil moisture of sub-surface piped drainage as compared with natural surface drainage only, in ploughed and unploughed soil, a factorial systematic design with four replicated blocks was used on a calcareous clay (45%) of the Wicken series. The effects of drainage treatments were found to be temporary, except a crusting effect during the drying of the unploughed surface drained soil. The soil strength at 7.5-22.5 cm depth was linearly related to the matric suction with the range of -3 to 20 cm water. In the ploughed soil, compaction data established the beneficial effects of subsurface drainage in reducing damage from tractor trafffic, decreasing wheel sinkage. It was concluded that for clay soils having a temporary excess moisture, draining the water table below 50-60 cm depth should be recommended as a precautionary measure to minimize structural damage.

1167

STEPHENS, J.C. Peat and muck drainage problems. J. Irrig. Drain. Div. ASCE, 1969, 95:285-305.

The paper discusses that drainage of organic soils results in lowering of the surface (subsidence) shrinkage due to drying, loss of the bouyant force of groundwater, compaction, wind erosion, burning and biochemical oxidation, the relative effect of each of these factors being dependent on soil origin, land management practices, and climate and the total subsidence rate being proportionate to depth of drainage; the lower the water table, the greater the soil loss; studies the Florida Everglades peats where largest losses (subsidence) have been due to biochemical oxidation which would make organic soils too shallow for agriculture by the year 2000. The paper explains recent water management program (1945-1968) ameliorating the consequences of overdrainage; inform that subsidence of organic soils from drainage is worldwide with observed rates upto 3.5 m annually; suggests to drain only those organic lands that are potentially arable, place drained lands into productive use quickly, and hold water tables as high as crop and human requirements permit.

1168

STEPHENS, J.C.; SPEIR, W.H. Subsidence of organic soils in the U.S.A. Pub. Int. Ass. Scient. Hydrol., Symp., Tokyo, 1970, 89:523-34.

Shrinkage rate of organic soils was proportionate to drainage depth; the lower the water table, the greater the subsidence. In the Florida Everglades, subsidence between 1913 and 1968 was rapid initially, mainly from shrinkage and compaction, then declined to a continuous steady rate, primarily from oxidation, until underlying mineral material was reached. Under similar drainage, organic soils subside faster in warm climates and lowmoor peats subside faster than highmoor peats. To conserve organic soils, water tables should be kept as high as crop and field requirements permit.

1169

STINGA, N. Effect of drainage on fertility of waterlogged soils used as crop lands. In Tara Birsei. Anal. Inst. Cerc. Imb. Func. Pedol., Ser. Pedol., 1968, 35:271-92.

Twenty-two methods that included the use of open ditches and earthenware tiles were tested on a humic gley over a period of seven years. The efficiency of the five most suitable methods followed the decreasing order: drainage with 20-cm bunds of brushwood at 1 m depth and 7.5 and 15 m distance, ¨ 10 cm concrete pipes at 1.5 m, depth and 15.0 m distance, ¨ 10 cm concrete pipes at 0.6 m depth and 3.0 m and 5.0 m distance. Total porosity was decreased by 7-13 percent and humus content (0-20cm) from 19.3 to 15.9 percent: NPK contents showed insignificant changes. Crops gave higher yields and more plants reached maturity.

1170

STINGA, N. Improvement of productivity of marshy soils in the Tara Birsei region in relation to drainage conditions. Probl. Agric., 1963, 15: 7:39-47.

The most promising spacings were 7.5-15 m with faggot-bundle drains at 100 cm depth and 5.3 m with concrete drains at 60 cm (or 22-15 m at 150 cm). The efficiency of tile drains was low, and drainage by open ditches was ineffective.

1171

STISEN,, S.; MORTENSEN, J.R. Artesian water drainage in relation to drainage in Denmark. In Land Drainage (Gardiner, M.J.; Ed) A.A. Balkema, Rotterdam, Netherlands, 1982, 175-87.

The use of vertical drains in landscapes affected by artesian water is described.

1172

STOJSIC, M.; RADEVSKI, A. Improvement and modernisation of the old drainage and irrigation systems in Yugoslavia. 11th Congr. Irrig. Drain, Glenoble, 1981, 2:773-87.

The projected improvement of drainage efficiency of reconstruction and additional construction of a network of ditches in the case of light soils (alluvial plains of big rivers) and by increased capacity of discharge structure at higher land levels. In heavy soils a system of horizontal drains will be installed.

1173

STOJSIC, M.; SEVIC, D. Hydraulic chareteristics of PVC corrugated pipes. 10th Congr. Irrig. Drain. Athens, 1978, V: 4:285-93.

Water flow at low velocities through corrugated

PVC pipes results in the formation of a border layer of water between the ribs which is at rest, the roughness values being similar to those in smooth pipes. At high velocities, the ribs function as a factor of roughness. The water flow to the drain through the opening is best when occurring in smooth pipes with long narrow slots. In corrugated pipes, the performations of ribs on the outside have the greatest effect per 10 mm^2, though their flow per opening is the weakest. The effect of Type 2 opening (transversal oblong 1 opening per rib) is very good individually, but is the poorest per 10 mm^2.

1174

STORCH, W.V.; TAYLOR, R.L. Some environmental effects of drainage in Florida. J. Irrig. Drain. Div. ASCE, 1969, 95:139-51.

The paper describes intensive agricultural development started in 1959 on a 2021.8 sq. km. region lying east of lake Okeechobee in south Florida, where effective drainage to coastal waters has been the key to agricultural expansion. By 1967, approximately 20% of the region was planted in citrus. Changes in the natural environment within the region took place almost immediately after drainage was initiated and indications are that the natural environment in adjacent areas can be affected as well. The effect of drainage on the immediate economic environment of the region has been favourable to-date. It has been found, however, that natural environmental changes have produced tensions, with economic overtons, in the social environment resulting in conflict between agricultural and conservationist interests, that the proper function of public government is to evaluate the possible environmental effect of drainage programs and to use its financial, persuasive and regulatory powers to guide total environmental change in accordance with its estimate of the best public interest.

1175

STORSBERGEN, C.; BOS, M.G. Drainage of a coastal plain in Java. <u>11th Congr. Irrig. Drain</u>. Glenoble, 1981, 3:521-32.

The coastal areas of Java's north coast are frequently flooded due to lack of maintenance of the drainage system, the topographical conditions, and uncontrolled sedimentation caused by heavy erosion in the upper catchment area. Layouts for coastal drainage schemes are given, together with a simplified calculation of the drainage requirements. The basic data needed are discussed. Economic factors play an important part in the drainage possibilities as the requirements are determined by a comparison of the cost of the drainage system versus the prevention of crop yield reduction. As criterion it is taken that once every 5 years there will be a crop damage as a result of too high water levels. In general, there will be no yield reduction if the period during which the rise of the water level exceeds 10 cm is less than 3 days and the average rise is less than 20 cm.

1176

STRYK, F.G. VON; BOLTON, E.F. Atrazine residues in tile-drainwater from corn plots as affected by cropping practices and fertility levels. <u>Can. J. Soil Sci</u>., 1977, 57:249-53.

Atrazine was applied pre-emergence to corn plots at Woodslee, Ontario which were drained by separate drains(average depth 70 cm). Water samples from the tile outlets were extracted with chloroform and the atrazine content determined by gas chromatography. The discharge of atrazine residues monitored over a 4 yr period ranged from 2.75 to 8.69 ug/ha annually. The concentration varied between 0.5 and 3.50 g/L of water. Losses of atrazine were greater from plots under a continuous corn crop and a high fertility level than from plots under rotation corn and low fertility.

1177

STUFF, R.G.; DALE, R.F. Soil water tables under corn on tile-drained Chalmers silt loam. Proc. Indian Acad. Sci., 1974, 83:454-64.

In trials in 1971-73, soil moisture content and water-table depth were measured in early-or later-sown maize on tile-drained Chalmers silt loam. Upward flux of water into the root zone was estimated from the relationship between available soil moisture and the drop in the water-table depth. Upward flux varied from 1 percent of actual evapotranspiration in the early sowing in the wet 1972 season to 11 % in the early sowing in the 1973 season which included long dry periods. It was concluded that water from the water table could furnish a significant part of the water requirement during dry periods.

1178

STURDY, R.D. The heavy soils of Essex; their problems and potentials. Soil Water, 1979, 7:8-11.

Some properties of the heavy soils of Essex are described. These soils are characterized by a large clay content and slow permeability of the subsoil, leading to long periods of waterlogging. The problems of farming on heavy land are discussed. Marked improvement of the water regime by drainage is limited in such soils.

1179

STUYT, L.C.P.M. A new approach in research into drainage materials. Tech. Bull. ICW(New series), 1982, 16, 25p.

A research project started recently at the Institute for Land and Water Management Research is described. The aim is to define criteria for the manufacture of synthetic drainage envelope materials. Laboratory apparatus is developed for testing existing envelope materials under condition constrained to narrow limits to ensure reproducibility. As a first step, the occurrence of chemical phenomena has been excluded from the research. Particle size and shape are introduced as design parameters. In due course, mathematical modelling of the filtering process in drainage envelope materials is planned. Testing has only recently started so no experimental data are yet available.

1180

SUSSER, P.;SCHWERTMANN, U. Iron oxide mineralogy of ochreous deposits in drain pipes and ditches.Z. Kult. Flurb., 1983, 24:38695.

Ten samples of ochreous precipitates from drain pipes or ditches contained poorly crystalline ferrihydrite (5 $Fe_2O_3 9H_2O$) as the dominant or sole Fe(III) oxide, as shown by X-ray diffraction or differential X-ray diffraction (DXRD). One sample may have contained some feroxyhite(FeOOH). The surface area varied between 90 and 279 m^2/g, and increased after H_2O_2 treatment to 190/400m^2/g. The C_t content was higher(5-21%) and the Si_o content was lower than in other natural ferrihydrites. The ferrihydriites are considred to be the product of rapid (baterial)oxidation of Fe(II) in the drain water.

1181

SUZUKI,T.; SHIGEMI,T. Drainage facilities at Kanbara plain(Mainly Kameda-Go Area). 7th Cong. Irrig. Drain Mexico, 1969, Q 25 R 18: 237-54.

Kanbara plain is a low-lying flat area occupying about 70,000 ha and located in the central part of the Japanese mainland Honshu on the Japan sea side. The area was marshy and in some parts there exist peats which hamper not only the farming operations but also the productivity of the land. The Kameda-Go area is one of the typical areas of this type. With the installment of the Kurinoki drainage pumping facilities unification of irrigation organizations, separation of irrigation system from drainage system, adjustment of partitions of agricultural land and drying of the field were carried out. The drying of fields completed within a short period of 15-16 years, compared with 40 years elsewhere, helped considerably in saving of labour owing to the introduction of farm machinery but created problems for the farmer who had to share proportionate financial burden which exceeded his capacity. This also resulted in decrease of yield temporarily. Since the irrigation and drainage improvements, rice yields have increased and it is possible to raise two crops in a year. However, each farming family is now free to control the water in each plot of its field as it likes. This causes over induction of water from the main or the branch irrigation canals. This, besides the late farming methods requiring use of more water for higher yields, forces progressively increase operation of the Kurinoki drainage pumping station.

1182

SVIHIA, V. Assessing the tendency of the soil to cause clogging in drain lines. Arch. Acker. Pflan. Bodenk., 1978, 22:169-74.

The tendency of a soil to clog drains was assessed,, for non-cohesive soils, on the basis of the size of the smallest pores, and for cohesive soils on the computation of indexes based on (1)moisture at the flow limit divided by moisture at pF_2 and (2)clay content($<$0.002mm) divided by content of 0.01-9.1 mm particles.

1183

SVIKLIS, P.B. Connection between the drainage network and the yield of agricultural crops on derno-carbonate soils in the Latvian SSR. Tr.Latv. Nauch. issled. Inst. Gidrotekh. Melior., 1963, 2:102-25.

Average crop yields increased with increasing

drain depth from 90 cm to 120-150 cm, but yield was not connected with drain runoff because of the great capacity of the ground above the drains for accumulating water.

1184

SVIKLIS, P.B.; NOVIKOVSKY, V.E. Soviet experience on the use of synthetic materials in irrigation and drainage. 8th Cong. Irrig. Drain. Varna, 1972, Q28R 12: 171-80.

The paper reviews the use of synthetic materials, existing and proposed for use during irrigation and drainage in the Soviet Union. In regard to drainage, the paper outlines drainage development of over wetted lands, the use of buried drainage composed of plastic pipes and the possibility of application of polymers for fabrication of drainage pipes. The description is given of drainage pipe design and of their shaped part conditions as related to their application and silting control. Plastic drainage efficiency and convenience of construction are evaluated.

1185

SVOBODOVA, V. Calculation of probable time period for tile drain clogging by iron compounds and the need for maintencnce. Ved. Pr. Vyzl. Ust. Melior. Ve. Zb., 1973, 13:47-60.

It was calculated that drains 70 m long spaced 20 m apart and constructed from 6.5 cm internal diameter tiles laid on a 4 percent slope would become clogged in 9 years and the gaps between tiles would be blocked in 3 months if the soil water contained 5 ug Fe/L. Deposition from soil water containing 3 ug/Fe/L would block the gaps between tiles in 4 - 6 years . Both calculations were based on an annual drainage outflow of 500 mm.

1186

SVOBODOVA, V. Clogging of drain tiles influenced by

the iron content in drainage waters on surface-waterlogged soils. Melliorace, 1970, 6:27-36.

The iron content in drainage water had been examined to relate it with the clogging problems of subsurface drainage systems. It is concluded that it gives a good estimate of clogging potential.

1187

SVOBODOVA, V. Decomposition of organic drain filters in soils. Scientia. Agric. Bohemoslovaca, 1981, 13:107-18.

Data from model experiments are presented on (a) the decomposition of wheat straw, spruce sawdust, spruce bark and oak bark and (b) the release of phenolic compounds during the decomposition of these materials and spruce branches.

1188

SVOBODOVA. V. Effect of drainage methods on N and P concentrations in drainage waters. Rostl. Vyroba. 1978, 24: 337-46.

Mole, cross and tile drainage combined with subsoiling led to greater leaching of N from chernozem-smonitza, pseudogley, gley and acid brown soils, compared with tile drainage alone. Leaching of P was negligible in chernozem but marked in acid brown soils after drainage and ferilization. Ammonium sulphate was less rapidly leached than nitrate and urea. Leaching of N and to some extent P, was dependent on quantity of drainage water and precipitation.

1189

SVOBODOVA, V. The practical use of theoretical and empirical data relating to ochre formation in drains.Arch. Acker. Pflan. Bodenk., 1977, 21: 821-6.

The probable times required for ochre formation in drains designed according to usual Czechoslovakian criteria are tabulated for experimentally determined discharges from the drains.

1190

SVOBODOVA, V. Water balance study on a drained heavy Redzine soil under winter wheat. Trans. 10th Cong. Int. Soc. Soil. Sci., Moscow, 10:292-300.

Drainage studies on a Redzine soil, clay marls of poor permeability (<8 cm/day) are reported. Excess moisture was better controlled on experimental plots where tile drains were combined with a filter. Among the slag and brick crush filter, performance of brick crush filter was superior in controlling moisture and hence the winter wheat yields. Soil water balance in the upper 1m of soil profile for the seven treatments are presented.

1191

SVOBODOVA,, V.; NINSOVA, M. The decomposition of filter materials in soil- a model experiment. Z. Kult.Flurb., 1979, 20: 42-50.

In a model experiment, the rate of decomposition of and the sorption of soil N by , the following drain filters followed the order: wheat straw> spruce sawdust > spruce bark > oak bark. The rates of decomposition and N sorption were higher in a heavy clay(chernozem-smonitza) rich in nutrients than they were in a moderately heavy loam(acid Braunerde) poor in nutrients, and in both soils they were higher at the maximum moisture capacity(MMC) than at 50% MMC. A high N content in the soil and a high NO_3 content in the drainage water resulted in rapid decomposition of the filters.

1192

SWARTZENDRUBER, D. Approximate water flow rates for tube drains in stratified soils. J. Geophys. Res., 1962, 67:2395-2402.

A general theoretical method of approximating the steady-state, two- dimensional seepage rate in stratified porous media is applied to tube drainage in layered soils. For drain tubes in a surface soil layer of given permeability, overlying a deep layer of different permeability, the approximate flow rate is within 8% of the exact flow rate.

1193

SZWAJ, S. Stress and strain distribution in the soil and its influence on the trajectory of motion of drainage equipment. Zs. Probl. Post. Nauk. Roln., 1978, 4:201:15-28.

A theoretical analysis of the conditions occurring during mole drainage.

1194

TABET, A. Analysis of experimental results of drains in heavy clay soils. 10th Cong. Irrig. Drain. Athens, 1978, 311-52.

A four-year study was conducted on a 12 ha plot in the Gharb plain on heavy swelling clay soils which develop deep(over 1 m) cracks in the summer. Extensive data on drain flow and water table drawdown have been obtained. Calculated values of hydraulic conductivity have been compared with values found by the auger hole method. Calculation of the drainage porosity revealed the influence of rain or irrigation. Some results are presented in detail. Recommendations are given on the importance of the residual charge in the calculation of drain spacing; the importance of flows during the irrigation period for designing collector drain systems; and the enormous water losses through cracks occurring with furrow or flood irrigation.

1195

TABUCHI, T. Studies on drainage in clayey paddy

fields. (1)-A hypothesis of drainage and the classification of drainage in clayey paddy fields. Trans. Agric. Engng. Soc. Tokyo, 1966, 18:7-11.

The paper brings out the importance of the cracks in the rice fields and classifies the drainage system into the crack, and the non-crack types.

1196

TAJINO, N. PVC pipes as a new material for underdrainage. 8th Congr. Irrig. Drain. Varna, 1972, Q 28 R 13: 181-98.

Drainage rates expressed by the ratio of atual discharge to theo-retical discharge for bamboo, brushwood, clay and plastic pipes had been compared to wire netting drain. The brushwood drain drainage rate is 34% whereas for clay pipe it was 100% and bamboo pipe it was 90%. PVC pipe on the other hand had a drainage rate of 91%. The comparison of the presence and absence of enveloping materials for underdrain protection yielded drainage rate of 13 and 100% suggesting the importance of enveloping materials. The maintenance procedure of new materials for underdrainage has been outlined.

1197

TAJINO, N. Subsurface drainage (in Japan). VIII. Determination of drain spacing. Tech. Bull. Fac. Agric. Kagawa Univ., 1979, 30: 197-213.

Several experimental and theoretical methods were investigated and some modifications were made to these methods. Data for the lowering of the groundwater table by underdrainage is presented and for this purpose a new coefficient, a seepage coefficient "u" is proposed, since "f" (soil porosity, drainable pore space) has certain drawbacks. Drain spacings calculated by 9 different methods ranged from 10.8 to 12.2 m. Methods of calculating the required spacings under specified hydraulic conditions are shown.

1198

TAJINO, N. Under drainage. IX. The spacing ratio for underdrainage. Kag. Daig. Nogak. Gakuz. Hok., 1979, 31:49-58.

Equations are given for the calculation of tile spacings for a range of soil depths.

1199

TAJINO, N. Underdrainage investigations. X. Underdrainage capacity. Kag. Daig. Nogak. Gakuz. Hok., 1980, 31:137-60.

In this mathematical expression of underdrainage capacity, under-drainage efficiency and the underdrainage discharge ratio were used. Good agreement was obtained between theoretical values and actual values obtained on two experimental farms.

1200

TAJINO, N.; IYANAGA, K. Disadvantages of PVC pipes as materials for underdrains and countermeasures for overcoming such disadvantages. 10th Congr. Irrig. Drain. Athens, 1978, 4:439-53.

Disadvantages of PVC pipes encountered during their use in Japan include breakage, faulty connection, uplifting due to buoyance in high water table areas, clogging of drain filter materials, and lack of drying of soil surrounding the drain due to rapid installation. Measures to overcome or avoid these problems are described.

1201

TAKAHASHI, T. Fundamental studies on the non-steady state tile drainage. I.R.C. network and the non-steady state drainage. Trans. Jap. Soc. Irrig. Drain. Recl. Engng., 1969, 30:40-6.

Resistance capacitance network has been suggested

to study the non-steady state drainage problems. A model based on this is presented and advantages over other models are discussed.

1202

TAKAHASHI, T. Fundamental studies on the non-steady state tile drainage. II-Experimental considerations on the non-steady state tile drainage. Trans. Jap. Soc. Irrig.Drain. Recl.Engng., 1969, 30:47-54.

The R.C. model developed has been used to test the non-steady state drainage design problems under different sets of conditions.

1203

TAKAHASHI, T.; BISHAY, B.G.; NAGAHORI, K. Theoretical and field studies of the ground water table movement under submerged drain outlets. Trans. Japanese Soc. Irrig. Drain. Reclam. Engng., 1978, 75:47-53.

A theoretical analysis of groundwater movement under submerged drain outlets in polder soils in relation to rainfall intensity, hydraulic conductivity of the soil, drainable porosity, drain spacing, pressure head inside the drain pipe and thickness of Hooghoudt's "equivalent layer" is presented. In a field study of five-polder soils ranging in age from 13 to 260 years after reclamation, a close agreement was obtained between experimental and observed groundwater table depths. The rate of groundwater table drawdown was strongly correlated with hydraulic conductivity and drainable porosity, and negatively correlated with back presesures inside the drain pipe in excess of 10 cm. Mole drains annd porous drain envelopes accelerated hydraulic conductivity and lowering of the groundwater table.

1204

TAKAHASHI, T.; NARITA, H.; TANAKA, T.; TONOSAKI, T.; KANAYA, H. Studies on the growth of grapevines in a drained paddy field. I. The influence of the water table on vine growth and fruit quality. Bull. Aomori Field Crops Hor.Expt. Stn. Aomori, Japan, 1976, 1:53-62.

Groundwater levels in vineyards next to irrigated rice fields were extremely high and root growth, number of fruiting branches per unit length of main branch, and brix were all low and root development was poor. An underground drainage system, dep ploughing to break the plough sole, use of large planting holes, application of an organic conditioner and fused phosphate all improved growth.

1205

TALMAN, A.J. Interactions between iron ochre and pile slots- an experimental study. Tech. Rep. Field Drainage Experimental Unit, U.K., 1978, 78/2, 13p.

This paper describes an experiment that was run in the laboratory to investigate the blocking of simulated drainage pipe slots with iron ochre. Both clay and plastic materials were studied, using slots of different widths. Water was passed through ochrous soil and then through the slots for a period of 4 months. The percentage of blocking was calculated using visual observations. The results indicated that slot width, the thickness of the material (pipe) and treatment with permeable fill, all played significant roles in determining the degree of blockage of the slots.

1206

TALSMA, T. The effect of soil physical conditions on reclamation of saline land. Trans. 6th Congr. Irrig. Drain. New Delhi, 1966, 2:Q 19 R 7: 83-91.

The steady state diffusion mechanism, describing evaporation from the water table as dependent on the meteriological conditions, soil properties and water-table depth, was found to be suitable in understanding and predicting chloride salinization from saline groundwater. The main soil property, involved in the steady state transfer of moisture and salt to the soil surface is the capillary conductivity-moisture suction relationship. Two of such relationships, for a loam and clay soil respectively, are shown. For prevention of serious salinization or resalinisation after reclamation, it is argued that the water table should be at or below that depth where evaporation is reduced to about 1 mm/ day.

1207

TALSMA, T. Leaching of the tile-drained saline soils. Aust. J. Soil Res., 1967, 5:37-46.

Leaching experiments were conducted on three tile-drained saline soils, each soil having a different average hydraulic conductivity. The experimental plots consisted of the area between two parallel lateral drains and the main disposal drain. Fresh irrigation water was alternately ponded to a depth of 15 cm and drained to slightly above tile-drain depth. During the ponding stage, desalinization proceeds more rapidly near the drains than midway between, while during the falling water table stage desalinisation is more even over the whole area. Comparison of continuous ponding with alternate ponding and drainage shows that in the latter case complete desalinization is achieved with considerably less leaching water. The leaching efficiency is not very high under conditions of continuous ponding.

1208

TALSMA, T. Techincal and economic considerations on drainage of irrigated land, in relation to soil properties and land-use planning. Trans. Int. Congr. Agric. Engng. Lausanne, 1964, 6: 2:215-23.

Where saline irrigation water is used, the design discharge coefficient is related to the leaching requirement and the leaching efficiency of the soil; this leads to discharge coefficients generally smaller than those used in humid regions. With an un-stable saline water table, the drainage depth is related to the moisture-conduction properties of the soil profile; this depth is 120-200cm and is greater than in humid areas. An example is given of setting up tile-drainage criteria for the Murrumbidgee Irrigation Areas.

1209

TAYLOR, G.S. Drainable porosity evaluation from outflow measurements and its use in drawdown equations. Soil Sci,, 1960, 90: 338-43.

A procedure for determining drainable porosity (volume fraction of pore-space water drainable from soil) at equilibrium conditions, and for estimating equivalent drainable porosity for non-equillibrium conditions from fluid outflow and water-table mea-surements is described, and the use of drainable porosity in drawdown equations and its evaluation from field drainage data are discussed.

1210

TAYLOR, G.S.; GOINS, T. Field evaluation of tile drain filters in a humid region soil. Reds. Circ. Ohio Agric. Res. Dev. Cent., 1967, 154, 18p.

Various filter materials, corn-cobs, wheat straw, crushed stone and vermiculite, were placed over recently installed tile drains in a Crosby silt loam soil which has developed from calcareous loam till. Other tile drains received no filter materials. After eight years, only negligible amounts of sediment were found in any of the tile drains. The soil adjacent to the drains was porous, regardless of filter treatment; it had water-stable aggregates and did not contribute significantly to siltation of the drains or to severe soil settlement around the drains. Filter materials are not recommended for agricultural drains in silty and clay-textured soils of the humid region.

1211

TAYLOR, G.S.; FAUSEY, N.R. Backfill alteration and drainage of clay soils. 4th Natn. Drain. Symp. Proc. ASAE, 1982, 12-82:139-48.

Backfill alteration studies on a Toledo silty clay which contains approximately 50% clay in the plow layer and 60% in the B-horizon indicated that as long as percentage ponding exceeded 20-30%, flow rates from drains with altered backfill were 2-4 times greater than those with unaltered backfill. Backfilling the trench with soil doubled drain flow rates compared to drains with no backfill alterations. The data indicated that there has been significant decline in drain flow rate in case of trenching and subsoiling. In case of trenching and refilling there was reduction in the flow rates after one or two years when the flow rates can be expected to be similar to unaltered backfill within another one or two years.

1212

TAYLOR, G.S.; HUNDAL, S.S.; FAUSEY, N.R. Permeable backfill effects on water removed by subsurface drains in clay soil. Trans. ASAE, 1980, 23:104-8.

A highly permeable backfill was created by trenching and back-filling directly over drains installed 6 year earlier in a lake-bed clay soil. The backfill increased drain flow rates by a factor of two or four during periods of ponding but had no effect on flow rates after ponding ceased. The full potential of the backfill could not be determined because of the limited hydraulic conductivity of the drains. The backfill had no effect on water table drawdown, apparently because of excellent surface drainage that diverted ponded water from untrenched drains to adjacent trenched ones. After one year, flowrates remained the same where the permeable backfill extended to within 5 cm of the drain but were reduced 20 to 25% where the backfill extended to drain.

1213

TAYLOR., G.S.; HUNDAL, S.S.; FAUSEY, N.R.; SCHWAB, G.O. Backfill alteration effects on pipe drainage of a clay soil. Soil Sci. Soc. Am. J., 1981, 45:611-16.

A permeable backfill may improve pipe drainage for soils that have a slowly permeable subsoil. Permeable backfills were created directly over drains installed 8 years earlier by replacing soil with gravel and by trenching and refilling. The permeable zone extended from the bottom of the plow layer to the drain. The research was conducted on a poorly drained, cultivated, lakebed clay soil in Ohio. The addition of gravel increased drain flow rates by a factor of 2 to 3 during periods of ponding but had no effect on flow rates under non-ponded conditions nor on water table drawdown rates. The ponded flow rates for drains with gravel have not changed after 3 years. During the first year, trenching and refilling had the same effect on drain flow rates as gravel. During the second year, the ponded flow rates declined to only 50 or 60% of those with gravel. The permeable backfill functioned as a hydraulic connector and permitted ponded water to bypass the relatively impermeable subsoil in flowing to the drain. Sediment concentrations in the drain water were unaffected by backfill alterations.

1214

TAYLOR, J.C.; MATTOS, C.C.L.; BELTRAME, L.F.S. Effects of soil drainage on the number of days available for planting summer crops. Rev. Bras. Ciencia Sol., 1981, 5:139-44.

A field experiment with tilled and untilled plots in drained and undrained areas was carried out in a Planosol of the Vaccai mapping unit, typical of the rice growing flood plains of Rio Grande do Sul, with the purpose of evaluating the effect of drainage on the number of days available for planting soybeans and maize under suitable moisture conditions. Criteria for counting the number of suitable days were established, based on correlation between soil moisture content and penetration resistance and visual assessment of the soil workability. The lower limit of plasticity was used as an alternative criterion. In the drained area, during the period from 24/07 to 18/12 of 1979 there were 61 and 46 days suitable for field work in the untilled and tilled plots, respectively. In the undrained area the equivalent available time was less, being 35 and 23 days, respectively.

1215

TAYLOR, G.S.; WORSTELL, R.V. Ponded water flow in layered soils. Trans. 7th Congr. Soil Sci., Madison, 1960, 1:480-4.

An analog study of ponded flow into drains was conducted in layered soils. When the drains are located in a layer whose conductivity is low compared to the other two layers, a high per centage of the potential is dissipated very close to the drain. Without exception, the stream lines intersect the ground surface at smaller horizontal distances from the drain center line in the layered soils than in the homogenous soils. For a particular conductivity profile, an increase in drain depth causes a streamline to intersect the ground surface at greater horizontal distances from the drain. When the conductivity ratio of the two adjacent layers is less than 5, then placement of the drain in the less permeable layer does not significantly reduce the drain flow. However, it is reduced significantly if it exceeds five and the drain is placed in the less permeable layer.

1216

TAYLOR, P.A.; WATTS, D.G. An empirical model for predicting drainage system performance. Trans. ASAE, 1976, 10:723-6, 729.

A subsurface drainage system with 10 cm diam clay

tile laterals placed at a depth of 90 cm and spacing of 9, 18 and 27 m has been described. The observations were simulated by a computer model and effect of carrying capacity of mainlines, tile spacing and depths were found out. Long term water table records were simulated. It was observed that a spacing of 27 m at 60 cm depth and a 1.25 cm capacity of mains would provide better drainage for all return periods. At 30 cm depth 36 or 45 m spacing with a 2.54 cm capacity main would provide better drainage for events occurring once in ten years. From practical point of view a less costly system at larger spacing with a slightly higher capacity of the mains (1.9-2.5 cm) was recommended for the area.

1217

TAYLOR, W.R.; WILLARDSON, L.S. Target grade control for trenching machines. Proc.Natn. Drain. Symp, Illinois, ASAE, 1971, 47-9.

The grade control by the target method is described and its advantages and dis-advantages over other methods reported.

1218

TEJIMA, S.; OGINO, Y.; YABE, K.; MURASHIMA. K. Experimental studies on new subsurface drainage method by the use of negative pressure. Bull. Univ. Csaka Prefexture, Japan, 1980, 32:11-22.

A new subsurface drainage method is proposed in which negative pressure is applied to the water column in porous pipe buried in the soil. Experiments on the fundamental principles of this method were carried out using ceramic pipe. The physical proper-ties of various pipe materials were compared. Although ceramic pipe properties were comparable with other materials, it was inferior in practice. After removal of gravitational water by traditional tile drains, a further portion of soil water was removed by the negative pressure drainage technique.

1219

TERVONEN, M.; ISSAKAINEN, J. Effect of ditch spacing and fertilization on the revival of radial growth of Scots pine on shallow-peat bog of the small-edge type. Folia Forestalia, Inst. Forstable Fenniae, 1980, 444, 14p.

Radial increment was measured in autumn 1979 on trial plots established in 1972 on drained bog at Muhos, N. Central Finland. Four ditch spacings (10, 20, 30 and 40 m) and four fertilizer treatments (Nil; $P_{96}K_{60}$; $N_{46}K_{96}K_{60}$; and $N_{46}P_{48}K_{30}$) were compared. Differences between the fertilizer treatments were greatest at 20-m ditch spacing. $N_{46}P_{48}K_{30}$ increased radial growth most. $N_{46}P_{96}K_{60}$ improved growth more than PK immediately after fertilizing, but the effect started to weaken after four years. The effect of PK was significant only at 20-m spacing. Refertilization of $N_{46}P_{48}K_{30}$ plots in 1975 clearly improved growth. Radial growth increased with decreasing ditch spacing. At 10-m spacing, the trees grew equally well throughout the area, whereas at wider spacings the growth was best in the trees nearest to the ditches.

1220

THANOPOULOS, GHR. Existing drainage schemes and need of its completion in Larissa plain. Bull. G.C.I.D., 1964, 8:14-15.

The paper gives a brief description of the actual conditions in Larisa Plain (the largest one in Central Greece), the existing drainage schemes and the anticipated benefits on their completion.

1221

THEOBALD, G.H. Methods and machines for tile and other tube drainage. FAO, Rome, 1963, 104 p.

The book contains 8 chapters such as the need for

drainage, methods of draining, under-drainage layout, materials, tools and machines, drain installation, junctions and outfalls, mole drainage. Mostly practical aspects of drainage are presented.

1222

THIEL, T.J.; BORNSTEIN, J. Tile drainage of a sloping fragipan soil. Trans. ASAE, 1965, 8:555-7.

An electric resistance analog was used to study the problem of locating the drains in a sloping fragipen soil. The results of the electrical model study have substantiated field data indicating that, for the most part, water generally moves downslope parallel to the soil horizon boundaries. The analog results showed that the influence of subsurface drains on flow direction under the conditions of this study is restricted to the region within about 1.8 m on either side of the drain. An analysis of the hydraulic potentials showed that, when the drain was located in fragipan material, water flowed both above and below the drain and moved on downslope to the lower drain. However, when the drain was backfilled with A_{2g} soil material it intercepted all downslope flow, and the water divide occurred slightly downslope from the drain. Although drain outflow was increased about 50 percent by changing the drain depth from 45 to 75 cm increases of nearly 300 percent were obtained by backfilling the drain with the more permeable A_{2g} soil material.

1223

THIEL, T.J.; DIXON, R.M.; PETERSON, A.E. Drainage of a peat soil overlying an artesian aquifer. II Analog studies of tile and well drainage. Res. Rep. Coll. Agric. Univ. Wis., 1966, 21:7-11.

An electrical resistance network was used to study the removal of excess artesian water and rainfall from a deep peat soil overlying a thick silt stratum. Reduction of the aquifer hydraulic head (as may be accomplished by artesian wells and tile drains in the peat layer) was simulated with the network. 10 cm drains placed 1.25 m deep and a steady rainfall of 2.5 cm per day were used as design criteria. Drain spacings of 4 to 30 m and aquifer hydraulic heads from 1.5 m above to 0.75 m below the soil surface were simulated. Reduction of hydraulic head by artesian relief wells had little effect on the position of the water table in the peat layer, or on the drain spacing necessary to control the water table. Lowering the hydraulic head of the aquifer (1.5 m above to 0.75 m below the soil surface) lowered the midplane water table only 7.5 cm for all drain spacings. A drain spacing of 12 m would keep the water table at or slightly below, the soil surface during a steady rainfall of 2.5 cm per day. Drain spacings obtained were significantly closer than those obtained with a modification of the nomographic method and with the ellipse equation.

1224

THIEL, T.J.; TAYLOR, G.S. Utilizing a sand tank model to study some moisture flow problems in drainage. <u>Trans. Int. Congr. Soil. Sci</u>., Madison, 1960, 1:47-9.

A small tank model is described for studying subsurface drainage problems. Provisions are made for varying drain depth and spacing. Glycerine-sand flow systems are used. An example of the type of problem which can be studied is illustrated by some experimental findings for drawdown in homogeneous media.

1225

THIHATMER, J. Relations between ground water table and plant cover in different sandy substrates. <u>Arch. Acker Pflan. Bodenk</u>, 1978, 22:397-403.

In lysimeter experiments using 4 sands of different textures, a mixture of perennial ryegrass, meadow fescue, Poa trivialis and red clover was grown with water tables at 40, 60, 80 or 100 cm depth. At all water levels, yields decreased as the coarseness of the sand increased. The yields were highest with the water table at 40 cm and decreased sharply when it fell below 60 cm. Plant cover decreased from 100% on fine or medium fine sand with the water table at 40 cm, or on fine sand with the water table at 60 cm, to 30% on coarse sand with the water table at 100 cm. The limiting water table depths for good plant growth were 40, 40-50, 60-70 and ≫ 100 cm in coarse, medium coarse, medium fine and fine sand, respectively.

1226

THOMASSOM, A.J.(Ed). Soil and Field Drainage. Soil Survey Tech. Monograph-7. Harpenden U.K., 1975, 79 p.

Technical monograph with 7 chapters and three appendices provides informations on soil maps and their influence on drainage design. Drainage design in relation to soil series and soil properties affecting drainage design have been discussed.

1227

THORBURN, A. Ochre formation in soils. Rep. Welsh Soils Disc. Gr.U.K., 1977, 18:81-90.

A localised but often severe problems in the drainage of agricultural land is that of ochre formation causing blockage and failure of pipe drain systems. This paper presents picture of the nature and formative processes of ochre deposits. Possible control methods are discussed, and the problem is illustrated by results from investigations on an affected East Angian fen soil.

1228

TLAPAK, V. Enhancement of soil moisture conditions in tile-drained soil under drought. <u>Acta Univ. Agric. Fac. Agron. Brno</u> (A), 1975, 25:15-25.

Certain relationships governing the moisture conditions and water dynamics of soil under less favourable climatic conditions, in particular, low rates and unfavourable distribution of precipitation, were studied. Precipitation frequencies and moisture occurrences were studied. The soil moisture dynamics as affeted by atmospheric precipitation were assessed on the basis of pF curves. Tile-drained soil gave better water utilization and good water economy even with low precipitation. Due to their moiste-ning action, tile drains appreciably increased the moisture content of the depper soil horizons.

1229

TOKSOZ, S.; KIRKHAM, D. Steady drainage of layered soils. I. Theory. J. Irrig. Drain. Div. ASCE, 1971, 97:1-18.

The problems of steady drainage of two-layer and three-layer soils have been solved. Two sets of flow nets have been given for each problem. Solutions of steady drainage problems for soils with more than three layers have been deduced from the two-layered and three-layered solutions.

1230

TOKSOZ, S.; KIRKHAM, D. Steady drainage of Layered soils. II. Nomographs, J. Irrig. Drain. Div. ASCE., 1971, 97:19-37.

Drain spacing equations for the steady drainage of two and three layered soils have been given. A set of 26 nomographs has been prepared for determining explicitly the drain spacing for the two-layered problem.

1231

*TOMN, E.D. A scientific basis for the trenchless laying of a drainage system. Vestn. S-Kh. Nauk. Mosk., 1977, 4:78-91.

Closed drain can be laid without trenching in irrigated soil whose groundwater is not more than 1.0-1.5 m below the soil surface by means of a machine whose cutting component is a passive vertical knife with step-like incisions. The drainage line (pipes and filters) must be continuous, and excavation for and laying of, the drains is a simultaneous operation. The flow of water to the drains is calculated for different conditions. Nomograms describing drawbar resistance are established on the basis of calculated and experimentally determined resistances.

1232

TOMITA, O.; SAVU, P. The efficiency of a drainage network at reduced depth on salinized soils in the northern district of the Albita-Falciu embanked enclosure. In Lucrari Stiintinfice, institutul Agronomic "Ion Ionescu de la Bard" Horticulture, 1978, 9-12.

The efficiency of a drainage network at 0.8 m depth on drainage and salt leaching of a clayey salinised soil, with and without ploughing, has been examined. With loosened soil, the drainage rate was higher and salt removal from the drained layer following gypsum application and leaching was adequate.

1233

TOTH, A.; KOSCONDI, J. Role of combined drainage in the improvement of pseudogley brown forest soils. 10th Congr.Irrig. Drain. Athens, 1978, 4:159-64.

Long-term studies on the improvement of pseudogley brown forest soils in western Hungary using tile or plastic drains at double the usual spacing combined with subsoiling, mole drainage and liming are outlined. The drainage system was laid out according to calculations based on the requirement to drain excess water from the upper 70-80 cm within two days. The formula for this calculation needs further refinement.

1234

TOVEY, R. Alfalfa growth as influenced by static an fluctuating water tables. Trans. ASAE, 1964, 7:310-2.

Alfalfa can tolerate water table upto 60 cm belo ground level. In case of fluctuating water tabl the excess above 60 cm must be removed within days to insure optimum crop production. After days the root system deterioration became progres sively worse.

1235

TRAFFORD, B.D. Agricultural land drainage. Report, Welsh Soils Disc. Group, Ministry of Agriculture, Fisheries and Food, UK, 1971, 12:68-84.

Non-variable design factors (climatic conditions, soil forming method) and variable design factors, steady state and non-steady state drainage designs in present use in the UK and their limitations are discussed. The best design for a given situation may be one which is less than perfect from the purely drainage engineering standpoint. Soil series provide an excellent basis for broad policy decisions.

1236

TRAFFORD, B.D. The analysis of inflow into some experimental agricultural drains. Bull. Int. Assocn. Hydrol. Sci., 1973, 17: 431-40.

Irrigation equipment was used to create strong flows in agricultural drains of clayware and plastic pipes, the latter in both smooth and corrugated conformations. The piezometric heads at 10 m intervals along drains were analysed to show that the inflow per unit length varied greatly. Both inflows and outflow were found, some drains had substantial inflows at the origin. For given conditions the flow rates are as much as 17% less than flow rates determined for similar conditions in the laboratory.

1237

TRAFFORD, B.D. The application of synthetic materials to field drainage techniques and practices. 8th Congr. Irrig. Drain. Varna, 1972, Q 28 R 3: 33-57.

Results of survey indicated that 25% of agricultural land in U.K. would benefit by drainage. Traditional materials for drainage viz., synthetics are compared and their merits and demerits discussed. The hydraulic and hydrologic qualities of various syn-thetic products including plastic pipes are examined and in general shown to have some technical advantage over the traditional materials specially baked clay ware. The cost structure of whole underdrainage job is examined and it is shown that cost can be reduced through mechanization though material cost is more.

1238

TRAFFORD, B.D. Field drainage. J.Roy. Agric. Soc., 1970, 131: 129-52.

A review, with 54 references, of the history of field drainage, drainage theory (soil-water relations), experimental results, and present drainage trends.

1239

TRAFFORD, B.D. Improving the design of practical field drainage. Soil Sci., 1975, 119:334-8.

There exists mathematical models to determine the optimum depth and spacing of field drains from measurements of hydraulic conductivity, or indirect estimates of this. The results are reasonably useful as long as the hydraulic conductivity exceeds perhaps 1-2 m/day. For heavier, less permeable soils, it is necessary either to use very cheap "pipes" (mole drains) at the calculated spacing or normal pipes at wider spacing after increasing the hydraulic conductivity by subsoiling.

1240

TRRAFFORD, B.D. The interactions between field drainage and nitrogen. Tech. Bull., ADAS, Field Drainage Experimental Unit, 1976, 76-8, 13p.

Some of the literature on the role of field drainage in the loss of nitrogenous fertilisers in drainage water is reviewed. It is suggested that some of the seemingly contradictory results may be explained by considering the likely flow path of the water through the soil and hence likely transit times. A method for making rough estimates of transit time is presented. Attention is drawn to the situations in clay soils without effective under-drainage. Here drainage will take place either as surface runoff or as lateral seepage in the cultivated layer. Some evidence is presented which suggests that this might lead to higher nitrate losses than would be experienced with effective underdrainage. The need for comprehensive field experiments is stressed and some indications of the necessary observations are given.

1241

TRAFFORD, B.D. Recent progress in field drainage: Part I. J.Roy Agric. Soc. England, 1977, 138:27-42.

The aspects considered are: the present state of field drainage in England and Wales; drainage theory; drainage benefits.

1242

TRAFFORD, B.D. Recent progress in field drainage. Part II. J. Roy. Agric. Soc. England, 1978, 139:31-42.

Recent field and laboratory studies concerning subsoiling and mole drainage, pan busting, drain spacing, drainage materials, grassland poaching, the interaction of drainage with cultivation and N fertilizer, planning drainage, installation through a growing crop, sub-irrigation, and iron ochre are discussed.

1243

TRAFFORD, B.D. Soil water regimes--what is known, the work which is in hand and suggestions for progres. Tech. Bull. Field Drainage Experimental Unit, Ministry of Agriculture, Fisheries and Food, U.K., 1974, 74/13, 19p.

Data are presented for the response of various crops to constant water-table levels at different depths; the optimal level for most crops was 80-100 cm below ground level and at this level, damage to soil by cattle or cultivation is minimized. The effects of short periods of waterlogging on different crops is reviewed and, in general, crops were most susceptible to damage at germination and at flowering. The importance of soil water regime in the basic drainage status and its future use in studies of drainage design and economics are discussed.

1244

TRAFFORD, B.D. Some conclusions from experimental drainage work on clayland. Soil Water, 1976, 4:7-9.

Mole drainage appears to be the most effective and cheapest way of changing the soil water conditions in the subsoil of clay soils. Drains alone have relatively little effect on soil water regime or crop yields.

1245

TRAFFORD, B.D.; OLIPHANT, H.M. The effect of different drainage systems on soil conditions and crop yield of a heavy clay soil. Exptl. Husb., 1977, 32:75-85.

In trials in 1971, 73 and 74, winter wheat was grown on a heavy clay soil with (a) natural drainage (b) 15-m spaced drains, (c) 30 m spaced drains with permeable fill or (d) 60-m spaced drains with permeable fill with treatment (b), (c) and (d) being super-imposed by no subsoiling, deep subsoiling or moledraiange. Averaged over type of drainage, grain yield was highest (4.70 t/ ha) where moling was superimposed over drainage. Averaged over type of superimposed treatment, grain yield was highest at 4.81 t/ha with (b) compared with only 3.25 t/ha with (a) and no subsoiling. It was concluded that efficient drainage could increase winter wheat yield by about 1 t/ha.

1246

TRAFFORD, B.D.; RYCROFT, D.W. Observations on the soil-water regimes in a drained clay soil. J. Soil Sci., 1973, 24:380-91.

Details of soil-water profiles between drains and the associated drain-flow data are presented on a clay soil of the Ragdale series formed in drift over Lias clay in Midlands of England. Mole drainage is shown to modify the drainage performance consi-derably by providing a quicker disposal of water and significantly lower water table with much smaller fluctuations in level. Where drains have been laid in natural subsoil without disturbing, the water-table level is frequently close to cultivation level and there appears to be little movement in the subsoil. It is concluded that for effective drainage, soil of this type requires deep cultivation to modify the subsoil and allow the water movement at greater depths.

1247

TREWHELLA, W.; WEBSTER, A. Salinity control and drainage in the Shepparton and Kerang regions. In Hydrogeology of the Riverine Plain of South East Australia, (R.R. Storrier, I.D. Kelley, Ed.), Aust. Soc. Soil. Sci. Inc, 1978, 169-95.

The Shepparton and Kerang Regions with a gross area of 0.83 mha= about half of which is irrigated, are situated on the Riverine Plain in northern Victoria. In the Kerang Region, watertable levels have been within two m of the surface almost everywhere, and large areas salinized, for several decades. No new drainage schemes are at present in hand, or proposed, because of severe effuent disposal problems and prohibitory costs. The possibilities for on-farm disposal of surface run-off and groundwater effluent including re-use are being examined. In the Shepparton Region watertable levels are within two metres of the surface under at least 100,000 ha and are still rising under other areas. Salinity symptoms in fruit trees, hitherto confined to a few pear orchards, appeared in more than 3000 ha of pears in 1975. Salinization of low lying pastures is increasingly apparent. To control water table levels and salinity the installation of 650 groundwater pumps is propsoed, together with extension of the surface drainage network. Plans for the safe disposal of the effluent are outlined. Physical and economic constraints to these proposals are discussed.

1248

TRUE, G.M. On-farm drainage design procedures used by ADS, Inc. In Water Management for Irrigation and Drainage. Proc. Spec. Conf. ASCE Irrig. Drain. Div. ASCE, New York, 1977, 1:14-21.

The design concepts and procedures found useful in solving on-farm drainage problem for irrigated lands by Advanced Drainage Systems Inc (ADS Inc) are presented. Drainage coefficient can be reduced as much as by 30% if area irrrigated is 10% or less of the total irrigated area. Minimum depth of water table between drains may be 122-153 cm depending upon the texture of the soil in case underground water is saline. For non-saline water it may be around 107 cm. Several aspects of sump design, selection of filter material and economics of the drainage systems are described.

1249

TSUTSUI, H. Water management in waterlogged paddy

fields with reference to drainage improvement. <u>ICID Bull</u>., 1963, 12:62-5.

Surface and subsurface drainage needs of rice are discussed in respect of waterlogging of these soils. Effect of drainage on soil properties specifically the permeability of soils is highlighted. It is suggested that to economise on water, the irrigation of rice fields should be developed as if for an upland irrigation system. The problems encountered in any such system are discussed.

1250

TYAGI, N.K.; LENKA, D.; ACHARYA, N. Waterlogginng in Hirakud Ayacut. Symp. Waterlogging-Causes and Measures for Prevention. CBIP New Delhi,. 1972, 118, 1:1-6.

In Hirakud Ayacut the problem of waterlogging since 1958, after the introduction of canal irrigation, is causing concern. The water table has gone up by 5 m and nearly 60000 ha of land is suffering from waterlogging. Two types of causes i.e., physiographic and man-made have been identified. Irrigation water management, lining of canals and distributaries, land management and provision of field drainage channels may help in alleviating the problem to a greater extent.

1251

TZIMOPOULOS, C.; TERZIDIS, G. Unsteady flow of water through soils drained by parallel ditches. <u>J. Hydrol.,</u>1975, 27:73-93.

Water movement through soils drained by parallel ditches is generally unsteady and can be described by the non-linear equation of Boussinesq. Numerical solutions of this equation by means of implicit finite differences of the type described by Crank-Nicolson, Laasonen and Douglas agreed well with experimental data obtained with a Hele-Shaw model.

1252

U.A.R. MINISTRY OF IRRIGATION. Drainage of irrigated lands in U.A.R. National Committee of ICID, Cairo, 1968, 44p.

The paper gives brief account of the history of land drainage in UAR, source and type of water for irrigation, subsurface water effect, drainage system design and practices, drainage economics, research in drainage and research project plans.

1253

UNHANAND, K.; KADIR, T.N. A theory of the combined mole-tile system. Water Resour. Res., 1975, 11:111-9.

A theory of water movement in the combined mole-tile drain system based on the transient state condition was developed. Two general equations were derived to describe the height of the water table at any location at any elapsed time after the drainage process begins; one is applicable for the stage in which the water table is above the mole drains, and other for when the water table falls below the drains. The two general equations were simplified for the point located at midpoint between the tile drains and mole drains. In the derivation, assumptions regarding the flow, condition of groundwater and shape of the water table profile at certain boundaries, were made. Field experiments were then conducted and the data were used in verifying the equation for the first stage.

1254

UNHANAND, K.; TUAMSANGIEM, K. performance and evaluation of combined mole-tile drain system in heavy soils. ICID Bull., 1975, 25:36-52.

A field experiment on the investigation of performance and cost of a combined mole-tile drain system in heavy soils in comparison with a similar tile drain system is reported. Three experimental plots consisted of a tile drained plot, a combined (single and double mole) drained plots. They were constructed with a tile drain spacing of 37 m and mole drain spacing of 1.83 m. The experimental results show that the combined system was more effective than the drain system in lowering the water table. The difference in the effectiveness of the single mole drain over double mole drain was not distinctly apparent. No deterioration of the mole drains during the period of two years tested was detected. The cost analysis indicated that the annual cost of the combined system is always less than that of an equivalent tile drain system even if the mole drains in the combined system have to be redrawn every year.

1255

UNIITED STATES DEPARTMENT OF AGRICULTURE-SOIL CONSERVATION SERVICE. Principles of drainage-SCS National Engineering Handbook. Section 16, Chapter 1. Washington, 1961, 331 p.

The paper presents basic principles and theories concerning the removal of excess water from agricultural lands and reviews methods of subsurface flow as related to solution of drainage problems. Their limitations have been discussed.

1256

UPADHYA, G.S., DATTA, N.P.; DEB, D.L. Note on the effect of selected drainage practices on yield of rice and the efficiency of nitrogen use. Indian J. Agric. Sci.,1973, 43:888-9.

Grain yield was not affected by various management practices, but straw yield was slightly increased by mid-term drainage for two or ten days, as compared to continuous flooding. Yields were not significantly affected by N source (urea or ammonium sulphate). Utilisation of fertilizer N (labelled with ^{15}N) was increased by the drainage treatments, and by application at the promoridial initiation state rather than transplanting.

1257

UPPAL, H.L. Serious waterlogging in Punjab and Haryana-How cured and measures to prevent its recurrence. **Symp. Waterlogging causes and Measures for Prevention** Cent. Bd. Irrig. Pwr. New Delhi, 118, 1972, 2:1-31.

Rise of water-table resulting in waterlogging occurred in all the canal tracts in 12-15 years after the commissioning of the canals. Salinity followed in the wake of waterlogging. In the Punjab, serious waterlogging occured. Eight pilot projects (each over 0.4 lakh ha) were set up in different parts of the state for studying the hydrology, geology and geophysics of the tract and also the effect of different ameliorative measures adopted. Two lakh tubewells were installed from 1966 to 1972. The water table which had its peak in 1962 had been going down due to the combined effect of surface drainage, flood and choe control and installation of tubewells as well as drought experienced during the period. The water table at 3.5-4.5 m should be maintained for effectiveness to irrigated agriculture. Hydrological changes occuring in the state must be taken into consideration. Research must be continued.

1258

UPPAL, H.L.; SINGH G., DHAND, A.D. Effect of release pipes in the bed of surface drains in heavy soils on withdrawal of ground water. J. Irrig. Pwr., 1964, 21:419-26.

The paper describes that surface-cum-seepage drains do not function well in heavy soils as the groundwater does not easily drain out from such soils on account of their high clay contents and high pH value. In order that these drains may function as seepage drains and suggests the introduction of release pipes in or near the bed of surface drains in clayey soils which would not only act as storm drains but also as seepage drains.

1259

UPPAL, H.L.; MANGAT, N.S. Waterlogging in Muktsar town and adjacent areas; its causes and remedies. J. Inst. Engrs. India, 1981, 62:193-200.

Factual presentation of the problem of waterlogging in the Mukatsar town in Faridkot district of south-west Punjab is presented where water table rise is around 0.61 m/year. The rise of water table in the tract is mainly due to the advance of depth to water table with a rate of 0.29 km/year from the north-west punjab towards the town owing to the charge of ground water by Satluj river. This is helped by the advance of depth to water table from north-east Punjab to south-west also. Some of the measures including lining of unlined pond in the town have been suggested to remedy the situation.

1260

UPPAL, H.L.; SEHGAL, S.R. Design and spacing of shallow open drains as an anti-waterlogging measure. J. Inst. Engrs. India, 1964, 44:777-9.

The paper describes that in area of high water table where sub-soil water has nearly come to the surface, as shallow open drain is an effective measure to lower it to limits considered safe and gives design of a shallow open type drain, where there is a source of seepage at some finite distance from the projected location of the drain.

1261

UPPAL, H.L.; SEHGAL, S.R. Drainage requirements of the Punjab state. ICID Bull., 1965, 14:23-31.

Flooding of an area can occur either because of spills from rivers or from runoff resulting from excessive rainfall. For the latter type of flodding, surface drainage has to be provided to dispose of expeditiously the large runoff of the tract. It is the disposal of this type of flooding termed as 'drainage requirement' that is the subject matter of discussion of this paper with particular reference to various canal irrigated tracts of the Punjab state.

1262

UPPAL, H.L.; SINGH, G.; DHAND, A.D. Investigation of a suitable subsurface drainage system for anti-waterlogging. Indian J. Pwr. River Valley. Dev., 1964, 14:19-24.

To evolve a suitable design of sub-surface drainage system, examined the functioning of different types of sub-surface drain; observed that flat brick tiles drains functioned best and the next best was sub-surface drain made out of porous cement concerte tiles; in view of the difficulties involved in keeping the flat brick tiles sub-surface drains (without base) well flushed and clean, suggest that porous cement concrete or ordinary cement-sand tiles may be adopted for land drainage and these may be shrouded with 7.5 to 10 cm thick gravel pack all-around.

1263

URVOY, C.; FAUCHER, A.; URVOY, C. Developments in sub-surface drainage. 10th Cong. Irrig. Drain. Athens, 1978, 4:27-32.

Technological developments in materials, machinery and control systems have had a major impact on agricultural drainage schemes.

1264

USHAKOV, B.A. Effect of drainage on adjacent dry valley stands. Lesn. Khoz., 1979, 3:22-4.

Contradictory Russian reports as to the effect of agricultural and forest drainage schemes on the conditions and growth of stands on adjacent dry areas are discussed. Some observers have reported a deleterious effect of drainage in these conditions, whereas others have found no effect or even a beneficial effect. In 1974, investigations were made in the Volyn region (NW Ukraine) in the vicinity of a forest drainage scheme installed in 1962; the drainage channels were 0.9-1 m deep and 200-250 m apart. The radial increment of Scots pine was determined before and after the drainage scheme, and in control areas well away from the drainage scheme. The results indicate that in these conditions at least, drainage had no significant beneficial effect on adjacent dry-valley stands.

1265

USSERY, L.R. Salinity control through tile drains in Hidalgo clay loam. J. Rio. Grande Valley Hort. Soc., 1960, 14:200-4.

A leaching test is described. The drainage system (15 cm concrete tile; 2 to 2.25 m below the surface; 60 m apart) removed in the first year 75 ton salts per ha from the top 2 m. Soil salinity was reduced in the top 1 m to a level permitting the growth of crops of low salt tolerance. A fluctuating highly saline water table, previously high enough to restrict crop root development, was lowered to permit leaching of salts from the soil profile by irrigation and high intensity rainfall.

1266

VAIGNEUR, H.O.; JOHNSON, H.P. Drainage design based on moisture balance. Trans. ASAE, 1966, 9:764-7.

Precipitation records have been used to predict water table fluctuations for the designated soil characteristics and predetermined drainage facilities. A tile depth of 1.2 m, a tile of 15.0 cm and a depth to the impervious layer below the drain of 1.2 m were used in the calculations. A transient-flow equation was used to calculate daily water tables for 16 treatments combination of soil conditions and tile spacings. It was found that tile spacing of 16.5, 33,64.5, 115.5 m are feasible for hydraulic conductivities of 30, 60, 120 and 240 cm/day respectively for the criteria of a 90 cm water table for a duration of atleast one day. An increase in this duration by one day allows the spacing increase by 3.6 m.

1267

VALK, G.G.M. VAN DER; SCHONEVELD, J.A. Effect of groundwater level on the production of some horiticultural crops on clay & loamy soils (1 & 2). Meded. Dir. Tuinb.,1963, 26:296-300.

Gladiolus, cauli-flower, early red beet, savoy cabbage and onions were grown on heavy clay with sticky-clay or light fine sandy-clay subsoils and on light fine sandy-clay profiles. Groundwater levels varied from 20 to 160 cm. On heavy clay topsoils cultivation was late and unaffected by groundwater levels of 30-160 cm. The fields on heavy clay topsoils were not affected by the different subsoils. Ground-water levels above 50-60 cm depressed yields with the exception of gladiolus, the yield of which was not unfavourably affected even with a level of 30 cm. Earlier cultivation was possible on light fine sandy clay where crop establishment was not affected by weather. On these soils production of all crops reacted to ground-water level.

1268

VAN AART, R. Waterlogging and salinity problems in the Indus Plain (Pakistan). Rep. Int. Inst. Land Reclam. Improv., 1973, 45-53.

As a result of seepage from irrigation canals, the ground-water table in the Indus Basin has risen and in some places is $<$ 1.5 m below the surface. In areas where the aquifer water is of good quality, tubewell drainage is preferable to horizontal drainage since in addition to the drainage function, it provides good irrigation water. In areas with saline groundwater, horizontal drainage produces less contamination and is preferred. Pipe drains are preferred to open drains even though flushing may be necessary every ten years.

1269

VAN BEERS, W.F.J. Some nomographs for the calculations of drain spacings. Bull. 8. Int. Inst. Land Reclam. Improve., Wageningen, Netherlands, 48 p.

The bulletin reports some of the nomographs of drain spacing equation for use by field practitioners.

1270

VAN DEN BERG, C. General considerations of drainage problems. F.A.O. Seminar on Waterlogging in relation to irrigation and salinity problems, Lahore, (1964), 1965, 133-9.

The paper deals with hydrological conditions in a river system and the effect of introducing irrigation in such areas; gives formulae for working out depth and spacing of surface drains together with a table calculated by Hooghoudt formula.

1271

VAN DER BEKEN, A. Research on the hydraulics of drain pipes. 8th Cong. Irrig. Drain. Varna, 1972, Q 27 R 23: 349-62.

Recent developments in the analysis of hydraulic resistence of pipes can be applied sucessfully to drain pipes. The effect of joints and perforations are evaluated rationally. The corrugated plastic pipes proved to be not fully hydraulically rough. When drain pipes are runing full, the piezometric height above the pipe will interact with the lateral inflow. A formula is presented which gives the relationship between the lateral inflow and the piezometric height.

1272

VAN DER BEKEN, A.; TROCH, F.D.; SOMER, M.D.; ZUIDEMA, F.C. The lateral inflow into submerged drains. Bull. Int. Assocn. Hydrol. Sci., 1972, 17:273-89.

An attempt is made to explain the field measurements of piezometric height and discharge rate in a submerged drain system. The lateral inflow into the drain pipe is not necessarily uniformly distributed as is usually assumed. A mathematical treatment is presented resulting in a general formula for calculating the actual distribution of the lateral inflow and a practical application is discussed.

1273

VAN DER MEER, M. Pre-drainage research in land consolidation areas. Proc. Int. Drain. Workshop, Wageningen, Netherlands, (1978) 1979, 25:136-49.

The requirements of pre-drainage research are presented and suggested that cut on expenditure on pre-drainage measurements can be made by investigations of soil maps and profile structures. Soil maps can be used as a starting point for any drainage investigation.

1274

VAN HOORN, J.W.; RISSEEUM, J.; JURADO PRIETO, J.M. Desalinisation of highly saline soil-prediction and field observation. In Managing saline water for irrigation. (H.D. Dregne, Ed.) Texas Tech. Univ. Lubbock, Texas, 1977, 558-74.

The clay soil had EC_2 values increasing from 12 dS/m at 0-25 cm to 34 dS/m at 150-200 cm depth. The hydraulic conductivity was low, 0.05 to 0.1 m/day. Below the depth of 1.5 m, the soil was impermeable. For two cases viz. drains located (1) well above an impermeable layer, and (2) on an impermeable layer, desalinisation has been calculated. For tile drains spaced at 10,20 and 40 m, for a period of 4 year, quantity of salt leached, amount of drainage water and average salinity of drainage water, have been worked out. In reality the desalinisation process occurs in a way that differs from the assumed for the calculation. This may be ascribed to differences in vertical and in horizontal flow through the soil profile, causing a decrease of salinity towards the drain, and to capillary rise, causing an increase of salinity in summer. However, for spacings of 10 and 20 m the calculation assuming a decrease of percolation water with depth, a reasonable approximation of the total amount of salt removed, can be obtained.

1275

VAN SCHAIK, J.C.; MILNE, R.A. Reclamation of a saline-sodic soil with shallow tile drainage. Can. J. Soil Sci., 1962, 42:43-8.

Leaching studies showed that soils which have become saline-sodic can be reclaimed with tile drains installed at a depth of 75 cm with a 9 m spacing. The glacial till at the experimental site was at a depth of 60-90 cm below the surface and was slowly permeable. Only after the application of 177.5 cm of water, the saline-sodic soil was reclaimed. There was no difference in salt movement with respect to proximity to the tile drains.

1276

VARUNTSIYAN, E.S. Some experience on rapid irreversible desalting of saline soils and ground water. ICID Bull., 1962, 11:13-7.

Jafarkhan tile drainage system installed in 1930 is described wherein tiles are laid at 3-4 m depth and 350-844 m spacing. In the course of 5 years since the drains began functioning, the soils were completely leached of the salts and it is estimated that some salt has been leached even from below the tile depth layers. The drain discharge during leaching was observed to be 2 L/sec per km of the drain length. It is also verified that drains installed at a depth of 3-3.5 m and spaced at 220 m can enhance the leaching process and complete the desalinization within one season.

1277

VARUNTSIYAN, E.S. Some experience of rapid irreversible desalting of soils and ground water. Annuc. Bull. ICID, 1962, 13-7.

The work conducted at Mugan Experimental Land Reclamation Station located in the eastern part of the Azerbaijan lowland formed in place of an ancient bay of the Caspian sea, has been described. The soils and ground water contain salts to the extent of 1 to 3%. Ground-water mineralisation of the upper layer averages 30 g/L and is often as high as 80-100 and 115 g/L. The drains made of tile and concrete, 0.75-0.8 m long, 200-250 mm in diam were placed in gravel. Drains laid 3-4 m deep, spaced 350-844 m apart, removed large amounts of water and salt. During the first 6 years of operation the amount of salt removed averaged 120 t/ha a year, while during other 30 years the amount removed annually was 1700 to 3500 t/ha. In the course of 5 years since the installment of drains, salt was also removed from the stratum occurring below the drain bed. The depth of complete desalting, of soils and ground water average 8 to 9 m below the ground surface, which continued to increase showing thereby achievement of irreversible desalting. The drainage system should include permanent drains and temporary deep drains laid in the water bearing stratum. Blind pipe should be used as permanent ones. The operations were economical.

1278

VENETS, C. Estimation of unconfined ground water flow to ditches.
J. Hydrol., 1971, 13:338-42.

The flow from unconfined aquifers to drainage

ditches is mathematically examined and formulae are derived for a step by step computation of the ground water flow corresponding to a certain sequence of infiltration with constant input within time units appropriately chosen.

1279

VIEIRA, D.B.; PISSARRA, A.E.; POMBO, J.G.; PEREIRA, D.S. Drainage and desalinization studies and design in Lezina Grande, Portugal. In Polders of the World. Vol. 1. Polder projects-Land and water management aspects. Int. Inst. Land Reclam. Improve., Wageningen, 1982, 384-95.

Studies were carried out to improve drainage and desalinization conditions in the mostly heavy saline and alkaline soils of Leziria Grande, a flat and alluvial area, protected against floods and tides, in the Tejo Valley near Lisbon. This type of study was recommended as it enabled drainage criteria to be established from which conclusions could be reached as to drainage design.

1280

VIETS, F.G. JY.; MUCKEL, D.C. Reclamation of tidal soils. Trans. 4th Congr. Irrig. Drain., 1960, Q 11: 1-16.

Levees are usually required to prevent inundation during high tides. Main and lateral drain ditches are required for the removal of excess surface and subsoil water, and outlets are required for the disposal of excess water. The chemical and fertility problems involved in reclamation after drainage and water control which have been established are: removal of salts, replacement of exchangeable sodium, correction of acidity, and adequate supplies of plant nutrients. Removal of salts by leaching and the displacement of exchangeable sodium by leaching using gypsum when needed, do not pose any unusual problems. But careful evaluation of the problem of excess sulfides which may lead to extreme acidity when the soil becomes aerated, must be made.

1281

VIMOKE, B.S.; TAYLOR, G.S. Apprroximate equations for fluid flow into buried drains. <u>Soil Sci. Soc. Am. Proc.</u>, 1960, 24:418-9.

Approximate equations for four cases viz. (1) multiple drains equally spaced over an impervious layer, (2) a single drain over an impermeable layer, (3) multiple spaced drains in a homogeneous soils and (4) a single drain in a homogeneous soil are presented. The results are comparable with equations available in the literature and calculation time is considerably reduced specially with the equation of first case.

1282

VIMOKE, B.S.; TYRA, T.D.; THIEL, T.J.; TAYLOR, G.S. Improvements in construction and use of resistance networks for studying drainage problem. <u>Soil Sci. Soc. Am. Proc.</u>, 1962, 26:203-7.

A resistance network analog is described which incorporates some improved features in flexibility, accuracy and ease of operation. The network consists of 575 variable resistance units, plug in type and the mesh size is smaller on 1/4th of the board. Improved equations are presented for calculating network resistances adjacent to the drain. These equations utilize a logarithmic expression rather than a linear one. Comparisons are made of drain flow rates in homogenous medium as evaluated by network and by analytical solutions, the network data generally deviates $<$ 2% if the logarithmic expression is used to calculate drain resistors whereas it is as high as 30% if a linear relationship is used.

1283

VISSER, W.C. Formation of nitrogen in the soil and

depth of field drainage. Stikstof, 1964, 43:231-37.

The growth equation $(1-{}^{q}Q)$ $(1-{}^{q}{}_{rx})$ $(1-{}^{q}{}_{sy})$ =D in which Q=maximum yield, q=yield attained and r, s and D are plant factors as modified by incorporating the relations between ground-water depth, aeration and response of plant and micro-organisms. The usual deep drainage in the heavy clay area is apparently done mainly to increase available N by stimulating N fixation. Without N application, the plant responds only to the release of fixed N, and micro-organisms respond favourably to a drainage of 2 m or more. With sufficient added N, the yield response depends on the sensitiveness of the plant to aeration, which in this case is already satisfied at a drainage depth of 50 cm.

1284

VISSER, W.C. Rules of transfer of water management experience, with special reference to the assessment of drainage design constants. <u>Tech. Bull.</u> Inst. Land Wat. Mgmt. Res., 1968, 59:90-149.

A solution of water-management problems is presented which is usable when only simple research means are available. Curve-fitting techniques, adjustment of constants and methods of graphical solution are dealt with. The elaboration of alternative solutions was done by varying the parameter values of those properties which were subject to change by technical methods. The dependence of water management studies on formulae for plant response, discharge, evaporation, soil moisture profile, desorption curve, and rainfall frequency was demonstrated.

1285

VITHAYATHIL, F.L.; BOAST, C.W.; HOENER, D.M.; COMMONER, B. A systems approach to evaluate nutrient and pesticide transport from artificially drained cropland. <u>Environ.Mgmt.</u>, 1979, 3:123-32.

An inventory of artificially drained croplands in the USA is presented. The results of studies to determine losses of nutrients, sediment and pesticides in drainage water are summarised. Losses are extremely variable. Average losses of nitrate through subsurface drainage and surface runoff are 17.6 and 11.6 kg/ha/year, respectively. Total phosphorous is lost mainly through surface runoff (average 1.3 kg/ha/year). Sediment yields average 505 kg/ha/year. Losses of the herbicides atrazine and dicamba average 5% of aamounts applied, while losses of the insecticides aldrin and heptachlor are less than 0.01% of amounts applied. An interdisciplinary systems approach, utilizing a computer simulation model for predicting nutrient and pesticides load in sediments and water from artificially drained cropland, is described.

1286

VLAHINIC, M.; RESULOVIC, H. The effect of draining pseudogley soil on wheat yield. Trans. 10th Congr. Irrig. Drain., Mos., 1:310-7.

The investigations of the influence of the system and intensity of drainage on pseudogley on the wheat yields were carried out. At the one experimental site wheat yields were maximum with crosswise drainage (1969-70) and mole drainage (1970-71) whereas at the other, wheat yields were maximum with tile drainage (10m spacing) in both the years. It is, however, reported that considering other factors crosswise and mole drainage will be most expedient in those conditions. The yield responses of 15 to 210%, were obtained.

1287

VLUGTER, IR.H. Irrigation-cum-drainage. Trans. 6th Congr. Irrig. Drain., New Delhi, 1966, Q 19 R 9: 105-11.

Proposes a method of drainage combining irrigation and drainage channels for flat irrigated areas typical of West Pakistan, describes that the principal advantages of the combined system are that with the frequent interception of the drainage water under gravity flow, pumping plants are not required and also no canal lining is necessary; considers that this method is economical on account of the dual purposes of the channels, results in lesser earthwork and proper functioning of the drains.

1288

VOMPERSKAYA, M.I. The effect of a shallow drainage network on the nutrient regime and growth of spruce plantations on boggy soils. Lesoved., 1980, 5:52-61.

An account is given of investigations in 1963-76 in the Yaroslavl region (N. Central Russia) on peatlands carrying low-grade birch. Drainage ditches 25-30 cm or 40-50 cm deep were made at various spacings, and Norway spruce seadlings were planted. Data are presented on the foliar N contents and growth of the spruce. Drainage improved the N nutrition of the spruce, but did not always create optimum conditions. In wet years, N uptake was clearly better on the drained plots but in the dry year the foliar N content was virtually the same as on the control plots. The deeper drainage ditches were superior to the shallow drainage furrows in regulating the water regime. Liming the soil did not improve the N nutrition and growth of the spruce.

1289

VOTRUBA, J. A combination of measures for improving the water and air content of heavy clay soils of Czechoslovakia. Arch. Acker. Pflan Bodenk., 1979, 23: 337-43.

Combined drainage (drainage by pipes together with mole drains or deep loosening of the soil) is used as well as traditional pipe drainage in some areas. Regulatory drainage (which removes or supplies water to the water-table in order to control the height of the latter) is under investigation.

Improvement of soil structure by slag from steelworks, fly ash, material mobilizing water thermal stablization, and response to a magnetic field is also being investigated.

1290

VRIES, C.A. DE; VAN BEEK, B.C.P.H. Drainage of Agricultural Land- A biblography. Int. Inst. Land Recl. Improv., Wageningen Bib. 5, 1966, 28 p.

This bibliography contains selected references to the literature on the drainage of agricultural land with emphasis on subsurface drainage, mostly published between 1950 and 1966.

1291

VVEDENSKAYA, A.E.; RUSINOV, I.F. The hydrological regime and water balance of drained lands of the forest zone of West Siberia. IAHS Pub., 1975, 105:313-20.

The paper gives the results of three years of study of drainage flow on the first experimental site in West Siberia where plastic drains were used in thick peatlands. Observations todate of runoff, groundwater, evaporation and soil moisture enable both quantitative and qualitative statements to be made about the drainage flow. Detailed studies of the dynamics of spring drainage have shown that it is similar to the surface flow from experimental plots. The data obtained have confirmed the feasibility of drainage under the severe conditions found in Siberia and may serve as preliminary recommendations for the design of plastic drains in the forest zone of West Siberia.

1292

WALKER, P.N. Pipe stiffness of permanently deflected corrugated plastic drain tubing. Trans. ASAE, 1979, 22:1313-6.

Thirty cm lengths of corrugated polyethylene drain

tubing were permanently deflected from 0 to 60% in the laboratory using a parallel plate device and heat treatment. The heat treatment itself did not affect pipe stiffness of the tubing. The 250 test samples were loaded with a parallel plate device to determine the force, F, required to deflect the tubing additional amounts, ΔY. The ratio of $F/\Delta Y$, defined as pipe stiffness, was found to increase with an increase in permanent deflection, starting at approximately 10% permanent deflection for 102, 127 and 152 mm perforated helical-corrugated and 102 mm non-perforated ring-corrugated tubing. The pipe stiffness of 102 mm perforated ring-corrugated tubing decreased with permanent deflection over the entire range tested.

1293

WALKER, P.N.; SCARBOROUGH, J.N.; WAUTHIER, D.S. Deflection of buried predeflected plastic drain tubing due to wheel loads. <u>Trans. ASAE</u>, 1982, 25:322-6,332.

Tubing samples were permanently deflected into ellipses and installed at depths of 0.38, 0.54 and 0.71 m. After a combin repeated passes, deflections were noted. It was observed that tubing which has even a small initial deflection (5%) is less able to withstand additional deflection than tubing which is circular after installation. The results are intercepted to method of drain tile installation. The recommendation that the minimum depth of cover for 150 mm diam tubing should be 0.6 m is confirmed by this study.

1294

WALKER, P.N.; THRONE, M.D.; BENHAM, E.C.; SIPP, S.K. Yield response of corn and soybeans to irrigation and drainage on a claypan soil. <u>Trans. ASAE</u>, 1982, 25:1617-21.

Five yr of corn and 2 yr of soybean yield data from forty field plots with combinations of irrigation and drainage treatment are presented. The irrigation treatments were sprinkler, furrow, and no irrigation. The drainage treatments were surface, subsurface, surface plus subsurface and no drainage. The plots were located on a claypan soil in south central Illinois. The plots with surface drainage had a slope of 0.5 percent; the others were graded level. Subsurface drainage was provided with plastic tubing on 6 m centers. The data indicate average corn yield increases of 0.8 and 2.4 t/ha due to drainage and irrigation, respectively. Together, irrigation and drainage act synergistically to produce an average yield incrse of 4.8 t/ha.

1295

WALKER, P.N.; WELLS, J.D. Pipe drainage of heavy soils: a field evaluation. Trans. ASAE, 1983, 26:112-15, 122.

Five heavy Illinois soils were studied at sites with subsurface drainage installed previous to the study. Plots were irrigated to bring the water level to the surface and drainage performance till water table reached 300 mm was studied. The drain spacing necessary to drop the water table midway between drain lines to a depth of 200 mm within one day were: Bryce, 24.2 m, Bonnie, 13.9 m; Wabash, 12.9 m; Darvine, 10.8 m; and Piopolis, 13.1 m.

1296

WALTER, M.F.; BLACK, R.D.; ZWERMAN, P.J. Tile flow response in a layered soil. Trans. ASAE, 1979, 22:577-81.

The dranage data were collected and analysed for a 16 yr period on 12 plots of Luscustrain silt loam, with a shallow restricting layer. Av. annual tile flow corresponds to 7.7 cm, or 7.5% of the av. annual precipitation; extremes of 28 cm and 1.8 cm were recorded. The av. annual precipitation was found to be an unreliable indicator of tileflow. The correlation coefficient for the 16 yr period being 0.60 and when the data from one extremely wet year were omitted was 0.20. The highest av. monthly tile flow occurred during March and April, and were typically 10 times as large as those during the driest months of July to September. 84% of the average annual tile flow took place between November and April, and only 16% during the growing season (May to October). The correlation coeff. for av. monthly tile flow versus precipitation was only 0.09. Correlation between monthly potential evaporation and tile flow revealed a hysteresis effect, reflecting seasonal evaporation and corresponding soil moisture changes.

1297

WANG, F.C.; OVERMAN, A.R. Impact of surface drainage on ground water hydraulics. Water Resourc. Bull., 1981, 17:971-7.

The paper presents an analytical technique and a numerical model in quantifying the difference of surface and subsurface runoff before and after the construction of drainage canals and for delineating the effects of drains on channel level and regional water table in adjacent areas in south Florida. The surface runoff model is formulated on the climatic water balance technique and the ground water model is treated as a one dimensional transient phenomenon that forms a nonlinear flow problem. Analytical solutions are derived through problem linearization. These two models are coupled to estimate the impact of drainage canals on adjacent water table drawdown.

1298

WAITE, P.J. Soil salinity, water table modelling. Rep. Hydraulic Research Station, Wallingford, Oxon, 1981, No. OD 36: 10 + ix p.

The model described is an approximation to the movement of the water table in a two layer soil, drained by open ditches within the upper layer. The two dimensional model is developed using the Dupuit-Forchieimer approximation and a form of equivalent transmissivity for the two layer soil. The equation is linearised and solved using the finite element method for the spatial derivatives and a Crank-Nicolson approximation of the time derivative. Solutions for different depths and spacings of the field drains are used to check earlier calculations on drain spacing for an experimental site at Zewia, in the northern Nile Delta, Egypt.

1299

WALKER, W.R.; SKOGERBOE, G.V.; EVANS, R.G. Optimization of salinity control in Grand Valley. J. Irrig. Drain. Div.ASCE, 1979, 115: 15-28.

A mathematical simulation of the coat effectiveness relationship for various agricultural and desalinization structural alternatives for reducing salt pickup by irrigation return flows was developed and applied in the Grand Valley of Western Colorado. Cost effectiveness functions developed from the model were then optimized using a least cost criterion to determine the effective strategies for salinity control in the valley. The analysis indicates that lining small conveyance channels to reduce seepage and increase the efficiency of the existing furrow irrigation systems should be given the highest priority. Canal lining and desalination of return flows are costly alternatives. Some structural alternatives such as field and interceptor drainage exhibited more than an order of magnitude higher costs and should not be considered.

1300

WANG,H.C.; TU,K.H.; CHOU, C.T.; KU, K.A.; CHAI, M.T.; TAI, C.T. Effectiveness of mole drainage in heavy clay soil. Acta Pedol.Sin., 1978, 15: 194-200.

On a lowland farm, Huochiou County, Anhwei Province on heavy clay soils, crops (mainly wheat, barley) were seriously injured by waterlogging during the rainy season. This condition has been greatly improved by widespread application of mole drainage. The drains were laid at intervals of 10 m, at 60-70 cm depth, and it was found that, when combined with surface drainage by shallow ditches laid at 50-100m intervals, the system could cope with concentrated rainfall of moderate intensity. The beneficial effect of underground drainage on crop growth, the operation of farm machines and crop yield was also shown in experiments on soybean and wheat crops. However, the increase in crop yield depends on the severity of deluge rainfall effects during the growing season.

1301

WANG, H.C.; ZHOU, Z.D.; DU, G.H.; DAI, C.D. Study on the durability of mole channels. Acta Pedol. Sin., 1980, 17:187-92.

Data on the effect of soil texture on the stability and duration of functioning of mole channels are given and photographs are presented showing various states of collapse of the wall structure of mole channels in some clayey soils. The effectiveness of different shapes of mole channel is also considered.

1302

WANG, P.L.; YANG, S.J.; CHANG, Y.T. Effect of subsurface drainage on sugarcane growth and yield in fine-textured soils. Rep. Taiwan Sugar Research Institute, 1980, 88:19-34.

In trials during 3 crops years on fine-textured low humic gley soils, subsurface drainage was laid at 0.8 or 1.2m below the soil surface with drains 15,20 or 25m apart. A surface -drained plot was used as the control. After heavy rain water table level on the drained plots decreased much faster than on the control plot. The sprouting, development and nutrient uptake of ratoon cane were appreciably better on the drained plots. Average yield increases on the drained plots of plant cane, first ratoon and second ratoon were 17,40 amd 37per cent, respectively, over the control. Differences between the 2 depths and 3 spacings were small.

1303

WARRICK, A.W. Determination of equivalent raddi for rectangular drains. Soil Sci. Soc. Am. Proc., 1973, 37:809-11.

A relationship is developed giving the circular drain size which is equivalent to a rectangular drain. A plot of r_e versus °a.b' is given as a function of a/b with r_e as the circular drain radius and a and b are the semi-width and semi-height of the rectangular drain. The minimum value of re/ab is 1.18 and corresponds to a square.

1304

WARRICK, A.W.; KIRKHAM, D. Determination of equivalent raddi for half-tube and whole -tube drains in contact with an impermeable barrier. Soil Sci.Soc. Am. Proc., 1968, 32:449-51.

For soil drainage problems, it is shown that a halftube of radius, a, the diameter of which is coincident with a plane barrier, is equivalent to a whole-tube drain of a smaller radius, b= 2a/n = 0.6366 a, installed with its circumference touching the same plane barrier. The value 0.6366 °a' lies between the limits 0.5 a and 1.0 a found by Kirkham. The result is aplicable by analogy to all types of potential problems, such as those of electricity and heat conduction.

1305

WARRICK, A.W.; KIRRHAM, D. Two-dimensional seepage of ponded water to full ditch drains. Wat. Resourc. Res., 1969, 5:685-93.

A mathematical treatment of situations in which an impermeable barrier may or may not be present and in which the ditches are a finite or an infinite distance apart.

1306

WATERS, P. The effect of drainage on soil temperature at Drayton EHF. Tech.Bull., Ministry of Agriculture, Fisheries and Food, U.K. 1977, 77/6, 16p.

The influence of drainage status on the soil temperature regime in a heavy Lias clay of the Denchworth series was investigated in drained and undrained plots at the Drayton EHF. At any given depth, a graph of soil temperature readings versus time revealed a diurnal waveform which could be described as a sine wave. Results illustrated that the wave properties of mean temperature, amplitude and phase were influenced by factors which included season, weather, drainage status and depth of temperature measurements in the soil. Analysis of a drying period, 26 March-9 April, 1974 showed that above 8 cm depth the drained plot was warmer than the undrained plot for longer periods as the drying period progressed; but that below 8 cm the mean temperature, amplitude and phase values were practically identical for both plots. It was also shown that a potential 3 day advantage in seedling emergence of spring wheat drilled at 4 cm depth was available on the drained plot, and that enhanced bacterial activity would occur with increased soil temperature.

1307

WATSON, K.K. Some applications of unsaturated flow theory. In Drainage for Agriculture (Agronomy No.17, Schilfgaarde, J. Van, Ed.),Am. Soc. Agron. Wisconsin, 1974, 359-400.

The unsaturated flow theory is applied to drainage situations using numerical methods. Hysteresis behaviour and one-and-two dimensional systems are considered. A method for determining water table hydrographs for tile drains systems is described and the current status of field investigations in unsaturated soils discussed.

1308

WEERD, B. VAN DER. A registration unit for drain outflow, groundwater depth and precipitation. J. Hydrol., 1977, 34:383-88.

A description is given of an operational field registration unit for open and submerged drain outflow, groundwater depth and rainfall. The drain outflow meter has a measuring range from 20 to 400cm^3/s, with a mean error in recorded sediment-free outflow of 15 per cent at flow rates below 60 cm^3/s, and 1.5 per cent at higher flow rates. The ground-water level meter has a measuring range of 1.25 m but it can be adapted to deeper levels. Its accuracy is 0.5 cm when a float of 4.5 cm in a 5 cm diam piezometer tube is used. Larger floats give higher accuracies. The rain gauge is a commercial pluviograph changed from mechanical to electrical registration. Recorder, power suply and scanning system are built into a dust and waterproof box. The maximum distance between meters and recorder is about 150 m.

1309

WENDTE, L.W.; DRABLOS, C.J.W.; LEMBKE, W.D. The timeliness benefits of subsurface drainage. Trans. ASAE, 1978, 21:484-8.

A simulation model was developed to predict the available spring work days and earliest possible maize sowing dates for poorly drained fields on Drummer silty clay loam or Elliott silt loam with a subsurface drainage system. According to the model, weather rather than drainage limited the number of work days in March. In 80 per cent of yr, 2 days would be gained in April, 13 in May and 19 in June by applying the recommended drainage system. A closer drain spacing than that recommended (24-30 m at 91 cm deep for Drummer soil and 18-27 m for Elliott soil) would not generally allow significantly earlier sowing. The optimum drain spacing based on the timeliness benefit to earlier maize sowing was 30-61 m for drummer soil and 24-37 m for Elliott soil. The average annual benefit was $ 37-156/ha.

1310

WERTZ, G. The determination of drain spacing with particular reference to the area near the drains. Arch. Acker. Pflan. Bodenk., 1979, 23:371-5.

The determination of drain spacing by means of criteria varied for the zone near the drains is based on a mathematical treatment of laboratory and field data. Calculation of the contact resistance (or discharge resistance), which occurs on passing from lower to higher permeabilities in the zone near the drain pipes permits solutions for different drainage methods and special conditions (e.g. the presence of envelope material) in the zone near the drain pipes.

1311

WERTZ, G.; HILLIGER, D.Damming-up and infiltration for water regulation on deep groundwater sites. Arch. Acker. Pflan Bodenk., 1980, 24:209-14.

A combination of techniques and measures were used to improve the permeability of the drainage regulation zone. Rothe and Hooghoudt's formulae were used to determine drain spacing. Slope was a direct factor considered. Drainage discharge was controlled by adjusting the position of the drains in relation to slope.

1312

WESSELING, J. A comparison of the steady state drain spacing formulae of Hooghoudt and Kirkham in connection with design practice. J. Hydrol., 1964, 2:25-32.

It has been proved that the steady drainage equations proposed by Hooghoudt and Kirkham give results which differ less than 5 per cent in cases where the flow in the layer above the drains is neglected. Where a higher water table midway between drain occurs and particularly in cases where the hydraulic conductivity of the upper layer is large in comparison with that of deeper layers, solution of Hooghoudt must be preferred.

1313

WESSELING, J. The effect of using continuously submerged drains on drain spacing. J.Hydrol., 1964, 2:33-43.

The mathematical treatment of Kirkham has been extended for submerged drains; some calculation examples show that under certain favourable conditions a considerable increase of drain spacing can be obtained by using deep, submerged drains.

1314

WESSELING, J. The entrance resistance of drains as a factor in design. Proc. Int. Drain. Workshop, Wageningen, Netherlands, (1978), 1979, 25:354-65.

In design of drainage systems, the entrance resistance of the pipe system is generally neglected, although in material research quite some emphasis is laid on this aspect. Small diameters of pipes and high entrance resistance lead to submerging of drainage systems. With the aid of solutions for the flow problem in drainage it is explained what effect submerging may have on the drainage system applying Ernst's theory. It is proved that by submerging, the radial resistance of the system decreases. Arguing that a decrease in radial resistance can be used to allow a certain entrance resistance some simple formulae have been developed to take into account the entrance resistance. The results shown for entrance resistance of pipes with filters are generally such that they have no effect on the drain spacing. This does not hold for very thin filters, especially when smooth drainage pipes are used.

1315

WESSELING, J. Hydrology, soil properties, crop growth and land drainage. "Soil-water Plant" <u>Proc. Inf. Commun. Hydrol. Res.,</u> 1969, 15:9-27.

Subjects discussed compriseul: water in soil, drainage equations , and the effects of watertable depth and of some other factors (flooding, soil temperature) on crop yields.

1316

WESSELING, J. (Ed.) Proceedings of the International Drainage Workshop. Int. Inst. Land Reclam. Improve., Wageningen, Netherlands, (1978), 1979, 25, 731 p.

The proceedings is divided into two groups. Section one contains the synthesis of the discussions and contribution in four fields of drainage. The four areas covered are: (1) design and research, (2) drainage materials, (39 installation methods, and (4) drainage of irrigated lands. Section two contains contributory papers on these subjects. In total 53 papers are included covering the four areas.

1317

WESSELING, J. Report to the Government of Israel on field drainage research. F.A.O. Rome(1963), 1964. Rep. No.1752, 1-25.

The drainage problems of Israel are discussed in relation to the research necessary for the adaptation of the system of parallel drains and interception drainage to local condition; some preliminary results of an experimental field on a heavy clay soil, and urgent further research to be done.

1318

WESSELING, J. Storage factor as a drainage criterion. <u>Meded. Inst. Cult. Tech. Wat-Huish</u>, 1969, 118, 8p.

Relationships between discharge, drain depth, storage coefficient and frequency of occurrence of a specified water-table height are given in mathematical and diagramatic form.

1319

WESSELING, J.; HOMMA, F. Entrance resistance of plastic drain tubes. <u>Neth. J. Agric. Sci.</u>, 1967, 15:170-82.

The entrance resistance of plastic drain pipes with various amounts of longitudinal slits and covered with certain filter materials has been measured in a vertical sand tank model. It became apparent that the influence of an increasing amount of perforations on the lowering of the entrance resistance lessens at high perforation amounts. In general, a decrease of the entrance resistance can be better obtained by means of larger pipe diameter than by increasing the amount of perforations. For practical applications, a 4 cm pipe with 1 sheet of glass fibre requires about 450 mm^2 of perforation per running meter. Increasing the thickness of the filter (assuming it will retain its high permeability under field conditions) will be more effective in decreasing the entrance resistance than increasing the pipe diameter. Some examples from practice show that, with the use of 4 cm pipes having 2 rows or 450 mm^2 of longitudinal slit perforations and covered with one sheet of glass fibre, the extra loss of head due to the entrance resistance is in most cases negligible. Only when the hydraulic conductivity of the backfill of the trench is low (about 10 cm/day) this will not be the case. In such circumstances it is advisable to use a thicker filter or a larger pipe diameter.

1320

WHEELER, W.B.; MANSELL, R.S.; COLVERT, D.V.; STEWART, E.H. Movement of 2,4 D in drainage waters from a citrus grove in a Florida Flatwood soil. <u>Proc. Soil Crop Sci. Soc. Fl.</u>, 1978, 37:180-3.

The concentration and total discharge of 2,4D were measured in subsurface drain water for a 13-day period, after application at a rate of 3.7 kg/ha to a field site at Ft. Pierce. Outflow water volumes were highest in surface tillage (ST) plots with deep tillage plus heavy lime plots (DTL) being lower and deep tillage plots(DT) being the lowest. Concentrations of 2,4D were low, generally in the range of 20-25 ppb. (Parts/1000 million), the exception was in water from an open-outlet drain in the DTL plot which had a peak concentration of 75 ppb. The maximum hourly discharge ranged from approximately zero in the DT open-outlet drain water to 500 mg/ha in the DTL open-outlet drain. The soil tillage treatments appeared to influence 2,4D levels in drainage water. The DTL plot with a relatively alkaline soil had less than half the maximum water flow rate as compared to the ST plots, but produced higher total 2,4-D outlflow. The drain-outlet treacments (open and submerged outlets) also appeared to influence total pesticides outflow; the open outlet drains for ST and DTL discharged more than twice as much 2,4D, as did the submerged outlet drains. Considering that the total quantity of herbicide discharged ranged from 0 to 0.5 per cent of that applied, 2,4D does not appear to represent a significant environmental hazard when used in this manner.

1321

WIDMOSER, P. Study of flow in the immediate vicinity of drains. <u>8th Congr. Irrig. Drain. Varna</u>,1972, Q 28.1 R 2: 11-31.

The paper gives an easily understandable introduction to the aspects of approach flow (filters, compactions, influence of pipe opening etc.) of water in the vicinity of drains. The entrance resistance of various types of pipes are compared.

1322

WIEN, C.; LAL, R.; PULVER, E.L. Effects of transient flooding on growth and yield of some tropical. <u>In</u> Soil Physical Properties and Crop Production in the Tropics. (Lal, R.; Green-land, D.J. Eds.).John Wiley and Sons, New York, 1979, 235-45.

Effect of flooding at two growth stages of cowpea, soybean and maize for two soil groups viz. loamy sand and clay loam are presented. In each case yields decreased; two floodings giving a substantial decrease compared to one flooding.

1323

WIJK, A.L.M. VAN; FEDDES, R.A. A model approach to the evaluation of drainage effects. <u>Tech. Bull</u>. ICW (New Series), 1982, 15, 19 p.

A model is described for computing the effect of drainage on different soil types over a number of years.

1324

WIKLANDER, L. Leaching of plant nutrients. I. The contents in drainage water. <u>Grundforbattring</u>, 1970, 23:117-41.

EC, pH, Ca, Mg, K, Na, Mn, Fe, Cl, N, S and P were determined in 358 samples of drainage water from cultivated Swedish soils. The data were distributed so that about 60-70 per cent of the values were less than the means. Cl and Na were positively and closely correlated, and so were Mg and Na. The effects of calcareous soil and underlying calcareous rock and of S-rich soils on the contents of Ca and S in the water were obvious. The effect of sea salts on the Na and Cl content of the drainage water in western Sweden was also obvious. Normal rates of fertilizers will not cause eutrophication of lakes and rivers. Leaching of K in soils of not too low adsorptive capacity is usually less than or about the same as, atmospheric deposition of K. Loss of $H_2PO_4^-$ and HPO_4^{2-} is negligible. Leaching of NO_3^- may cause dangerous pollution if rates of NO^{-3} application are high.

1325

WIKLANDER, L.; HALLGREN, G. Leaching of plant nutrients. II. In cultivated soil at Marsta and Gammalstorp. Grundforbattring, 1971, 24:69-75.

The nutrients in drainage waters from 2 field sites were estimated. The respective losses of Ca, Mg, Na, Cu, Cl, S and P from a clay overlying calcareous clay with drain depth of 110 cm were: 89, 13.2, 1.9, 7.6, 0.017, 5.5, 3.0 and 0.038 kg/ha/yr; and from a silty loam overlying non-calcareous clay with a drain depth of 75 cm were: 45.0, 14.2, 14.9, 0.014, 8.5 26.3 and 0.044 kg/ha/yr.

1326

WILCOX, L.V.; RESH, W.F. Salt balance and leaching requirement in irrigated lands. USDA ARS Tech. Bull. 1963, 1290, 23 p.

Salt balance and leaching requirement were studied in the Rio Grande project in New Mexico and Texas. The finding supports the conclusions that (1) salt balance is a reliable and useful indicator of year-to-year trends in salinity conditions on irrigation projects. Salt balance data can be interpreted in terms of the average increment of salt added to the soil extracts and (2) leaching requirement is a reliable criterion for anticipating the leaching that would be necessary in order to prevent the salinity in the soil from exceeding a specified level.

1327

WILLARDSON, L.S. Drainage coefficients. Trans. ASAE, 1982, 25:1251-3.

Drainage coefficients used in humid and arid regions are developed by different methods. The procedure by which the data are handled and the design equations used, steady state or transient, can affect calculated drain spacing. A tensor product Lagrange interpolant is presented that can be used to adjust design water table heights to get equivalent drain spacing using either design method.

1328

WILLARDSON, L.S. Plow installation of plasic drain tubing. Trans. ASAE, 1974, 17:172-3.

A discussion on ploughs modified or designed for installing corrugated plastic drains.

1329

WILLARDSON, L.S. Practical installation of clay, concrete and corrugated plastic subsurface drains. 8th Congr. Irrig. Drain. Varna, 1972, Q 28 R 8:129-37.

Installation of subsurface drains in Imperial Valley of California has been discussed. Drains are installed at depths of 1.8 m or more with coarse sand envelope or filler at least 8 cm thick around the drain. The cost of installation of 10 cm diam pipe with envelope is less than $5.28 per m for all currently used materials. A plow developed for installing corrugated plastic drains is discussed which can install at speed six times faster that of a trencher. Competition between contractors is believed to be one of the reasons of reduced cost of drainage.

1330

WILLARDSON, L.S. Synthetic drain envelope materials. Proc. Int. Drain. Workshop, Wageningen, Netherlands, (1978), 1979, 25:297-305.

Synthetic drain envelopes are suitable for use in protecting subsurface drains provided the hydraulic failure gradient of the soil is not exceeded in the installation. Tests for drain envelopes based on tensile strength, abrasion and chemical resistance, permeability, opening size of the fabric soil, hydraulic failure and subsurface drainage performance are discussed.

1331

WILLARDSON, L.S.; ARELLANO, M.T. and BECKER, E.R. Comparative laboratory tests of new drainage conduits. Trans. ASAE, 1966, 9:352-4, 358.

Drainage conduits were tested for characteristic flow rate and for load carrying capacity. The materials tested include 10 cm diameter concrete draintile 30 cm long with butt joints, 10 cm diam perforated bituminized fibre pipe;7.5 cm diam. PVC mole liner with 0.375 mm wall thickness; 10 cm diam semirigid polyethylene tubing with 1.95 mm wall thickness; 7.5 cm diam steel reinforced fibre glass wrapped plastic; semi-circular galvanized steel with a corrugated roof in 18 gauge and 28 gauge; enamelled steel tubing with .125 mm and .20 mm wall thickness. The results indicate the necessity of further research on both the hydraulic and physical strength properties. The position and number of openings in conduits need to be studied, and further tests are needed to establish the bedding conditions required to fully develop the strength of flexible conduits. There is also a need for development of standard testing procedures for drainage materials so that they can be evaluated for field conditions.

1332

WILLARDSON, L.S.; BOLES, A.J.; BOUWER, H. Interceptor drain recovery of canal seepage. Trans. ASAE, 1971, 14:738-41.

Results of a analog computer model have been presented for design recommendations for canal seepage recovery systems. The results indicate that to reduce the system losses to a minimum, the drain should be located away from the canal and as deep as possible.

1333

WILLARDSON, L.S.; DAVIS, D.; MULDER, D.; SPENCER. J.R. Drain envelope response to field treatments. Trans. ASAE, 1975, 18:134-6.

In field experiments with 12.5 cm internal diameter, concrete pipes laid in a fine sand as little as 5 per cent soil in the drain envelope seriously impaired drain envelope performance. Flooding the trenches to consolidate the backfill material did not significantly affect envelope performance.High pressure jet cleaning of the drains improved envelope performance.

1334

WILLARDSON, L.S.; FOUSS, J.L.; REEVE, R.C.; FAUSEY, N.R. Entry velocity control limits drain sedimentation. J. Irrig. Drain. Div. ASCE, 1968, 94:455-63.

The paper shows that control of the velocity of water entering a subsurface drain will effectively prevent sedimentation; describes in detail the anti-sedimentation concept based on the Stokes Law settling velocity of particles; impresses that the essential element for implementation of this concept is an entrance flow path or channel to the drain in which both the magnitude and direction of the inflow velocity of water are controlled; includes laboratory tests for different sized particles, shapes of opening, and cross-sectional areas of flow; suggest tentative design criteria and recognises that the prevention of sediment entry into drains by this method is not by a filtering process.

1335

WILLARDSON, L.S.; MEEK, B.D.; HUBER, M.J. A flow path ground water sampler. Soil. Soc. Sci. Am. Proc., 1972, 36:965-6.

A sampling system is described that enables collection of water samples from different flow paths in the ground-water system around a tile line. The sampler can be installed as regular part of a tile drainage system.

1336

WILLARDSON, L.S; MEEK, B.D.; GRASS, L.B.; DICKEY, G.L.; BAILEY, J.W. Nitrate reduction with submerged drains. Trans. ASAE, 1972, 15:84-5, 90.

To study the performance of submerged drains as a means of dissipating nitrates, subsurface drainage system was installed in a Panoche silty clay, deep fine textured and slowly permeable soil. Laboratory and field studies have shown that denitrification occurs in saturated soil where there is ample organic carbon for bacterial metabolism and a shortage of oxygen. Denitrification and dilution of high nitrate ground water were accomplished in the field.

1337

WILLARDSON, L.S.; SHELDON, J.M.; DONNAN, W.W. Drainage protection of the Imperial Valley resource. ICID Tech. Mem., 1972, 1:1-14.

Irrigation in Imperial Valley of California which began in 1901 resulted in deterioration of 40,469 ha by 1919. Open drainage system installed in 1921 did not solve the problem. Tile drainage system completed in 40,469 ha by 1949 resulted in a favourable salt balance in the area. New drainage problems such as clogging by chemical deposition and ground water pollution being investigated at present are discussed.

1338

WILLARDSON, L.S.; WALKER, R.E. Synthetic drain envelop soil interactions. J. Irrig. Drain Div. ASCE, 1979, 105:367-73.

Synthetic drain envelope materials currently

available commercially showed equal effectiveness. The materials function primarily as mechanical support for the soil drain opening interface. Soils have a limited characteristic resistance to internal hydraulic gradients. Flows causing gradients in excess of the hydraulic failure gradient of the particular soil will result in failure of the soil structural skeleton and the drain envelope.

1339

WILLIAMS, R.J.B. The chemical composition of water from land drains at Saxmundham and Woburn, and the influence of rainfall upon nutrient losses. Rep., Rothamsted Exptl. Stn., 1971, 2:36-67.

In a 2-yr study of flow rates and chemical composition of land drainage water, drainage at Saxmundham (boulder clay over relatively impervious sub-soil) varied widely and depended on short periods with much rain. At Woburn (deeper and more permeable soil from lower Greensand), drainage flow was more constant and depended on longer periods of dry or wet weather. Sharp maxima in NO_3-N concentration in spring at Saxmundham were related to recently applied fertilizer N. There was no marked NO_3-N maximum in drainage from Greensand soils at Woburn. The relationships between NO_3-N concentration and other ions are discussed and the mean annual concentration of ions in drainage water from both sites given. Soil type influenced drainage composition at Woburn. Water collected from a pond at Saxmundham and a lake at Woburn had different compositions from the drainage that fed them; Ca, NO_3-N, SO_4-S were less, and NH_4-N, PO_4-P and Cl were more in the pond and lake. Surface waters from shallow pools left after rain in boulder clay soil contained less Mg and Na than drainage water, but more of other ions contained in fertilizers. Estimates show that large amounts of nutrients, especially lime and N, were removed by drainage.

1340

WILLIAMS, R.J.B. The chemical composition of water from land drainage at Saxmundham and Woburn, 1970-75. Rep., Rothamsted Exptl. Stn., 1975, 37-62.

The flows of land drainage from a field of 3 ha at Saxmundham on a soil of Beccles series were measured and compared with an old system, under arable and harbage crop and from Flitwick Rowsham and Evesham series at Woburn. Flow rates from the new system were larger and more persistent than from the old drains and annual losses of nutrients were more in wet years. More nutrients were lost under arable conditions than under leys of cut grass given fertilizer N and lucerne receiving none. Under arable conditions from the old and new system the relative magnitude of annual losses in kg/ha were Ca > SO_4-S > Cl > Na > NO_3-N > Mg > K > NH_4-N > PO_4-P. Under herbage crops, losses of chloride were more than SO_4-S and magnesium more than nitrate N. Larger and more continuous flows at Woburn compared with Saxmundham resulted in larger losses of nutrients partaricully from two sources associated with natural springs. The order of magnitude of the estimated losses as kg/year at Woburn were similar except that nitrate exceeded sodium, a reflection of the distance of Woburn from the sea. Silicon losses were similar to those for potassium.

1341

WILLIAMSON, R.E.; CARREKER, J.R. Effect of water-table levels on evapotranspiration and crop yield. Trans. ASAE, 1970, 13:168-70, 176.

Effect of water table on crop yield and evapotranspiration were measured on several crops in lysimeters filled with Norfolk fine-sandy loam and Bayboro loam soil. There was complex ralationships between depth to water table, soil properties and method of watering in regard to these effects on crop yield. Drought effects were greater with deeper water tables. The generally favourable crop response and reduced ET with the greater depths to water table under the conditions of this study, indicate that water tables should be maintained at 60 cm or more below the soil surface.

1342

WILIAMSSON, R.E.; KRIZ, G.J. Response of agricultural crops to flooding, depth of water table and soil gaseous composition. Trans. ASAE, 1970, 13:216-20.

Effect of depth to water table, short term flooding and soil gaseous composition on crop yields is revised and results from various studies are presented in tabular form.

1343

WILLIAMSON, R.E.; SCHILFGAARDE, J. VAN. Studies of crop response to drainage: II. Lysimeters. Trans. ASAE, 1965, 8:98-100, 102.

Sheltered and non-sheltered field lysimeters have been found useful for studies of the drainage requirements of crops and preferable to growth chambers in some respects. Two sets of lysimeters with automatically moving rain shelters have been built, differing only in dimensions. Salinity problems, which can develop quickly with subsurface watering, can be controlled readily with careful management. Since plant response under wet conditions is primarily a function of the aeration status of the soil, the measurement of soil oxygen, either with bare platinum electrodes or with a membrane-covered platinum electrode, was found to be one of the most effective indices of adequacy of drainage. The field data obtained indicated that some deep-rooted crops, such as corn, require a deeper water table than do shallow-rooted crops, such as soybeans. The water-table depths required for maximum growth for most crops used in these experiments were the same for both sheltered and non-sheltered lysimeters.

1344

WILLIS, G.H.; ROGERS, R.L.; SOUTHWICK, L.M. Losses of diuron, linuron, fenac and triflurain in surface drainage water. J. Environ. Qual., 1975, 4:399-402.

Diuron at 0.84 kg/ha, linuron, 2.24 kg/ha, chlorfenac 3.36 kg/ha and triflurain 1.4 kg/ha were applied to various annual field crops grown in silty clay loam soil on plots designed for sampling runoff. During the 3 year study highest concentrations in runoff, were usually associated with rainstorms that occurred soon after application. Maximum seasonal losses were 0.12, 0.3, 2.9 and 0.05 per cent of the applied diuron, linuron, chlorfenac and triflurain, respectively. Proper use on agricultural land in the lower Mississippi River valley of these herbicides pose little threat to adjacent aquatic areas.

1345

WIND, G.P. Application of analog and numerical models to investigate the influence of drainage on workability in Spring. Neth. J. Agric. Sci., 1976, 24: 155-72.

Model of non-steady unsaturated moisture flow were used to calculate moisture conditions in a top soil from natural rain and evaporation data for 23 years. The calculations were made with 5 drainage depths and 3 drainage intensities using an analog model. The number of days when the soil was workable and the first workable day showed large variations over the years. Drainage depth had a pronounced effect on workability but drainage intensity had hardly any effect.

1346

WIND, G.P. Application of analog models to investigate optimal drainage conditions. In Physical Factors of Soil Environment I. Panstwowe Wydawnictwo Naukowe, 1979, 391-9.

A model investigation of the effect of drainage on crop yield is reported. It is reported that whereas a yield reduction of 36 per cent in spring beans and 12 per cent in potatoes and sugar beets at a drain depth of 70 cm the reduction is only 8 per cent and 4 per cent for the two cases respectively at drain depth of 150 cm.

1347

WIND, G.P.; BUITENDIJK, J. Simulation over 35 years of the moisture content of a topsoil with an electric analog for 3 drain depths and 3 drain spacings. Proc. Int. Drain. Workshop, Wageningen, Netherlands, (1978), 1979, 25:214-9.

With an electronic analog model the moisture content of the top 10 cm of a schematized silt loam soil was calculated over 35 years. Drain depths of 70, 100 and 130 cm below surface were used, each with three drainage intensities. To obtain a low number of days with a very wet soil, a high drainage intensity appeared to be important, the drainage, however, should not be shallow. To obtain a high number of workable days, deep drainage is important, and drainage intensity has little significance.

1348

WINGER, R.J. Jr. Construction and maintenance techniques for subsurface pipe drainage systems in irrigated lands. Proc. Int. Drain. Workshop. Inst., Wageningen, Netherlands, (1978), 1979, 25:645-63.

Installation of subsurface drains in irrigatecd lands requires specialized trenching machines capable of excavating a suitable trench to depths of at least 3.7 m. Trenching machines should also be capable of installing a designed, well-graded gravel envelope which will provide adequate permeability for convergence, prevent fines in the base material from moving into the envelope and drain line and provide the required bedding support for corrugated plastic drain tubing. One of the more complex problems in drainage of irrigated lands is establishing a stable drain trench when saturated, unstable soils are encountered. Stabilization of drain trenches can be accomplished by using a coarse gravel bedding, installing a system of well points, or installing a temporary dewatering drain below the grade of the permanent drain. Back-filling the trench after the drain tubing has been installed must follow some type of backfilling techniques that will ensure consolidated backfill to the original ground surface. After the drain has been completed the effectiveness and quality of installation should be evaluated by flushing or pulling a torpedo shaped plug having a diameter 2.5 cm less than the drain diam through the drain in a manner that will not affect the installation or efficient functioning of the drain. A TV camera can be used after the drain is in place to observe obstructions, determine tubing deflection, and condition of the water inlet perforation. Periodic inspection and maintenance are essential for keeping the drains from plugging with roots and from bacteria. Cleaning equipment consists of high pressure jetting equipment and high speed auger type cutting heads. Copper sulfate can be used to kill roots and sulfur dioxide gas or safe sulfuric acid in pellet form can be used to reduce iron bacteria.

1349

WINGS, R.J.; LUTHIN, J.N. (Eds) Guide for investigation of sub-surface drainage problems on irrigated lands. Am. Soc. Agric. Engng. Michigen. Pub.No. SP-04-66, 1966, 57 p.

Pre and post-irrigation drainage investigations necessary in identifying the problem are presented. The report lists and describes the various tools and methods in accomplishing the investigations. Surface and sub-surface drainage design investigations and its analysis and interpretation techniques are given to arrive at a suitable solution of the problem. Field sheets for taking various records in these investigations are included.

350

'INGER, R.J.; RYAN, W.F. Gravel envelope for pipe-.rains design. Trans. ASAE, 1971, 14:471-4,479.

Mathematical model has been developed to calculate the minimum envelope permeability suitable for a given base material. The mathematical solution takes into account the effect of joint width, type, length and diameter of pipe, the maximum possible rate of inflow to the pipe considering the soil characteristics and drain spacing, the thickness of the envelope material, and use of a plastic or asphalt saturated felt cover over the top half of the pipe.

351

ISDOM, A.S. Land drainage. Sweet and Maxwell, London, 966, 312 p.

A reference and text book on land drainage, including varied aspects of the problem, usually needed by an agricultural engineer.

352

ISER, E.H.; WARD, R.C.; LINK, D.A. Optimized design f a subsurface drainage system. TRANS.ASAE, 1974, 7:175-8.

A technique for the evaluation of the optimum values of depth and spacing for a uniform subsurface drainage system, in terms of benefits exceeding costs, is presented.

353

OLSACK, J. Permanent regime of drainage in eterogeneous and anisotropic soil. ICID Bull., 1979, 8: 21-6.

The unconfined aquifer problem is considered in

the case of an anisotropic and vertically heterogeneous medium, without making the Dupuit assumptions. Using an auxilliary dependent variable, an equivalent problem is derived, which can be solved exactly for the steady state. Further on, an approximate solution of the initial problem can be deduced. The method is applied to drainge by open ditches or by tile drains laying upon an impervious layer, eventually with moling superimposed. An 'equivalent horizontal hydraulic conductivity' is derived which takes the profile heterogeneity in account. This parameter can be estimated in situ by pumping tests.

1354

WOODGYER, W. Clay or plastic for farm drainage. N.Z.J.Agric., 1980, 14:61.

On a silt loam soil (Keeble Farm, Massey University) there was a measureable increase in porosity when this soil type was drained; dry matter production of pasture was increased over the winter period by 30 per cent with drainage, but there was no apparent difference between the drainage performance of clay or plastic(Novaflo) pipe with mole drainage as secondary treatment.

1355

WOOLEY, J.T. Drainage requirement of plants. Proc. Drain for Efficient Crop Production Conf. Chicago, Illinois, 1965, 2-5.

Lack of information on drainage requirement of crops has been emphasized making the job of an design engineer difficult. Data on the growth and respiration with deficiency of oxygen or excess of carbon dioxide have been reported. Movement of oxygen to roots both through the soil and root and bio-chemical considerations of drainage to plants have been reported. It is emphasized that certain special tissues have special type of metabolism, knowledge of which might help in understanding the reaction of roots to drainage.

1356

WORLD BANK. Proceeding, Agricultural Sector Symposia CPS. Agric. and Rural Div. Deptt. and Pers.Mgt.Deptt. World Bank, Washington, USA, 1980, 563p.

The records of contribution to various aspects of agriculture are presented. The records pertaining to drainage of irrigated lands appear at three places under the common head "Problem of drainage related to irrigation projects".

1357

WRACHIEN, D. DE. Contribution to the design of tile drainage network by means of a three-dimensional mathematical model. Riv. Ingeg. Agrar., 1980, 11:19-23.

Some numerical methods are illustrated for the study of two dimensional steady state flow fields through porous media, in relation to determining the spacing of tile drainage networks. The practical necessity and suitability of three-dimensional mathematical models are analysed.

1358

YADAV,J.S.P. Reclamation of salt affected soils. Proc. Indo-Hungarian Sem. Mgmt. Salt Affected Soils Karnal, 1977, 134-59.

An exhaustive review paper, presents information on different methods of reclamation and management of salt affected soils in India and abroad. The chemical methods deal with use of gypsum, its dose, size of particles, depth of application and frequency of application; $CaCl_2$; sulphur; pyrites; phosphogypsum, sulphuric acid; lime; and pressmud. Agro- technical methods describe the cropping sequence, varieties and fertiliser application and mulches. Biological methods cover green manuring, addition of organic materials and plant growth. In the end, hydrotechnical methods have been discussed in three parts viz. leaching, drainage and deep ploughing.

1359

YADAV, J.S.P.;BANDYOPADHYA, A.K.; RAO, K.V.G.K.; SINHA, T.S.; BISWAS, C.R. Coastal saline soils of India. Bull. CSSRI Karnal, 1979, 5,34p.

In the coastal tracts of West Bengal,Orissa, Andhra Pradesh, Pondicherry, Tamilnadu, Kerala, Karnataka, Maharashtra, Gujarat and Goa, the saline soils occur to the extent of 2.1 m ha. Physical features of the coastal saline soils with respect to physiography and relief, river system, tidal effect, groundwater characteristics, climate and nature of soils salinity have been described. Reclamation measures have been discussed extensively which cover protective bunds, surface drainage, sub-surface drainage, leaching of soluble salts, evaluation of water quality for irrigation and water resources for irrigation, crops and their verietal improvements. Agronomic and cultural practices have been listed.

1360

YAMAZAKI, F. et al. The role of cracks in subsoil on under drainage in clayey reclaimed padddy fields. Trans. Agric. Engng. Soc. Tokyo, 1966, 16:1-8.

Reports results of the investigations carried out at two reclaimed rice fields in the polder along with Kojina Bay. The cracks in subsoil in newly-reclaimed fields, play the same role of under drainage, on clayey fields in a alluvial plain. In rice growing season, cracks formed on the surface soil due to intermediate drying of a field are connected with those of the sub-soil, and the water in the rice field is drained through these continuous cracks into the drain pipes. In winter cropping season, surface soil is ploughed and gets dried, the rain water seeps into the surface soil and is drained to the drain pipes through the cracks in subsoil. The field is well drained and the soil gets improved by weathering.

1361

YAMAZAKI, F.; TAKENAKA, H.; TABUCHI, T.; TADA, A. On the effect of cracks in subsoil upon tile drainage of clayey paddy field soil. J. Agric. Engng.Soc. Japan, 1964, 32:151-9.

The paper deals with the mechanism of drainage of the clayey rice field soil at Gyoda city, Saitama Pref. Japan and shows:(1) The investigated rice fields were severely waterlogged and it was difficult to plough them. The tile drains were settled and physical and chemical properties of soil were improved for plant growth and tillage. In non-waterlogging season the permeability of subsoil including cracks is very large. Excessive rain water flows into tile drains through cracks in mole. In irrigation period, the mole becomes impervious, thus cracks remain by and large permeable. By drying up rice fields in early August, cracks are created in mole, so it become very permeable. Therefore, consideration must be given to cracks in subsoil on establishing the design of drainage in such a clayey soil. Tile drains should be settled when cracks are developed in sub-soil by the action of open ditches. In design, the optimum tile depth for laterals is limited by the depth of cracks. It is of no use to settle tile drains in the layer below the cracks.

1362

YANG, S.J. WANG,, P.L. Soil aeration, nutrient uptake and yield of sugarcane as affected by tile drainage. In 17th Congr. Int. Soc. of Sugar cane Tech. Vol. 1(Lovez, M. B.; Maddrazo, C.M. Eds.) Manila, Phillippines, 1980, 754-66.

On a low-humic gley, tile drainage (compared with surface drainage) resulted in increases of 17,40 and 37 per cent in the yield of the plant crop and the first and second ratoon crops, respectively.

1363

YANG, S.J.; WANG, P.L. Reclamation of saline sugar cane plantations by a subsurface drainage system. Proc. 16th Congr. Int. Soc. Sugarcane Technol. Brazil, 1978,919-30.

Field experiments were conducted in three different soils to evaluate the effectiveness of subsurface drainage in reclamation of saline cane plantations. Drain characteristics and leaching of salts were studied separately. The outflow rate from each drain was very high, the peak discharge varying with hydraulic conductivity of the soil but all values being greater than 17 mm day. The water tables midway between the lines were lowered 1.5 m from the surface 7 days after drainage. A linear relationship was found between the discharge and the height of the water table at the mid-point between drains, the slope of the line increasing with increase in soil permeability. Five commonly used equations were compared for assessing the performance of a given scheme, and the predicted spacing from the Glover-Dumm equation found to be closest to field data and so best for use in design of a drainage system. Desalinization is rapid using a sub-surface drainage system, the ECe falling from 17 to below 2 dS/m over 16 months.

1364

YANG, S.J.; WANG, P.L.; CHANG, Y.T.; Effect of drain depth and spacing on performance of tile drainage system in fine-textured soils. Rep. Taiwan Sugar Research Institute, 1977, 77:1-11.

The effects of two depths and three spacings of drains on the tile discharge rates and drawdown in fine-textured low humic gley soil were evaluated in a field experiment. Drain depths were 1.2 and 0.8 m below the surface and the spacing were 15,20 and 25 m. The observations were taken immediately after heavy rainfall. For the same spacing drains at 1.2 m depth lowered the water table faster than those at 0.8 m and the narrower the drain spacing, the more effective was the drainage. Both peak discharge and total discharge increased with decreasing tile depth and spacing. A parabolic relation was found between drain discharge and water table level. Such results confirmed the unsteady state flow theory of Dumm when drains were located on or close to an impermeable layer. Satisfactory results were also obtained when comparison of the field data with unsteady state flow theory was made. It is concluded that the van Schilfgaardde equation should be adopted for design purposes.

1365

YANG,S.J.; WANG, P.L.; CHANG, Y.T.; YANG.T.C. The performance of subsurface drainage and its effect on sugarcane growth on fine-textured soils in Taiwan. Taiwan Sugar, 1977, 24:378-84.

In a fine-textured low humic gley soil, subsurface drainage at depths of 1.2 or 0.8 m and at intervals of 15,20 or 25 m increased cane yield by 15-19 per cent. Peak and total discharge of water increased with tile depth and decreasing spacing. In the rainy season O_2 content at 20 or 30 cm depth seldom dropped below 10 per cent in drained plots whereas it never exceeded 5 per cent in control plots.

1366

YEH,W.W.; SINGH, R. Transient flow between parallel drains. J.Hydrol., 1970, 11:301-12.

The transient-flow problem of subsurface drainage between parallel drains has been solved by the method of weighted residuals. The governing nonlinear partial differential equation is reduced to ordinary differential equations which are easily solvable on the computer. The solution is presented in two non-dimensional graphical forms for easy field applications.

1367

YEN, S.C. Natural leaching effect of salt-affected soils in Taiwan. II-Land with subsurface drainage system. <u>Rep</u>.Taiwan Sugar Res. Inst. , 1977, 77, 13-25.

Two sites of newly sub-drained salt-affected soils at Kan Hsi and Chi Ku were selected for study. The effectiveness of natural leaching was greatly improved by lowering the water table, impounding rainfall and mulching with cane trash. The total salt content within the profile depth of 180 cm soil was lowered from 20 to $<$5 d s/m. More than 100 t/ha of sugar cane were produced by the first crop following soil desalinization. The productivity of this desalinized soil was comparable with nearby normal soils. It has been suggested that the desalinization of salt-affected soils in Taiwan could be accomplished by sub-drainage and natural leaching without any additional irrigation water. Soil amendments such as filter cake and gypsum were beneficial on newly reclaimed saline soil. Application of these materials would increase sugarcane yield by 20-40 per cent. Natural leaching was less effective during the growing period. The soil salt balance depends largely on the design of the tile system and the management of the drainage unit.

1368

YEN, S.C.; FANG, Y.T. Natural leaching effect on salt affected soils in Taiwan. I. Land with open drainage system. <u>Rep.</u> Taiwan Sugar Res. Inst.,1976, 74:15-25.

The fluctuation of soil salinity in a 180 cm profile of a representative saline sugarcane field in the western coastal plain in each 3 day period after rainfall were recorded. From combined records of rainfall and groundwater it was concluded that the low leaching effect of high rainfall on the salt affected soils in Taiwan is due to high ground water table, high rainfall intensity and restricted external drainage. The most effective rainfall for leaching would be in the range of 40-100 mm for 3 day period and about 350 mm annually (25% of annual rainfall). The most effective measure for increasing natural leaching would be to lower the ground water table.

1369

YOUNG, C.E. Water table, soil moisture and oxygen diffusion relationship on two drained wetland forest cites. Soil Sci.,1969,107:220-2.

Relationship between soil moisture 15-45 cm below ground level and the water table,oxygen diffusion rate at 30.5 cm below ground level (Y_{OD}) and ground water level (X_1) are determined. The last relationship is linear of the form; $Y_{OD} = a + bx_1$.

1370

YOUNGS, E.G. Calculation of ponded water-drainage for flow regions of various geometries to demonstrate effect of disturbed soil zone shape on drain performance.J. Agric. Engng. Res., 1982, 27:441-54.

A general solution, obtained by conformal mapping, is presented for the two dimensional drainage problem of flow from a horizontal ponded surface to a cylinderical drain in a soil region of any shape that is symmetrical about the vertical plain through the drain. For each geometry of flow region the flow rate from the drain is maximum when the drain is located at some depth less than the maximum depth of soil. Specific solution to V-Shaped semi-circular and a rectangular cross-section of flow regions are discussed. From the calculated results, it is argued that better drain performance can be obtained if the zone of distrubance has a rectangular, rather than a V-shaped cross section.

1371

YOUNGS,E. G. A comparison of the performances of some plastic and tile drains. J. Agric. Engng. Res., 1965, 10:202-3.

Watertable heights obtained with plastic and tile drains in a hydraulic model sand tank were measured for various steady-state simulated rainfall rates. Results showed much higher watertables for given rainfall rates than occur in a similar system of completely permeable drain channels.

1372

YOUNGS, E.G. The contribution of physics to land drainage. J. Soil Sci., 1983, 34:1-21.

The physics of land drainage is develoved from the application of Darcy's law to the water movement produced as a result of the conditions at the boundaries of the groundwater region in a given drainage situation. The physical understanding of the groundwater problem so obtained allows land drainage to be considered as an engineering operation founded on rational scientific principles and allows recommendations to be made concerning the design or installations. However, for theory to be applied in many field situations, a greater understanding of the soil-water physics of heterogeneous soils, of fissured soils and of swelling soils is required.

1373

YOUNGS, E.G. Determination of the variation of hydraulic conductivity with depth in drained lands and the design of drainage installations. Agric. Wat. Mgmt., 1976, 1:57-66.

For lands drained by ditches dug to a horizontal impermeable floor, the variation in the hydraulic conductivity of the soils with depth may be obtained from the relationship between water-table height and drain-outflow rate. Some relationships obtained on an experimental plot on a clay soil, drained by tile drains with gravel backfill, and on another in the same field which was mole drained, were analysed to give the variation of hydraulic conductivity with depth by assuming that their performances approximated to that of ditches. For the tile-drained plot, the hydraulic conductivity value increased by three orders of magnitude near the bottom of the plough layer; this was reduced in a subsequent year when the field was uncultivated under grass with consequent higher water tables. The mole-drained soil was more permeable than the tile-drained soil at a lower depth, and hydraulic conductivity at this lower depth did not change in the subsequent year when the field was uncultivated. A general hydraulic approach to drainage design is suggested. Even if the drainage rate from an area is not measured, the water-table recession alone in an area drained by ditches may give sufficient information to design a drainage system on a rational physical basis.

1374

YOUNGS, E.G. Developments in the physics of land drainage J. Agric. Engng. Res., 1984, 29:167-75.

The historical development of the physical theory of soil water control by land drains is reviewed with special reference to the modifications that are required, for idealized theory to be applied to field situations with their complicating factors. The paper discusses the foundations of land drainage theory, solutions of drainage problem, complicating factors in the theory of land drainage which include three dimensional drainage, gappy nature of drainage channels, soil heterogeneity, variable specific yield; application of theory to drainage practices and unresolved physical problems. In the end, is presented a rational approach to land drainage which suggest that a knowledge of drainage theory is essential in any design of drainage system.

1375

YOUNGS, E.G. The hydraulic effect of filter materials around gappy non-ideal field drains. Agric. Wat. Mgmt., 1980,3:17-34.

The analysis of Widmoser and Kirkham for flow into gappy-non-ideal drain pipes with longitudinal slits and circumferential openings, respectively, were modified to take into account the hydraulic effect of filters when the boundary between filter material and soil is an equipotential surface. Upper and lower bounds were found for the effective drain radii of the drain pipes when surrounded by a cylinderical filter, by calculating the effective drain radii of the drain pipes with filter surrounds whose outer boundaries are the equipotential surfaces just contained within and without the outer cylinderical surface of the filter. The calculated bounds are close to each other for large filter thickness, but become further apart as the filter thickness is decreased. The results show that the maximum water-table height in drained land is considerably lowered in the presence of filter more permeable than the soil surrounding a gappy non-ideal drain, and that a thin filter of high hydraulic conductivity relative to the soil is hydraulically more beneficial than a thicker filter of hydraulic conductivity only slightly more than that of the soil. An exact analysis, useful for large filter thicknesses, is also given for the flow towards pipe drains with circumferential openings when surrounded by permeable material with hydraulic conductivity varying inversely with radial distance.

1376

YOUNGS, E.G. Physical aspects of land drainage. <u>Proc.N. Engl.Soils Discuss. Group.,</u> 1964, 1:25-9.

Unsaturated conditions of surface soil with its increased aeration produce a favourable environment for plant growth and increase the soi's mechanical strength. Hydraulic conductivity of soil in presence of drainage is discussed.

1377

YOUNGS, E.G. Water table height in homogeneous soils

drained by non-ideal drains. Soil Sci., 1974, 117:295-300.

The concept of the equivalent drain radius and entrance resistance of a non-ideal drain are used to obtain drainage formulae to describe the relationship between the maximum water-table height and average flux through the water table for parallel drainage installations where there is infinite depth of soil below the drain and when the drain is laid on the impermeable floor. For small equivalent drain radii or large entrance resistence, there is little difference in the two relationships. Experiments in a hydraulic model sand tank show that the proximity of an impermeable floor has little effect on the equivalent drain radius of a given drain, and that water-table heights are in fair agreement with the theoretical formulae.

1378

YOUNGS, E.G.; AL-NAJIM, M.A. Flow through unconfined aquifers containing interceptor drains. J. Hydrol., 1978, 37:339-48.

The flow into interceptor drains,located in an aquifer bounded by water standing at different levels in two parallel vertical ditches which extend in depth to a horizontal floor on which the aquifer rests, and the discharge into the downstream ditch are shown to be related to the water levels in the ditches by a modified form of Dupuits formula. For the case of an aquifer resting on an inclined impermeable floor, seepage analysis gives an approximate expression which is an extended form of that obtained without the inclusion of interceptor drains. Hele-Shaw analog results are used to give experimental confirmation of the theoretical relationships.

1379

YOUNG, T.C.; LIGON, J.T. Water table and soil moisture probabilities with tile drainage. Trans. ASAE, 1972,

15:448-51.

A method of soil moisture balance proposed by Ligon and transient flow equation by van Schilfg-garde were tested for use in a computer model of the environment, soil and drainage system. From the model output, relationships were determined between drain spacing, water table level, and return period for three sets of soil parameters. Soil moisture content was negatively related to depth to water table. An average decrease of 1.67 per cent of soil moisture in the upper 75 cm of soil was found for a 30 cm drop of the water table. The procedure is recommended for use at other places.

1380

YOUSEF, M.S.; AMER, M.H. Graphical solution of Amer and Luthin's spacing equations for tile drains. ICID Bull., 1972, 21:14-9.

Graphical solution of Amer and Luthin's equation of drain spacing and its use under given conditions is explained.

1381

YUN-SHENG,C. Drainage of paddy soils in Taihu lake region. Proc. Symp. Paddy Soils (Inst. Soil Acad. Sinica, Ed.), Springer-verlag, Berlin, 1981, 517-23.

Effect of tile drainage on paddy soils in relation to rice and barley cultivation are reported. Depth and spacing respectively of tiles in clayey soils is 1.1-12 and 6-10 m, in loamy 1.0-1.2 and 15 m and for sandy loam 1.0 and 20 m.

1382

ZAIDELMAN, F.R. Urgent problems of reclamation of mineral hydromorphic soils and their solutions. Trans.10th Int.Congr.Soil. Sci. Moscow, 1974,10:234-41.

A general paper on drainage discusses and suggests increased research into the causes of waterlogging; the degree of waterlogging; the properties and regimes of soils in the natural and ameliorated state and the influence exerted by local ameliorations on the hydrological regime and the ecological interconnections between the elements of a land scape as a whole, and the prognostication of the potential soil fertility. Principal ways of the solution of these problems are considered.

1383

ZAITSEV, B.D. Characteristics of the nutrient reserves in drained forest soils. Les. Prom., 1964, 490-3.

Reserves in the top 100 cm increased with increasing bogginess. Drainage accelerated decomposition of organic matter and increased the content of available nutrients, particularly in the rooting zone. The after effect of drainage can be forecast from the content of ash and of exchangeable Ca and Mg in the peat and from the mechanical composition of the mineral substrate of the lower horizons.

1384

ZAKS, V.G. Hydrological regime of waterlogged soil in the European Polesje areas and its transformation under the impact of drainage. Trans. 10th Int. Congr. Soil. Sci. Moscow, 1974, 10:275-81.

Water regime and expediency of draining podzolic non-gleyed, gleyed and peaty-gleyed soils in the Oka-Meshchera Polesja area are discussed as related to the influence of forest vegetation and crops, different weather conditions, and non-drained and adjacent to drained territories.

1385

ZAMPIGHI, G. Experimental monitoring of underground drainage installations. Irrigazione, 1981, 28:37-46.

A research programme to define and quantify the effects of underground drainage on 29 farms in the provinces of Ferrara, Bologna and Ravenna, Italy, is described. Hydrologic determinations include measurements of rainfall and drainage water volumes, comparisons of the depth of the phreatic layer in surface and subsurface drained areas, physico-chemical analysis of the drainage water, and periodic assessment of the efficiency of the underground drainage network. Variations in crop growth and in transmission of pests and diseases are noted. Increased crop production may result not only from improved distribution of fertilizers and herbicides and better aeration, but also from the larger cropping area due to the elimination of surface ditches. The economics of the underground drainage system and that of the surface system are compared.

1386

ZARADNY, H.; FEDDES, R.A. Calculation of non-steady flow towards a drain in saturated-unsaturated soil by finite elements. Agric. Wat. Mgmt., 1979, 2:37-53.

The problem of two-dimensional non-steady flow of water towards a drain in saturated-unsaturated soil is solved by a Galerkin-type finite element approach. Special emphasis is placed upon the type of boundary conditions to be used at the circumference of the drain, the hydraulic conductivity of the soil and the discretization scheme of the finite element network. Numerical solutions are compared with the Thiem-Dupuit well formula and with hodograph solutions.

1387

ZASLAVSKY, D. Definition of the drainage filter problem and a possible use of soil conditioners. Proc. Int. Drain. Workshop., Wageningen, Netherlands, (1978), 1979, 25:340-5.

The theory of piping or soil detachment by flowing water is developed. The soil parameters for this theory were measured by an erosimeter. There is a critical hydraulic gradient that would detach an aggregate of a given size or pass it through a certain sized filter pore. Soil conditioner added to the soil and soaked into the envelope material are useful.

1388

ZASLAVSKY, D. Drainage for salt leaching. <u>Proc.Int. Drain.Workshop,</u> Wageningen, Netherlands, (1978), 1979, 25: 664-85.

The water flow streamlines can serve as a guide foi a first approximation design of irrigation and drainage for salt leaching. The main guidelines are: (a) The different flow tubes should be as much as possible of the same cross-section and best should be parallel;(b) Stagnation areas cannot be leached; (c) To overcome stagnation areas the flow symmetry should be changed from time to time; (d) The most efficient leaching is by vertical infiltration into an unsaturated slightly more than the water held in the soil to be leached; (f) A drawdown of water table along with a possible drying up of the soil profile are necessary from time to time to facilitate leaching;(g)The depth of unsaturated soil should be sufficient to accept the amount of water held within the leached profile before the leaching. The classical formula for leaching requirement is valid only under very limited conditions. The new formula increases the leaching requirement, by a factor that is the ratio of field capacity to the dry state before irrigation. More sophisticated studies of solute movement in an active porous medium may be useful for a number of problems and for optimization of the water regimes in irrigated lands. However, the major conclusions in the study which are based on immiscible displacement are unchanged.

1389

ZASLAVSKY, D. Notes on the approach to drainage design. <u>Proc. Int. Drain. Workshop,</u> Wageningen,

Netherlands, (1978), 1979, 25:155-64.

The complexity of possible drainage solution is smaller than the input parameters that make a large number with many of them unknown and many of them of statistical nature. Much more good can be obtained by adaptation of existing knowledge to practice than by a new research. The emphasis should be put on a systematic approach for a fast and efficient elimination of design alternatives. Publication of and proper guidebook in this regard is suggested as the main solution to these problems.

1390

ZAYDEL'MAN, F.R.; ZAKS, V.G. Moisture regime and reclamation properties of light Polesian soils. Pochovedenie, 1972, 1:96-109.

Boggy light soils dry out deeply in the rooting zone in average and dry years, their profile is well aerated in wet years, and the level of their ground-water varies considerably. On undrained soils,the yields of agricultural crops and the productivity of meadows and forests vary with degree of bogginess, and drainage should not aim at producing a moisture regime such as occurs in non-boggey soils. Drainage of only the most boggy soils (mainly peaty soils) over a considerable area of a waterlogged tract consisting of mineral gleyed soils often produces a favourable or moderate decrease in soil moisture content and groundwater level.

1391

ZEIGLER, E.R.; WINGER, R.J. JR.; WILLARDSON, L.S. Friction, grade and alingnment studies for corrugated plastic drain tubing. In Water Management For Irrigation and Drainage. Proc. Spec. Conf. ASCE, Irrig. Drain. Div. ASCE, New York, 1977, 1:146-61.

Hydraulics of corrugated plastic drain tubing as a closed conduit or as open channel has been studied. Data on alignments are presented. Manning's n values have been reported for both the full tube and free surface flow. Movement of sediments at different velocities have been studied to find out the average flow velocity which moved the sediments. In the 0.3 m/s velocity range, a relatively small increase in water velocity produced a large increase in sand dune movement.

1392

ZEITS, J; TITZE, E. General information on subsurface drainage in the Netherlands. Proc. Int. Drain. Workshop, Wageningen, Netherlands, (1978), 1979, 25:481-511.

A review is given on geographic conditions and present state of art of drainage in the Netherlands. Various projects in which drainage play an important role are reviewed. Agriculture practices after drainage and institutes engaged in research on drainage are reported.

1393

ZEITZ, J.; TITZE, E.; WERIZ, G. Structure improvement and drainage of deep loam pseudogley soils in the north of the GDR. Arch. Acker-U. Pfla. Bodenk., 1982, 26:561-70.

Drainage with different drain spacings, and combinations of drainage, deep loosening, and envelope materials were investigated on a pseudogley soil with an impermeable layer (overlying marly till) at $\geq$ 0.6 m depth below the surface. Deep loosening increased the pore volume and air capacity. Entry resistance of the loosened layer was reduced by 25 to 82 per cent as compared to the compacted layer. However, recompaction of the soil was observed 28 months after deep loosening. The amelioration problem of the site could be solved by pipe drainage.

1394

ZERBI, G. Agronomic studies on drainage in the plain of the river Sele. Italia Agrocola, 1972, 110:361-74.

Plastic slitted tubes 12 m or 8 m apart improved the yield and quality of durum wheat as compared with open ditches. The soil moisture content in December-April was lower with plastic drains than with open ditches, and with the latter the moisture content exceeded the field capacity in January-February.

1395

ZERNOV, V.I.; DETSIK, T.A.; MIRONENKO, V.I.; KOTOVA, A.F. The effect of drainage on the water regime of adjacent dry valleys. Lesn. Khoz., 1977, 7:24-8.

A report is given on the effect of drainage operations on the water regime of adjacent areas and on the growth of pine (Pinus sylvestris) in Belorussia. The investigations were made in 1971-74 on areas where drainage had been done at least 5 years earlier, at distances of 20-950 m from the drains. Data are presented on current increment of the pine and on the water table. The results indicate that with drains of the normal depth (0.8-1m), the water table in dry valleys adjacent to the drained bogs is lowered by not more than 0.5 m, even in immediate proximity to the drain, and the increment of the pine is virtually unaffected by the drainage operations. If the water table is lowered by 0.5-1.5 m in the Pinetum myrtillosum and by more than 1 m in mossy pine types, the increment of the stands in the adjacent dry areas is reduced by about 15 per cent in the period after drainage.

1396

ZERYOUHI, I.; CARLIER, P. Impact of irrigation and research into a pumped drainage scheme-simulation by a mathematical model, using the groundwaters of Triffa and Garet (northeast Morocco) as examples. 10th Congr.

Irrig. Drain. Athens, 1978,4:187-209.

The development of irrigation in the plains of northeast Morocco had led to rise in groundwater levels for which existing drainage systems are inadequate. Schemes for drainage by pumping have been proposed, with possible reuse of the pumped water. A mathematical simulation is described and some results of the simulation in the plains of Triffa and Garet are presented, including maps of water table development under the influence of irrigation and pumped drainage.

1397

ZIVERT, A.A.; RIEKSTS, I.A.; SKINKIS, C.N. Investigation of runoff in drainage outlets and subsurface drained soils in the Latvian SSR. IAHS Pub., 1975, 105:117-27.

When designing drainage systems and drainage outlets, it is necessary to determine the expected hydrological regime during the operative life of the system. The procedure and technical requirements of the investigation of drainage system runoff are discussed in this paper and the main results of the observataion of drainage, surface and total runoff from subsurface drained areas are presented. The basic principles of forecasting the expected runoff in drainage outlets are investigated and characteristics of the changes in the runoff from some of the largest land reclamation schemes in the Latvian SSR (the Lielupe river, the Aiviekste river and Lake Lubanas) are given.

1398

ZUIDEMA, F.C.; SCHOLTEN, J. Maintenance of tile drainage system. 7th Congr. Irrig. Drain. Mexico, 1969, 5:25.310-25.319.

Mechanical and chemical means of cleaning tile drainage systems are discussed. Mechanical cleaning of little clogging or some plant roots in the drains can be done by means of piercing or syringing whereas larger amounts can be cleaned by low or high pressure engines. Chemical cleaning by means of SO_2 gas can be applied at serious joint-blockage and deposits of iron in the filter material. Use of larger drain diameter is an alternative which can be thought of at the design stage. In case of severe damage fields may have to be redrained with old or new drainage materials.

1399

ZUIDEMA, F.C.; SCHOLTEN, J. Model tests on drainage materials. Proc. Int. Drain. Workshop, Wageningen, Netherlands, (1978), 1979, 25:393-401.

The Scientific Department of the Ijsselmeer Polders Development Authority have presented the results which indicate that thin, woven or non-woven envelope materials are not advised for use in very fine sandy soils. In middle fine or coarse sandy soils, woven and non-woven envelope sheets may be applied. The Polypropylene fibre mat and plastic sheet envelope material has a low level of entrance resistance and can be used in fine sandy soils.

1400

ZVERKOV, Yu. Grasses on reclaimed peatlands. Zemledelie, 1976, 4:73.

In trials during 4 years in Kirov region, hay yields of brome grass, timothy and meadow fescue increased with increase in water table depth from 70-90 to 100-130 cm and in the peat layer of the soil from 30-50 to 60-90 cm. Brome grass gave the highest average hay yield of 5t/ha, followed by timothy with 4 t/ha.

1401

ZWERMAN, P.J.; CORPUZ, I.T. Nitrogen fertilization of crops to compensate for yield losses from poor drainage. Proc. Drain. for Efficient Crop Production Conf. Chicago, Illionois, 1965, 17-20.

The effect of water table on crops yields have been reviewed and reported that great reduction in yield of onion, potato, peppermint, carrot and corn were noted with water table at 40 cm. The yields for three water tables viz. 60, 80 and 100 cm, were similar. The examination of Dutch study indicated that for winter wheat, spring barley, spring wheat and caraway seed, addition of excess nitrogen at the rate of 40 kg N/ha resulted in significant increase in yield at 40 cm water table compared to 120 cm or 150 cm with the same additional nitrogen added. An index of drainage (SEW) is reported which indicates that cm days of water table is important in evaluating the drainage need of the crops. It is suggested that though nitrogen can compensate for yield decline to some extent more basic research is needed to tie together the theory and practice on this aspect.

1402

ZWERMAN, P.J.; GREWELING, T.; KLAUSNER, S.D. Nitrogen and phosphorus content of water from tile drains at two levels of management and fertilization. Soil Sci. Soc. Am. Proc., 1972, 36:134-7.

Average annual rates of 243.2 and 86 kg/ha N, and 31.5 and 12.2 kg/ha P as ammonium nitrate and superphosphate were applied to the soil. Concentrations of NO_3-N in the drainage water increased from 3.0 ppm to 51.1 ppm with increased fertilizer rate. During two 3-week test periods, NO_3-N delivered from the tile drains ranged from 0.225 to 2.75 kg/ha. Ammonium concentration were only slightly related to fertilizer rates. Orthophosphate concentrations calculated as P were 0.004 to 0.01 ppm with a weekly output range of 0.00016 to 0.00088.

SUPPLEMENT

ADDITIONAL ABSTRACTS

1403

ABELIOVICH, A. Avoiding ochre deposits in soil drainage pipes. Agric. Wat. Mgmt.,. 1985, 10:327-34.

Leptothrix sp. was identified as the causative agent producing ochre problems in an intensively irrigated area near the dead sea. By creating anaerobic or micro-aerophilic conditions in experimental drain pipes in the field with siphons installed at the outlet or in intermediate manhole, ochre production was avoided. Similarly clogged drain pipes were cleaned within weeks by raising the ground water level around the pipes.

1404

ABHANGE, G.V. Drainage problems in Maharashtra: past experiences and future challenges. Sinchan, 1986,4:50-80.

Drainage problems in Maharashtra State are described in relation to salinisation, alkalinisation and for shallow, medium and deep black soils. Past experiences of design according to a technical paper (TP-56) are illustrated. Data in respect of total extent of waterlogged areas, drainage schemes completed and sections of open and subsurface drains used in the state are provided.

1405

AGARWAL, R.S. Information regarding waterlogging and salinity in Kosi command area. Proc. Sem. Problems Waterlogging Salinity in Alluvial Soils, New Delhi, 1986 9. 1-9.23.

Out of gross command area of 7.36 lacs ha, 1.80 lacs ha is suffering from surface stagnation of water. Waterlogging through rise of watertable and soil salinity are developing in the area since the advent of Kosi irrigation water. It is because nearly 75 per cent of the water is lost to the ground water through seepage and deep percolation losses. Vertical drainage is being tried to overcome this problem.

1406

AMBEKAR, V.W. Status of waterlogging and soil salinity in Sharda Sahayak command area development project. Proc. Sem. Problems Waterlogging Salinity — Alluvial Soils, New Delhi, 1986, 10.1-10.13.

Development of waterlogging in Sharda Sahayak irrigation project is described. Out of 2.141 m ha of area surveyed in 35,184 ha the water table is below 1 m. In several villages in the command, area under rabi cultivation has reduced. Lining, perfect surface drainage, conjunctive use of ground and surface waters and a comprehensive plan for implementing drainage in the area are suggested.

1407

BAGARELLO, V.; SCIORTINO, A. Theoretical basis and criteria for the planning of drainage system under the pedoclimatic conditions of Italy. Irrigazione, 1985, 32:29-35.

Glover-Dumm equation for unsteady state flow conditions is discussed and its use in designing drainage system is explained.

1408

BAKER, J.L.; KANWAR, R.S.; AUSTIN, T.A. Impact of agricultural drainage wells on groundwater quality. J. Soil Wat. Conserv., 1985, 40: 516-20.

Agricultural drainage wells to remove surface and subsurface drainage inject water and are expected to contaminate the water supply wells in the vicinity. Nitrate-N was 10 mg/L in 85 per cent of the samples. Pesticides upto 1 ug/L were detected for subsurface and 80 ug/1 for surface runoff injection. Current pesticides water quality standards were not exceeded.

1409

BAL, A.R. A note on the effect of flood water inundation on two rice varieties. J.Indian. Soc. Coastal. Agric. Res., 1985, 3:149-50.

The two rice varieties (IR-8, Damodar) have been evaluated for their tolerance to water stagnation due to floods. Both varieties suffered yield reduction over control (no flood effect). However, var. Damodar was more tolerant and can be grown in areas where 7 days flood water inundation is expected.

1410

BANDYOPADHYAY, A.K. Studies on the effect of ground water table on salinisation of heavy textured soils of Sundarban and its effect on crop growth. J. Indian. Soc. Coast. Agric. Res., 1984, 2:26-8.

Critical depth of water table for soil salinisation for heavy textured soils of Sunderban is reported to be 110 cm from soil surface.

1411

BAZARAA, A.S.; ABDEL-DAYEM, M.S.; AMER, A.; WILLARDSON, L.S. Artesian and anisotropic effects on drain spacing. J. Irrig. Drain. Engng., 1986, 112:55-64.

Effect of artesian flow and anisotropy of the soils on subsurface drain spacing is discussed. For artesian conditions narrower than normal spacing is needed.For anisotropic conditions where horizontal hydraulic conductivity is larger than in vertical direction, neglecting anisotropy may lead to under design of the system.

1412

BENGSTON, R.L.; CARTER, C.E.; MORRIS, H.F.; KOWALCZUK, J.G. Reducing water pollution with subsurface drainage. Trans. ASAE, 1984, 27:80-3.

A subsurface drainage-runoff-erosion experiment on a loam alluvial soil near Baton Rouge, Louisiana, indicated that surface runoff and soil erosion reduced 29 and 16 per cent respectively in case of surface and subsurface drainage combination compared to surface drainage alone. Similarly potassium and phosphorus removed were also reduced 24 and 32 per cent respectively.

1413

BHATTACHARYA, A.K. Drainage for land reclamation. Sinchan, 1986, 4:101-7.

Extent of waterlogged and salt affected soils and measures for their amelioration are described.

1414

BOTTCHER, A.B.; MONKE, E.J.; BEASLEY, D.B.; HUGGINS, L.F. Sediment and nutrient losses from subsurface drainage. In Erosion and Conservation (Elswaify, S.A., Moldenhauer, W.C.; Lo, A., Eds.) Soil Cons. Soc. Am. Ankeny, U.S.A., 1985, 622-33.

Data on sediment, N and P losses over a five year period presented for a 17 ha field in Indiana.

1415

BRENNAN, P.J. Pumped field drainage. Booklet, ADAS, Ministry of Agric. Fisheries and Food, UK, 1985, 2514, 28p.

The design of a pumped field drainage scheme, pump selection, pumphouse maintenance and operation are some of the subjects covered.

1416

BUCKLAND, G.D.; HARKER, D.B.; SOMMERFELDT, T.G. Comparison of methods for determining satuarated hydraulic conductivity and drainable porosity of two southern Alberta soils. Can. J. Soil Sci., 1986, 66:249-59.

Saturated hydraulic conductivity (K_s) and drainable pore space (f) have been determined by various methods and by drain flow data. Auger hole and constant head well permeameter Ks were 140 and 110 per cent respectively, of Ks determined from subsurface drains. The drainable pore space, determined from soil characteristics curve (0-5 KPa) and drained flow data are close to each other.

1417

BUCKLAND, G.D.; HARKER, D.B.; SOMMERFELDT, T.G. The influence of drain depth on the rate of soil reclamation in irrigated areas of southern Alberta. Can. J. Soil Sci., 1986, 66:531-5.

Salinisation-resalinisation in subsurface drained plots (depth 0.76, 1.22 and 1.68 m) are studied. Leaching is more efficient with deeper depth of drains. Resalinisation was also quick and pronounced in fields with drains at 0.76 m depth. The efficiency of leaching was lowest for the 0.76 m depth drains at both sites.

1418

CANNELL, R.Q.; BELFORD, R.K.; BLACKWELL, P.S.; GOVI, G.; THOMSON, R.J. Effect of waterlogging on soil aeration and on root and shoot growth and yield of winter oats (Avena sativa L.) Pl. Soil, 1985, 85:361-73.

A lysimetric study on effect of flooding on winter oats; oxygen flux density decreases after flooding at the soil surface and in the upper 50 cm reduces to zero. Root growth is negligible in 20-50 cm zone and reaches as below as 95 cm. The total root length and shoot dry weight were 77 and 60 per cent of those in drained treatments. Leaf area index diminished and tillering was restricted. At harvest, total dry matter and grain yields were only 9 per cent less. It is attributed to less grains per panicle.

1419

CANNELL, R.Q.; CHRISTIAN, D.G.; HENDERSON, F.K.G. A study of mole drainage with simplified cultivation for autumn-sown crops on a clay soil. A comparison of direct drilling and mouldboard ploughing on drained and undrained land on root and shoot growth, nutrient uptake and yield. Soil Tillage Res., 1986, 7:251-72.

Effect of tillage and drainage intensity on cereal crops (wheat and oats) is examined. An interaction between the intensity of drainage and methods of tillage occurred only in the 3rd year. This is due to rainfall conditions in the four year study.

1420

CARTER, E.C.; IRVINE, J.E.; McDANIEL, V.; DUNCKELMAN, J.W. Yield response of sugar cane to stalk density and subsurface drainage treatments. Trans. ASAE, 1985, 28:172-8.

Four sugar cane planting patterns each with and without subsurface drainage were studied. Apart from increased yield due to new planting patterns suggested in the paper, it is recommended that subsurface drainage should be included in sugar cane production practices. It will reduce the adverse effects of high water table.

1421

CASTLE, D.A. A rationale for permeable backfill placed over pipe drains. Soil Use Mgmt., 1986, 2:58-61.

Technique and importance of permeable backfill in clayey, slowly permeable soils as a interconnector between mole drains, subsoiler fissures or permeable subsoil layers and the pipes are discussed.

1422

CASTLE, D.A. Mc CUNNALL, J.; TRING, I.M. Field Drainage, Principles and Practices. Batsford Academic and Educational, London, 1984, 250p.

The book on drainage has been divided into 19 chapters and discusses various aspects of surface and subsurface drainage. Design criteria and special problems such as drainage of peat soils, iron ochre, sodic/saline soils, and restoration of agricultural lands and drainage for fruit crops etc. have been discussed.

1423

CHENG, Y.S. Effect of drainage on the characteristics of paddy soils in China. In Organic Matter and Rice. Int. Rice Res. Inst. Los Banos, Philipines, 1984, 417-26.

Effect of drainage on physical properties of soils (Heavy paddy soils) is examined. Data on percolation rates, bulk density, hardness, O_2, CO_2 in soil air, rooting depth, root activity leaf colour, dry matter and yield are summarized. It is observed that drainage is limiting factor. Tile drainage followed by mole drainage are recommended.

1424

CULLEY, J.L.B.; COOTE, D.R. Water table regimes in an eastern Ontario soil with and without pipe drains. Can. Agric. Engng., 1984, 26:7-13.

Water table regimes in poorly drained clay soil were investigated with and without drainage. Pipe drains installed at 17 m spacing relieved the area greatly by lowering the water table. The water table rose to within 0.6 m of the soil surface only occasionally whereas in undrained it used to rise to the surface in the autumn. Mean drawdown rate due to drains was about 0.15 m/day.

1425

DAS, B. Surface drainage and runoff criteria for agricultural Areas. Proc. Natn. Sem. Estimation of Runoff Surface and Subsurface Drain., New Delhi, 1985, IBD1-IBD7.

Basic approach for planning surface drainage for an agricultural area has been indicated. Selection of design storms, fixation of loss rate and other parameters required for assessment of runoff have been given. Method of estimating the period of drainage and the volume of water to be drained has been outlined for tidal and non-tidal areas.

1426

DESMIER, R.E.; SCHRALE, G. Saline groundwater seepage into the pasture irrigation areas along the lower, Murray River. In: Hydrology and Water Resour. Symp. Barton, Aust., 1985, 25-8.

Salt and water balance at Long Flat has been studied for three years at twelve locations. It is reported that for every tonne of salt diverted from the river, 2.3 tonnes were returned. Alternative means of drainage disposal are outlined.

1427

DIRKS, F.J.H. Feasibility of vertical drainage in the Nile Valley. Proc. Sem. Problems Waterlogging Salinity in Alluvial Soils, New Delhi, 1986, 3. 1-3.27.

The technical and economic feasibility of vertical land drainage by pumping of groundwater for irrigation are investigated in a pilot area in the Nile Valley. The paper emphasizes the methodology of the investigations and the results of the forecasts of land drainage on present and future suitability of pumped ground water quality for irrigation. It is concluded that vertical drainage is hydrologically feasible and economically an attractive alternative.

1428

DURNFORD, D.S.; PODMORE, T.H.; RICHARDSON, E.V. Optimal drain design for arid, irrigated areas. <u>Trans. ASAE</u>, 1984, 27:110-5.

A methodology and computer model are presented for identifying optimum subsurface drainage systems in arid irrigated regions. A simple optimization scheme is used to determine the least cost drain layout. This optimization routine is part of a larger simulation model and evaluates the physical effects of installing the least cost drain system and predicts the resulting crop yield. By varying the design requirements and comparing drain costs with the predicted values of increased crop yields, an optimum level of drainage and the minimum cost drain layout is determined.

1429

ERONEN, J. Peatland drainage and wood production. <u>Suo</u>, 1984, 35:9-12.

A new classification of drained peatland for forest utilization is described. The classification stresses economic aspects of silviculture and logging operations.

1430

EVANS, E.J.; LEITCH, M.H., FAIRLEY, R.I., KING, J.A. Comparative studies on the growth of winter wheat on restored opencast and undisturbed land. <u>Reclam. Reveget. Res.</u>, 1986, 4:223-43.

The paper describes the soil properties of a progressively restored opencast site as influenced by drainage. The profile showed increased bulk density, reduced pore volume and an absence of continuous macro-pores. Installation of drainage largely improved the water movement. Drainage improved the yield of wheat but it was much less compared to a unworked soil.

1431

EVANS, R.O.; WESTERMAN, P.W.; OVERCASH, M.R. Subsurface drainage water quality from land application of swine lagoon effluent. Trans. ASAE, 1984, 27:473-80.

Subsurface flow volumes and nutrient concentrations in the effluent are measured for 20 months after a 6 year treatment period in which pig lagoon effluents were applied. Plots received 325,650 and 1300 Kg N/ha per annum. Nutrient mass balance is reported.

1432

FAUSEY, N.R. Drainage-tillage interaction on Clermont soil. Trans. ASAE, 1984, 27:403-6.

Interaction of subsurface and tillage operations is studied. Spring chiesel ploughing, no tillage and no-tillage on permanent beds were compared on a Clermout silt loam soil with subsurface drainage. Both population and stand of maize were reduced. Yield reduced with distance from drain as well. The tillage and drainage effects were independent. Surface drainage resulting from the raised beds had a highly significant effect on the performance of the no-tillage system.

1433

FRY, R.K.; SINAI, G.S.; SPOOR, G. machinery for installation of small diameter pipes at shallow depths. Agric. Wat. Mgmt., 1984, 9:23-36.

A method of installing 20-35 mm diameter drain tubes using low-cost machinery is described. Extensive soil loosening above and beside the drain is achieved using wing attachment and shallower leading tines. Top soil placement above the drain is incorporated in the same pass as the pipe is installed, to eliminate cavities, enable firm pipe bedding and good grade control.

1434

GOEL, A.C.; GUPTA, S.K. Drainage requirement of wheat in relation to its cultivation in Haryana. J. Res. Hissar,, 1986 16:30-6.

Analysis of data on performance of wheat crop under high water table, surface inundation and irrigation water quality is carried out. It is observed that whereas drainage measures are required in saline water belt comprising of Sonepat, Rohtak, Jind and some parts of Hissar and Sirsa districts, the good quality water zones mainly in Karnal, Kurukshetra, Ambala and Faridabad districts have been overdrained. The rainfall frequency analysis shows that surface inundation problem in eastern zone can occur with highest probability in the months of January and February. It is observed that marginal quality waters ($\leq$ 6dS/m) can be developed for irrigation of wheat to arrest the problem of rising water table in south western region.

1435

GROFFMAN, P.M.; BOX, J.E. Jr., TODD, R.L. Effects of artificial drainage on soil mineral nitrogen dynamics in winter wheat on the southern Piedmont. Commun. Soil Sci. Pl. Anal., 1984, 15:1051-63.

Artificial subsurface drainage to relieve excess soil moisture increased the rates of mineralization and nitrification and decreased rate of nitrification. This results in increased levels of soil mineral N in drained than undrained plots. Wheat yields were not increased with drainage as application of N (80 kg/ha) applied in previous spring compensated for the reduced mineral N levels in the undrained treatments.

1436

GUITJENS, J.C.; TSUI, P.S.; THRAN, D.F. Quantity and quality variations in subsurface drainage. Trans. ASAE, 1984, 27:425-8.

Fifteen parallel drains spaced at 37 m were selected to study spatial and temporal variability in drainage and to see whether such variations were induced by water management. Spatial variations were very pronounced in each parameter investigated. Each parameter produced a different spatial pattern for annual mean values. Management effects could not be separated from natural effects.

1437

GUPTA, S.K. Agricultural drainage methods for salt affected soils. Sinchan, 1986, 4:91-100.

Various definitions used in India for defining waterlogging are reported. The extent of waterlogging in India is reported to vary from 7 m ha to 20 m ha. Several experiments on subsurface drainage alongwith their salient features and conclusions are given. Crop yields from a case study are presented to show that subsurface drainage is essential to reclaim waterlogged saline soils. Crop tolerance tables for alkalinity as well as to saline irrigation waters are presented. The table can be usefully employed in taking decisions on the use of poor quality drainage effluents.

1438

GUPTA, S.K. Analysis of flood and rise in water table problems of Rohtak district. Curr. Agric., 1984, 8:16-25.

The analysis of physiographic position, probability analysis of rainfall and dry days and the rise in water table of low and high water table areas in Rohtak district has been done. Physiographically district is a flat terrain almost making a saucer shaped depression and hence is prone to floods. The rainfall in the area though erratic, has a probability of heavy storms of 15 cm once in 5 years (48 hr duration). Annual water table rises at the rate of 0.08 m/year and 0.6 m/year in already high water table and deep water table areas respectively. The paper lists several precautionary and curative measures to overcome the problem of floods.

1439

GUPTA, S.K. Physico-chemical characteristics of saline soils under reclamation. J. Agric. Engng., 1986, 23:53-60.

The physico-chemical characteristics of saline soils before and after reclamation are presented.

1440

GUPTA, S.K. Planning and managing irrigation projects for salinity control. Proc. Natn. Symp. Form. Appraisal of Irrigation Projects, Patna, 1985, VII.15-VII.25.

Drainage and leaching are two essential elements of any plan of irrigation to untapped salt affected lands. Leaching water is needed for initial reclamation and for salt balance. Agricultural considerations such as on-farm irrigation/drainage infra-structure, on farm land development and in-project training of farmers should be given due weightage in projects which are presently heavily biased in favour of engineering considerations.

1441

GUPTA, S.K. Subsurface drainage design - comparison of unsteady state equations. Proc. Natn. Sem. Estimation Runoff Surface and Subsurface Drain., New Delhi, 1985, IV SKG 1-IV SKG 7.

Drainage design equations for falling water table based on both D-F assumption and on potential theory have been compared. It is reported that equations based on potential theory give generally higher spacing than those based on D-F theory. Among the equations based on D-F theory van Shilfgaarde equation gives higher spacing because it accounts for flow from below and above the drains. The discussion on these equation is presented in the light of a case study carried out by the author at Sampla (India).

1442

GUPTA, S.K. Subsurface drainage for waterlogged saline soils. Irrig. Pwt. J., 1985, 42:335-44.

For the reclamation of waterlogged saline soils of Indo-Gangetic alluvial plains, subsurface drainage is feasible and essential. Drains at 50 m spacing and at 1.5 m depth can usefully be employed to drain these areas. By operating the drainage system when water table rises above 1 m depth, water table can be controlled and salinity reduced. There are considerable differences in salt status of drained and undrained areas resulting in differences in crop yields.

1443

GUPTA, S.K. Subsurface drainage strategies for waterlogged lands; a case study from Rohtak and Sonepat districts of Haryana. Proc. 53rd R & D Session, CBIP, New Delhi, 1986, 184:377-87.

The existing draft of ground water can take care of only 53.7 per cent of total average annual recharge of 20.6 cm. The draft can be increased by encouraging ground water development, installing subsurface (horizontal) drainage systems in critical areas (60 m spacing). The combination of two drainage methods will solve the menacing waterlogging and salinity problems. Ground water recharge and draft calculations for a sample block of Sampla, in Rohtak district are presented in an appendix.

1444

GUPTA, S.K. Subsurface drainage for waterlogging and salinity control: field experiments in India. Proc. 51st R & D Session, CBIP, New Delhi, 1984, 113-24.

Work on critical depth to water table for various crops in India has been reviewed. Review of experiments on subsurface drainage suggests that provision of drainage helps in maintaining aeration status of the soil during critical periods and thus helps in establishing better crops compared to undrained lands. The paper lists major areas of research activity needing attention particularly in respect of conditions prevailing in India.

1445

GUPTA, S.K.; GUPTA, I.C. Management of saline soils and Waters. Oxford and IBtl Publishing Co. New Delhi, 1986.

The book a comprehensive treatise on the subject of management of saline soils and waters, deals both with theory and practice. It contains eight chapters viz., Introduction, Quality of Irrigation water, Nature and properties of saline socils, crop Tolerance, Land Development and Leaching, Iirigation Practices, Drainage and Agronomic and cultural practices. The research work done in India in recent years has been recived very exhanstively. Crop to lerance tables with respect to EC of irrigation waters and ESP of soils have been given.

1446

HASSAN, A.A.; ANUKULARMPHAI, A. Design parameters of mole drainage system in Bangkok soil. In Agric. Mech. in Asia, Africa and Latin America. Irrigation and Water Mgmt., Directorate, Bangladesh, 1985, 16:49-54.

Mole drains at 40, 50 and 60 cm depth and spaced at 2, 3 and 4 m were drawn in a soil with 48 per cent clay content. Different combinations were tried. Mole drains constructed at 50 cm depth and at 3 m spacing were more stable and more effective in terms of extent of fissures and cracks and outflow behaviour.

1447

HATHOOT, H.M. Graphical design of tile and mole drainage systems. ICID Bull., 1985, 34:43-8.

An exact spacing design formula is established. By means of computer programme, numerical solutions for the design formula are provided and accordingly a number of non-dimensional spacing design charts covering a wide practical range are given. Numerical examples are given to illustrate their use.

1448

HEIKURAINEN, L. The maintenance and future of forest drainage areas. Suo, 35:13-8.

The paper supports the conventional drainage practices used in Finland for draining areas for forests.

1449

HEIM, S. Determination of parameters for establishing the dimensions and design of drainage systems on high groundwater soils. Arch. Acker. Pflan. Bodenk., 1984, 28:217-23.

The suitability of different methods for drainage design parameters evaluation has been investigated. The paper presents results in respect of hydraulic conductivity of silty clay black gley soils.

1450

HEIM, S. Geohydraulic effect of an experimental system for groundwater regulation on a heavy bottomland soil in the Unstrut-Helme region. Arch. Acker. Pflan. Bodenk., 1984 28:207-16.

A method of ground water regulation has been proposed and tested. Extensive survey on a Auenschluff-Auenton-Schwarzglay, silty clay, black clay bottomland soil showed large horizontal and vertical heterogeneity. Infiltration of water in the regulation zone with drain spacings of 5, 10, 20 and 40 m got established within 7-11 days. The technological suitability improved considerably due to drainage.

1451

HOVE, P. Drainage problem in peat soil. <u>7th Int. Peat Congr. Dublin</u>, Ireland, 1984, IV:230-6.

Problems in functioning of drains in peatland are discussed. Compaction by heavy machinery, accumulation of gas bubbles in soil pores and drain slope are some of the problems encountered.

1452

IRVIN, J.E.; RICHARD, C.A.; CARTER, C.E.; DUNCKELMAN, J.W. The effect of row spacing and subsurface drainage on sugar cane yields. <u>Sugar Cane</u>, 1984, 2:3-5.

Sugar cane was grown in drained and undrained fields both with four treatments of V furrow (control), wide furrow, 3 drills 0.45 cm apart and a V furrow on 0.6 m wide row. Yields of cane and sugar were significantly lower in undrained plots in the second ratoon and for a 3 year cycle. The 0.6 m-wide row had the highest sugar (165% of control) in the plant cane crop and 91 per cent in the second ratoon crop. Drainage or spacing had no significant effect on cane quality.

1453

JACOBS, T.C.; GILLIAM, J.W. Riparian losses of nitrate from agricultural drainage waters. <u>J. Environ. Qual.</u>, 1985, 14:472-8.

From a Middle-coastal Plain watershed 10-55 kg/ha per annum nitrate-N moved into subsurface drainage waters. Subsurface drainage moved more nitrogen into surface waters than fields without drainage. Effect of buffer strips of forested vegetation on nitrate pollution is described. In lower coastal plain watershed, the nitrate-N loss is 6-12 kg/ha per annum. These losses were mostly in surface runoff.

1454

JAVIER, R.R. Effect of flooding on seven species of tropical pasture legumes. Ann. Trop. Res., 1985, 5:91-101.

7 spp. of tropical pasture legume were grown in pots and flooding effect for 10 and 21 days was studied. The pots were flooded at flower initiation stage. *Macroptilium lathyroides* was superior in tolerance and *Casia rotundifolia* and *Vigna parkeri* were sensitive. Root growth was most affected. Nodulation was reduced in all except *M. Lathyroids.* High stomatal conductance was maintained in flood-tolerant spp. with increased duration of flooding, while high leaf diffusive resistance occurred in flood sensitive spp.

1455

JOSHI, P.K.; AGNIHOTRI, A.K.; SINGH, O.P. Investment on installation of tube drainage. Proc. Natn. Sem. Estimation Runoff Surface Subsurface Drain., New Delhi, 1985, IV JAS1-IV JAS8.

The paper presents cost-benefit of subsurface drainage systems for reclaiming barren saline soils in India. Cost of subsurface drainage varies from Rs. 8685/ha for 100 m spacing to Rs. 21912 for a spacing of 25 m. The cost-benefit ratio varies from 1.65 to 2.84 depending upon the crop. The paper also discusses other technical and socio-economic benefits of drainage such as generation of employment, inter-sectoral terms of trade, rain water management and environmental improvement.

1456

KANWAR, R.S.; JOHNSON, H.P. Sensitivity analysis of a

drainage simulation model. Water Resourc. Bull., 1984, 20:339-42.

A sensitivity analysis of a computer model simulating major water and nitrogen processes of a soil-water-plant climatic system on an annual basis, was conducted to determine how the model reaches to variations in selected hyrdrologic and nitrogen movement parameters. Two major output variables i.e. total subsurface drain volume and cumulative nitrate loss were selected for the sensitivity analysis. Model sensitivity analysis showed that the model is most sensitive to hydrologic parameters. The model is very sensitive to variations in the initial water content in the soil profile.

1457

KAPILA, B.K. Utility of shallow tubewell pumping to lower ground water table - a case study. Res. Mag. Irrig. Pwr. Res., Instt. Amritsar, 1985, 11:27-30.

Effect of shallow tubewells (68 No.) on ground water table were investigated in Tarn Taran tehsil of district Amritsar. It is brought out that 15 cm diam gravel packed shallow tubewells of design discharge 0.014 cumecs have not lowered the water table. Detailed investigations for installing deep tubewells of higher discharge are recommended.

1458

KOOL, Y.M.; SAHU, H.L. MEHTA, R.S.; YADAV, R.D. Rising water table trends and waterlogging at Barna Command Madhya Pradesh. Proc. Sem. Problems Waterlogging, Salinity in Alluvial Soils, New Delhi, 6.1-6.14.

Drainage investigations in the waterlogged soils of Barna and Tawa command areas in Madhya Pradesh are presented. A combination of preventive measures such as lining of surface water distribution system, vertical drainage through deep augmentation tubewells and subsurface horizontal drainage in specific areas are suggested to overcome the problem of waterlogging in these commands.

1459

KOWALD, R.; MARTINS, O. Investigation on the prevention of root penetration into drains. Z. Kult. Flurb., 1986, 27:130-4.

To reduce the root penetration problem openings should be positioned at the tops of the ridges. The drain openings can be protected by small rise in front of and behind the openings.

1460

KRUSEMAN, G.P. Waterlogging concepts and misconcepts. Proc. Sem. Problems of Waterlogging Salinity in Alluvial Soils, New Delhi, 1986, 1.1-1.13.

The cause of rising water table and eventual waterlogging is described as a disrupted equilibrium between recharge and discharge. It can seldom be prevented in areas irrigated with surface water. Remedial measures have to be taken either by installing pumping systems or pipe drains to lower the water table.

1461

KUMAR, R.; AGARWAL, M.C.; SINGH, J. Management of rising water table in semi-arid regions. Proc. Sem. Problem of Waterlogging Salinity in Alluvial Soils, New Delhi, 8.1-8.12.

Paper discusses three alternative drainage measure for waterlogged areas in Haryana i.e. conjunctive use for stabilising water table, skimming wells and drainage through horizontal tiles. Details of investigations and design of tile drainage system at the Haryana Agricultural University Hissar farm are given.

1462

KUNTZE, H.; SCHEFFER, B.; EILTS, F. Organic drain filter to counteract iron clogging. 4th report: Oak Wood sawdust as drain filter. Z. Kult. Flurb., 1984, 25: 308-12.

The effect of oak sawdust (15 kg/m) as a ditch filter is examined and reported that it is as effective as Trillo (acorn cupulae from a small Asiatic oak) in reducing iron clogging. The effect lasted for five years which was longer than that of Trillo. Phenols and oxidizable substances in the drainage effluent increased in the drainage effluents for the first 2-3 weeks. The use of synthetic filter material alongwith sawdust is suggested as latter is unable to stop silting up of the drains.

1463

KUNTZE, H.; WETZEN, F.A. Soil hydrological effects of a combined drainage in a bottomland soil. Z. Kult. Flurb., 1984, 25:360-9.

Results of a 7 year trial on heavy clay soil of the upper Middle-Weser-River are reported. The effects of conventional pipe drains (10 m interval) and pipe collecters at 32, 64 and 128 m with mole drains at 2 m interval were investigated. Mole drains lasted for 3 years. However, positive hydrological and economic effects were observed within 7 years. Mole drains of 16-32 m length (collectors 32, 64 m intervals) were best and provided 51 more field-working days and lowered the water table by about 30 cm. Average surplus grain yield of 2 dt/ha was sufficient to cover the cost of combined drainage (64 m collectors) within 4 years.

1464

LESAFFRE, B.; NORMAND, M.; VALENCIA, G. Water regime of a mole drainage experiment on a heavy clay soil of the sologne area, France. Int. Land Reclam. Improv.

Inst. Wageningen, Netherlands, 1984, 37:171-77.

Results of mole drainage on a clayey pseudogley soil are reported. Moisture content upto 1.6m depth and ratio of rainfall to drain outflow defined as outflow coefficients are given.

1465

LINDNER, H.; SCHAFFER, E. Model experiments on the efficiency of smooth-walled and grooved ceramic drainpipes. Arch. Acker. Pflan. Bodenk., 1984, 28:155-61.

Ceramic drainpipes with and without grooves have been compared for efficiency. Fibre glass fleece has been used as a envelope material in each case. The water outflow is 30 to 100 per cent more and the inflow resistance lower in case of grooved pipes. A 2 mm groove is inferior to grooves of ≥ 3 mm. Reduction or supression of deflection and aggregation of flow lines at the joints is attributed for the increased effectiveness of grooved pipes.

1466

LOVELL, C.J.; YOUNGS, E.G. A comparison of steady state land drainage equations. Agric. Wat. Mgmt., 1984, 9:1-21.

The physical assumptions and mathematical approximations leading to ten steady-state drainage equations for installation of parallel cylinderical drains are critically examined. Of the drainage equations considered, Hooghoudt's equivalent depth equation is the only one that gives results contained mainly within the known bounds that result from a consideration of the combination of equations.

1467

MANGAT, N.S.; UPPAL, H.L.; SINGH, T. Geohydrological imbalance in Ferozepur-Faridkot-Bhantinda area. J. Inst. Engineers (India), 1985, 65 (CI6):243-9.

The waterlogging problem in soutn-western Punjab is discussed. It is reported that 5500 km^2 of earstwhile arid land is now in the grip of severe waterlogging. Probable behaviour of ground water table in this area has been reported. Norms for ground water assessment in Punjab are described in an appendix.

1468

MOHAMMAD, F.S.; SKAGGS, R.W. Effect of drain tube openings on drainage and sub-irrigation. Trans. ASAE, 1984, 27:1455-62.

A laboratory soil tank model is used to study the effect of drain tube openings on drainage and sub-irrigation. There was an increase in drain flow, drawdown and subirrigation rates till the area increased to 38.5 cm^2/m. The perforation area, 79 cm^2/m increased the drawdown rate similar to a 5 cm gravel envelope. Location of the perforations on the drain tube had only a small effect on drain outflow, drawdown and sub-irrigation rates.

1469

MULLER, L.; SCHINDLER, V. Stability of mole drainage channels in bottomland clay soils. Arch. Acher. Pflan. Boden., 1985, 29:499-505.

The stability of mole drainage channels was tested in bottomland clay soils. The mole channels in these soils whose plasticity index exceed 0.25 should have their apex at a depth exceeding 0.55 m. The channels, for stability should not be overloaded during the formation and consolidation periods.

1470

MULQUEEN, J. The development of gravel mole drainage. J. agric. Engng. Res., 1985, 32:143-51.

The development and testing programme for a gravel

filled mole drain is described. Aggregates of 10-20 mm clean gravel were used to fill the mole drains. The results indicate that the mole drains become stable and for the last 17 years they are functioning.

1471

NATH, J.; SINGH, R. Effect of wter table depth and soil texture on evaporation rate - a comparison of different methods of prediction. J. Indian. Soc. Soil Sci., 1984, 32:409-14.

Evaporation rates for sand, sandy loam and sandy clay loam were evaluated with different depths to water table. The evaporation rates (maximum) were predicted using three equations. Gardener equation gave the closest agreement followed by those of Raats and Gardener and Campbell.

1472

NORRIS, E.R.; JUTRAS, P.J.; SCOTT, D.W.; BROUGHTON, R.S. A comparison of drainlaying times and delays in the operation of three subsurface drain installation systems. Can. Agric. Engng., 1984, 26:87-90.

A comparative study on time loss in various tile laying machines; trenchless plough either with a radio grade control or laser grade control was actually installing drain tube in 20 per cent of total field time whereas trencher with laser control was doing it for 40 per cent of time. Lost time is attributed to machine, field operation and laser control factors.

1473

PATRICIA, A.D.; ARMSTRONG, A. A field measurement of grassland poaching. J. Agric. Sci. Camb., 1986, 106:67-73.

Poaching which has been defined visually as a depression greater than 4 cm depth is studied for a period of 3 years over a new grass lay. Typically standard deviations along a poached, undrained transect were twice those of an un-poached, drained one. In this experiment drainage with pipes alone at 12 m spacing or mole drainage at 2 m with pipes at 40 m spacing controlled poaching though water table was not lower than in undrained.

1474

PEDERSEN, E.F. Investigations on drain water quality and leaching of nutrients from marsh and deep peat soils, 1971-84. Tidss. Plnteavl, 1985, 89:319-29.

Drain water quality of a marsh soil at Hojer and peat soil at Hedebybro in Skjerna river valley are reported. Sodium and chlorine were predominant ions in marsh soils with average concentrations of 293 and 476 mg/L. The peat soil which has a depth of 8-10 m, the drain water had a pH of 3.5 and average SO_4-S concentration was 578 mg/L. NH_4-N and iron contents were also high initially but decreased with time.

1475

PENNING-ROUSELL, E.C., PARKER, D.J.; HARDING, D.M. Floods and Drainage, British policies for hazard reduction, agricultural improvement and wetland conservation. Allen and Unwin, London, U.K., 1985, 199p.

Various aspects of flood alleviation and agricultural land drainage are discussed. These include institutional, hydrologic, hydraulic, economic, and environmental.

1476

PIDGEON, J.D.; THOROGOOD, P.J. Management of different soils for intensive winter cereal in central England: a farm case study. Soil Use Mgmt., 1985, 1:124-6.

Soil loosening and mole drainage increased wheat and barley yields (5.6-4.9 t/ha for 1979-81 to 7.4-6.8 t/ha for 1982-84) on a Ragdale series stagnogley soil. It is attributed to improved soil management.

1477

PRASHER, S.O.; RUSSELL, S.O.D.; PODMORE, T.H.; CHIENG, S.T. First and second order analysis of parameters uncertainty in drainage design. Trans. ASAE, 1984, 27:1081-6.

First and second order analysis is used in the design of a subsurface drainage systems to account for the variability of drainage design parameters. Both steady (Hooghoudt's) and unsteady (van Schilfgaarde and Boussinesq's) are used. Mean and variance of the design parameters, are incorporated in the design equations. The procedure is illustrated with examples to show how drainage with a specified probability of success can be designed.

1478

RAM, S.; CHAUHAN, H.S. Evolving a strategy for subsurface drainage. Proc. Sem. Problems Waterlogging in Alluvial Soils, New Delhi, 1986, 7.1-7.12.

The salient aspects of the problem of waterlogging and possible approach towards them have been discussed. The paper brings out numerous facts of this large problem so as to lead to an appropriate strategy for amelioration of waterlogged lands in the country.

1479

RAPTE, S.L. Drainage characteristics of vertisols. Sinchan, 1986, 4:108-10.

Drainage characteristics of vertisols are evaluated following installation of a brick subsurface drainage system. Cost of installation and benefits under rainfed and irrigated conditions are given. It is concluded that drainage system is feasible only when irrigation facility is available. The cost varies from Rs. 1678 (104 m spacing) to Rs. 13425 (13 m spacing).

1480

RAO, K.V.G.K.; SINGH, O.P.; GUPTA, R.K.; KAMRA, S.K.; PANDEY, R.S.; KUMBHARE, P.S.; Drainage investigations for salinity control in Haryana. Bull. Central Soil Salinity Research Inst. Karnal, 1986, 10, 95p.

Problem of rising water table in general and particularly in Haryana is reported. Investigations and results of drainage system installed at 3 places in Haryana are reported. The performance of drains is observed to be satisfactory with yields increasing from almost nil to those obtained in normal lands. General recommendations and precautions in the use of drainage are given.

1481

RAO, K.V.G.K.; SINGH, O.P.; KUMBHARE P.S. Drainage for reclamation of waterlogged saline soils - a case study. Proc. Sem. Problems Waterlogging Salinity in Alluvial Soils, New Delhi, 1986, 5.1-5.15.

Case studies of subsurface horizontal drainage at Sampla, Mundlana and Kailana Khas in the state of Haryana are presented. It is concluded that water table in the drained plots is lowered, salinity is reduced and crop performance is excellent. Subsurface drainage is recommended as an ameliorative measure for reclamation of waterlogged saline soils.

1482

REID, I.; PARKINSON, R.J. The nature of the tile drain outfall hydrograph in heavy clay soils. J. Hydrol., 1984, 72:289-305.

The drainflow hydrographs for subsurface drainage are presented over four seasons on a heavy clay soil. Seasonal differences are observed and are attributed to antecedent moisture and topographic conditions. The greatest risk of flooding in these areas is due to complex rather than simple storms. The hydrograph peaks in case of complex storm not only arrive early but are as large as 3 times the values in a simple storm having equivalent rainfall.

1483

ROBINSON, M. The extent of farm underdrainage in England and Wales, prior to 1939. Agric. History Rev., 1986, 34: 79-85.

History and areas under drainage prior to 1939 are reported.

1484

ROBINSON, M.; RYDER, E.L.; WARD, R.C. Influence of stream flow of field drainage in a small agricultural catchment. Agric. Wat. Mgmt., 1985, 10:145-8.

Mechanically induced fissures in fields with underdrainage/secondary treatments increased peak flow rates in a clay loam soil. The increase diminished with time because of intensive farming which eliminated fissures. Regular subsoiling is recommended.

1485

SCOTTER, D.R.; HORNE, D.J. The effect of mole drainage on soil temperatures under pastures. J. Soil Sci., 1985, 36:319-27.

Drainage has no observable effect on soil temperature inspite of differences in water table depth and soil water content. The details of thermal properties of soil are presented and it is reported that thermal diffusivity remained unchanged. Soil heat flux remains unaffected by drainage in spring. The predicted temperature differences due to drainage are of the order of 0.2°C.

1486

SHINDE, K.D. Drainage in Mula command, Sinchan, 1986, 4:129-35.

Causes of waterlogging and curative measures adopted for amelioration in the Mulla command area in Maharashtra are described. Inadequate field drainage, poor maintenance of natural drainage, excessive water application, seepage from canal system and lesser development of ground water are some of the important reasons of waterlogging in the command.

1487

SHIRKANDE, R.P. The most suitable plants to improve problematic lands with special reference to water drawn by eucaliptus plants. Sinchan, 1986, 4:116-8.

Fast growing plants which are suitable for waterlogged and waterlogged saline soils are reported. Whereas *Tamarix aphylla, Ailanthus altissima* and *Eucalyptus camaldulensis* are described in the first category, for the second category *E. hybrid, E. occidentalis* and *Emblica officinalis* are described. The drain spacing vis-a-vis discharge from a eucalyptus tree plantation are presented to highlight the importance of drainage through trees.

1488

SINGH, J.; SINGH, R,; KUMAR, R. An integrated control of rising saline water table. J. Res. Hissar, 1984, 14, 73-81.

Imbalance between recharge and withdrawal of ground water is associated with the rising water table in the inland alluvium basin of Haryana. The ground water quality is mostly poor. Vertical drainage is suggested for control. the pumped ground water can be used for cropping by mixing it with canal water or it can be evaporated to form salts.

1489

SINGH, J.; SINGH, R.; KUMAR, R. Groundwater studies for a shallow water table area. J.Res Hissar,. 1984, 14:334-41.

Ground water studies at government farm Hansi (Haryana) indicated that there is a ground water mound at the centre of the farm (675 ha). The water flow rate to the adjoining areas is estimated at 1.70 mm/day. To reduce this and to increase the ground water recharge during monsoon, it is suggested that water table at the farm should be depressed particularly in post monsoon season.

1490

SINGH, O.P.; PANDEY, R.S. Effect of subsurface drainage on hydrological aspect of saline soil reclamation and crop production. Proc. Natn. Sem. Estimation Runoff Surface and Subsurface Drain., New Delhi, 1985, IV SP 1-IV SP 6.

Hydrological effect of saline soil reclamation through subsurface drainage system is reported. Leaching of salts takes place as more amount of water flow occurs through the rootzone instead of surface runoff. The soils (barren) start yielding right in the first year though yield increases in each year following reclamation (upto 4 years).

1491

SINGH, O.P.; PANDEY, R.S. Operational performance of vertical subsurface drainage system in saline soils of Haryana. 53rd R&D Session, CBIP, New Delhi, 184:363-75.

General hydrogeological conditions of saline belt of Haryana indicate poor aquifer characteristics and their framework for installing a successful vertical drainage system. Based on a study, it is concluded that residual drawdown of a individual and a battery of 5 cavity wells is much less to be useful in a vertical drainage scheme.

1492

SINGLETON, P.L. Installation of PVC piping for soil water observations. N.Z. Soil Bur. Scientific Rep., 1984, 61, 12 p,

Procedures to install piezometers, observation wells and 20 mm diameter high pressure PVC pipes are described.

1493

SINHA, B.P.C.; BAHADUR, J. Environmental considerations in subsurface drainage of irrigated lands in a hard rock terrain. Proc. Natn. Sem. Estimation Runoff Surface Subsurface Drain., New Delhi, 1985, IV SB1-IV SB7.

Waterlogging and salinisation in irrigated agriculture in context of Indian conditions is described. A case history of predominantly hard rock terrain (Archean crystalline complex) is presented in which all inputs and outputs of the sytem have been quantified. An integrated approach in water-use planning is suggested to overcome the environmental degradation.

1494

SMEDEMA, L.K. Drainage coefficient for heavy land. J. Irrig. Drain. Engng., 1985, 111:101-12.

A new approach to determine drainage coefficients for soils with impeding nature of the subsoil is illustrated. The discharge, it is opined can be described by the linear reservcir type rainfall-discharge model. Procedures and guidelines are given for the use of this model. A comparative evaluation with existing methods and commonly used design standards shows the reliability and applicability of this method.

1495

SMEDEMA, L.K., POELMAN, A.; HAAN, W.DE. Use of Hooghoudt formula for drain spacing calculations in homogeneous- anisotropic soils. Agric. Wat. Mgmt., 1985, 10:283-92.

Cases of ground water flow in anisotropic soils can be transformed to equivalent isotropic cases, and then be solved with Laplace equation. On the basis of this principle, Hooghoudt formula is used to calculate drain spacing in aniosotropic soils. The comparison with numerical solution of the Laplace equation is used to confirm this.

1496

SOKOLOWSKI, J. Effect of filter materials on drain clogging. Rocz. Gleboz., 1984, 35:197-206.

Properties of filters which influence their durability and efficiency are discussed. A experiment with 15 filter materials and 3 soil types was used to find out the indices of clogging reduction which is also indicative of the effectiveness of filter materials. The water flow intensity index is defined as the ratio between the water flow in protected and unprotected drains.

1497

SOMMERFELDT, T.G. Performance of unlined and lined mole drains in a saline clay loam field. Can. Agric. Engng., 1982, 26:1-5.

Mole drains were tested with and without lining (5.0 m for the unlined and 7.6 m for the lined). Both drains functioned well for 10 years and lowered the water table effectively. The drain discharge was used to calculate the hydraulic conductivity and spacing was found commensurate with the evaluated values. The average discharge in case of unlined moles was only 24 per cent of the lined moles. Comparative values of hydraulic conductivity by various methods are presented.

1498

SOMMERFELDT, T.G., PAZIUK, N. Subsurface drainage with a plastic cut-off barrier to intercept canal seepage. Can. Agric. Engng., 1985, 27:111-4.

Performance of a intercepter drain to control water table along a seeping canal is evaluated. A subsurface drain with pea gravel envelope and a plastic barrier (to prevent bridging) was installed to intercept ground water flow. The drain performed well and lowered the water table from 1 m to about 2 m depth within one week.

1499

STONE, J.A. Poorly drained conditions and root development of eight intermediate soybean cultivars. Agron. J., 1985, 77:787-9.

Rate of tap root and lateral root extension in 21 day-old plants was significantly reduced when drainage conditions were poor. The top to root ratio was significantly higher in well drained compared to poorly drained lands because of greater reduction in root than in shoot dry weight.

1500

STRANDBERG, J.O. Flooding organic soils to control species of Pythium which attacks carrots and other vegetables. Proc. Flo. State Hort. Soc., (1984) 1985, 97:164-8.

Fallow flooding of organic soils reduced Pythium Spp. which are pathogenic to carrot roots. Oxygen rapidly decreased after flooding. Corresponding decrease was noticed in P. populations. Temperature effects are studied in a controlled environment which show that decrease is slow at temperatures > 20°C.

1501

STRYG, P.E. The economics of draining ferriferous soils. Volk. Hedes. Tides., 1984, 105:22-5.

Economic value of land drainage in ferriferous soils is examined. Significant increase in yields is reported. The apparent costs as demonstrated by the research are described. The recent estimates indicate that on an average expenditure on drainage in Denmark is 16000 kr/ha.

1502

SUDNITSYN, I.I.; SIDOROVA, M.A. Movement of moisture in soil with vertical vaccum drainage. Mosc. Univ. Soil Sci. Bull., 1984, 39:74-7.

Vaccum drainage process is described. The technique is suitable to drain areas around trees or shrubs or of micro-depressions.

1503

SWINFORD, J.M.; GEORGE, J.A. BENNINGA, W. Reclamation of a sodic soil at Pongola. Proc. 59th Ann. Congr. South Af., Sug. Techn. Assn. Durban and Mount Edge., 1985, 148-51.

The results of a reclamation project on sodic soils in Pongola are described. Subsurface drainage is installed to facilitate leaching. Amendments such as gypsum (26 t/ha), Sulphur (6 t/ha), filtercake (350 t/ha) and sulphuric acid (17 t/ha) were applied as various treatments. The cane yield in drained plots with amendments was consistently high compared to only the drained plots. The results, however, emphasize the reclamative effect of drainage alone on sodic soils reclamation. Amendments accelerate the reclamation process.

1504

TABUCHI, T. Underdrainage of lowland rice fields. In

Soil Physics and Rice. Int. Rice Res. Instt. Los Banos, Philippines, 1985, 147-59.

For farm mechanization of clay low-land rice fields rapid underdrainage is necessary. It can be achieved by levelling, making small shallow ditches, allowing ample time for drainage and backfilling with highly permeable material. The decision making on bund heights, precision of land levelling and drain spacing is explained. Data from various studies in Japan are presented.

1505

TOPP, G.C.; ZEBCHUK, W.D. The effect of auger hole diameter changes on hydraulic conductivity measurements. Can. Agric. Engng., 1985, 28:15-7.

Diameter changes of the auger holes have a profound and systematic effect on hydraulic conductivity values. Sensitivity analysis of the equations suggests that diameter changes should be taken into account as otherwise it will result in greater intensity of subsurface drainage.

1506

TOWNER, G.D.; YOUNGS, E.G. Application of drainage theory in the field. Soil Use Mgmt., 1986, 2:44-7.

The prediction of water table levels from groundwater theory agreed with data on water table heights upto 300 mm above the mean drain level. The drainage above that level was larger than predicted. Problems of measurements of water table levels and sample size in hydraulic conductivity measurements are discussed.

1507

TRAMPLER, R. A method of deep drainage requiring no form of support. Z. Kult. Flurb. 1985, 26:385-8.

A new sliding excavator claimed to be quicker, safer and cheaper is described. It can install drains upto 7 m depth.

1508

VUUREN, W. VAN, JORJAN, H. Evaluating subsurface drainage benefits: estimation from short-term cross-section data. Trans. ASAE, 1984, 27:1719-23.

A procedure has been described which estimates change in maize yield after subsurface drainage has been installed. Consequences of non-availability of long term data are discussed. Economic analysis based on short term data and conditions under which such procedures hold good have been outlined.

1509

WARHADE, M.S. Drainage of deep black soils around Mangalwedha Tal, Pandharpur, distt. Solapur. Sinchan, 1986, 4:124-8.

Steps used in drainage of black soils are described. The drainage is done by deep open drains or shallow widely spaced open drains with deep open wells. Neutralisation of alkali is done through burying a green manuring crop (Dhaincha or rice straw) and growing of rice in the crop rotation. The dimensions and costs of various drainage channels are reported. Pumping schedule of open wells is described.

1510

WESSELING, J.G.; WESSELING, J. The influence of seepage on the depth of water table in drainage. J.Hydrol., 1984, 73:289-97.

A theory based on two-dimensional flow is described to predict the depth of the water table in soils drained by a subsurface drainage system and underlain by artesian aquifers. Examples are given to illustrate the use of theory.

1511

WHEATLEY, R.E.; VAUGHAN, D.; ORD, B.G. Amelioration of the ochre problem in field drainage systems using coniferous bark. 7th Int. Peat Congr. Dublin, Ireland, 1984, IV:96-105.

It is reported that conifer bark provides an inexpensive method of ameliorating ochre formation.

1512

YOUNGS, E.G. A simple drainage equation for predicting water table drawdowns. J. Agric. Engng. Res., 1985, 31:321-8.

Based on Hooghoudt's equivalent depth steady state drainage equation and from the hodro, graph analysis for a parallel installation of drains of optimum size, a simple drainage equation is derived. The equation is useful in the analysis of falling water tables in drained land.

1513

YOUNGS, E.G. Characterization of hydrograph recession of land drains. J. Hydrol., 1985, 82:17-25.

A power-law drainage equation is used to derive a relationship between the drain flow rate and time. The relationship can be used to predict drainage parameters by superimposing the data on type curves. Use of graphical technique is illustrated by a typical example which reveals low values of specific yield for fine textured soils.

1514

YOUNGS, E.G. Water table heights in drained anisotropic homogeneous soils. Agric. Wat. Mgmt., 1986,11:1-11.

Theoretical study on water table heights in drained homogeneous anisotropic soils in which the vertical hydraulic conductivity differs from horizontal hydraulic conductivity.

Author's Index

Subject Index

GENERAL

(Numbers refer to Abstract numbers in the book)

E

Y

Subject Index

PLANTS CLASSIFIED

(Numbers reier to Abstract numbers in the book)

VEGETABLES

FRUITS

SPICES

TREES AND BUSHES